METHODS IN MOLECULAR BIOLOGY

Series Editor
John M. Walker
School of Life and Medical Sciences
University of Hertfordshire
Hatfield, Hertfordshire, UK

For further volumes:
http://www.springer.com/series/7651

For over 35 years, biological scientists have come to rely on the research protocols and methodologies in the critically acclaimed *Methods in Molecular Biology* series. The series was the first to introduce the step-by-step protocols approach that has become the standard in all biomedical protocol publishing. Each protocol is provided in readily-reproducible step-by-step fashion, opening with an introductory overview, a list of the materials and reagents needed to complete the experiment, and followed by a detailed procedure that is supported with a helpful notes section offering tips and tricks of the trade as well as troubleshooting advice. These hallmark features were introduced by series editor Dr. John Walker and constitute the key ingredient in each and every volume of the *Methods in Molecular Biology* series. Tested and trusted, comprehensive and reliable, all protocols from the series are indexed in Pub Med.

Comparative Embryo Culture

Methods and Protocols

Edited by

Jason R. Herrick

Department of Reproductive Sciences, Omaha's Henry Doorly Zoo and Aquarium, Omaha, NE, USA

Humana Press

Editor
Jason R. Herrick
Department of Reproductive Sciences
Omaha's Henry Doorly Zoo and Aquarium
Omaha, NE, USA

ISSN 1064-3745 ISSN 1940-6029 (electronic)
Methods in Molecular Biology
ISBN 978-1-4939-9565-3 ISBN 978-1-4939-9566-0 (eBook)
https://doi.org/10.1007/978-1-4939-9566-0

This Humana imprint is published by the registered company Springer Science+Business Media, LLC, part of Springer Nature.
The registered company address is: 233 Spring Street, New York, NY 10013, U.S.A.

Preface

The preimplantation period is a unique time during mammalian development when an embryo composed of between one and a few thousand cells survives in the absence of any maternal blood supply, relying entirely on the nutrients present in the fluids of the reproductive tract and/or limited stores of endogenous energy reserves. The relative autonomy of embryos at these early stages of development makes it possible to maintain viability during culture. This unique characteristic permits the evaluation of embryonic responses to specific manipulations of their environment, including changes in temperature, the gas composition of the atmosphere, pH, osmolarity, and the concentrations of a virtually endless list of specific molecules ranging from salts and nutrients to cytokines, hormones, and growth factors.

Although in vitro embryo culture was initially a purely academic pursuit, the application of this research to the treatment of human infertility and the propagation of superior livestock lineages has expanded interest in this field and led to significant advances in our understanding of preimplantation development. However, the applied, commercial value of embryo culture technologies often overshadows the utility of embryo culture for basic research. Studies of embryos in culture are essentially classic ablate and replace experiments. Initial studies focus on the identification of conditions that support the development of an embryo from one stage of development to the next, with the goal of achieving complete (zygote to hatched blastocyst) development in vitro. Subsequent refinements are then made not only to support survival and continued development but also to improve the quality of embryos and their ability to develop into healthy offspring, similar to embryos grown entirely in vivo. Importantly, as our understanding of the interactions between an embryo and its environment improves, we can provide more appropriate conditions for embryos during culture, and the in vitro model becomes increasingly useful.

Protocols for embryo culture are available for a wide variety of species, but the quantity and quality of embryos that can be produced differ dramatically. Some of these differences are due to the number of labs devoted to the species in question; some are due to differences in the culture requirements for each species. The mouse, human, and cow are the three most commonly studied species, the three species that are the most applicable to biomedical research, livestock production, and human medicine, and the three species with the most available funding, so it is perhaps not too surprising that ART protocols for these species are among the most successful. There are certainly other species with potential utility as research models (rats, hamsters, rabbits) or for commercial livestock production and biomedical purposes (pigs), but these species require different culture conditions than embryos from mice, humans, and cattle. Such unique culture requirements are often seen as a disadvantage of working with a particular species, but they are also valuable opportunities to study alternative developmental strategies that could have broad application. The historical note in this volume by Dr. Michael Kane describes his experience developing a medium for the culture of rabbit embryos. Unlike murine embryos that will form viable blastocyst in a simple medium lacking amino acids, amino acids are essential for rabbit embryos to form blastocysts. Although the relative importance of amino acids for development may be specific to the rabbit embryo, amino acids were later found to improve embryonic development and viability in a number of species and are now routinely used for the culture of embryos from most, if not all, species, including the mouse.

Although many labs focus on embryos from one species, this volume was assembled with the hope of inspiring embryologists to consider comparative studies. The species included in this volume represent a broad range of taxa, whose embryos have vastly different culture requirements and developmental characteristics. Among the species described in this volume, significant differences exist in the rates of cell division, the timing of the maternal to embryonic transition, the relative lipid content of the cytoplasm, the sensitivity of the embryo to specific environmental ions, the preferred nutrients and associated metabolic pathways used by the embryo, the timing and mechanisms of early lineage specification, the presence or absence of embryonic diapause, and the time from fertilization to implantation. The final chapter by Govindasamy and Bedzhov outlines a protocol for extended culture of embryos into peri-implantation stages. Given the diversity observed between species during the preimplantation period, extended culture will likely provide researchers with additional opportunities to discover novel developmental mechanisms. For whatever developmental phenomena studied, there is likely to be another species whose embryos have evolved a unique mechanism that could provide a valuable, comparative context for experimentation.

Omaha, NE, USA ***Jason R. Herrick***

Contents

Contributors

YUKSEL AGCA • *Department of Veterinary Pathobiology, College of Veterinary Medicine, University of Missouri, Columbia, MO, USA*

ABDULLAH N. ALOWAIMER • *Department of Animal Production, College of Food and Agricultural Sciences, King Saud University, Riyadh, Kingdom of Saudi Arabia*

JENNIFER P. BARFIELD • *Department of Biomedical Sciences, Animal Reproduction and Biotechnology Laboratory, Colorado State University, Fort Collins, CO, USA*

RIBRIO IVAN T. P. BATISTA • *Faculdade de Veterinária, Setor de Reprodução Animal, Universidade Federal Fluminense, Niteroi, RJ, Brazil*

IVAN BEDZHOV • *Embryonic Self-Organization Research Group, Max Planck Institute for Molecular Biomedicine, Münster, Germany*

DEBRA K. BERG • *AgResearch, Ltd., Ruakura Research Centre, Hamilton, New Zealand*

LUCIA BOCCIA • *Private Consulting, Salerno, Italy*

ELAINE M. CARNEVALE • *Equine Reproduction Laboratory, Department of Biomedical Sciences, Colorado State University, Fort Collins, CO, USA*

MARÍA-GRACIA CATALÁ • *Ciència Animal i dels Aliments, Universitat Autònoma de Barcelona, Barcelona, Spain*

ELIZA CURNOW • *Washington National Primate Research Center, Seattle, WA, USA*

SERENA DI FRANCESCO • *Private Consulting, Celbridge, Republic of Ireland*

JANE C. FENELON • *Centre de Recherché en Reproduction et Fertilité, Faculté de Médicine Vété rinaire, Université de Montréal, St-Hyacinthe, QC, Canada*

VICENTE JOSÉ F. FREITAS • *Laboratório de Fisiologia e Controle da Reprodução, Universidade Estadual do Ceará, Fortaleza, CE, Brazil*

DAVID K. GARDNER • *School of BioSciences, University of Melbourne, Parkville, VIC, Australia*

THADDEUS G. GOLOS • *Wisconsin National Primate Research Center, University of Wisconsin-Madison, Madison, WI, USA*

NIRAIMATHI GOVINDASAMY • *Embryonic Self-Organization Research Group, Max Planck Institute for Molecular Biomedicine, Münster, Germany*

P. S. P. GUPTA • *ICAR-National Institute of Animal Nutrition and Physiology, Bangalore, Karnataka, India*

CAROL HANNA • *Oregon National Primate Research Center, Oregon Health and Science University, Beaverton, OR, USA*

PETER J. HANSEN • *Department of Animal Sciences, University of Florida, Gainesville, FL, USA*

ERIC HAYES • *Washington National Primate Research Center, Seattle, WA, USA*

JASON R. HERRICK • *Department of Reproductive Sciences, Omaha's Henry Doorly Zoo and Aquarium, Omaha, NE, USA*

HIROYOSHI HOSHI • *Research Institute for the Functional Peptides, Higashine, Yamagata, Japan*

DOLORS IZQUIERDO • *Ciència Animal i dels Aliments, Universitat Autònoma de Barcelona, Barcelona, Spain*

ELIZABETH ANN JANNAMAN • *Department of Animal Sciences, University of Florida, Gainesville, FL, USA*

MICHAEL T. KANE • *Department of Physiology, National University of Ireland Galway, Galway, Ireland*
REBECCA L. KRISHER • *Colorado Center for Reproductive Medicine, Lone Tree, CO, USA*
LISA J. MACLELLAN • *Equine Reproduction Laboratory, Department of Biomedical Sciences, Colorado State University, Fort Collins, CO, USA*
PASCAL MERMILLOD • *Institut National de Recherche Agronomique (INRA), UMR7247, Physiologie de la Reproduction et des Comportements, Nouzilly, France*
A. MISHRA • *ICAR-National Institute of Animal Nutrition and Physiology, Bangalore, Karnataka, India*
TOMOMI MITO • *Research Institute for the Functional Peptides, Higashine, Yamagata, Japan*
S. MONDAL • *ICAR-National Institute of Animal Nutrition and Physiology, Bangalore, Karnataka, India; Animal Physiology Division, ICAR-National Institute of Animal Nutrition and Physiology, Bangalore, Karnataka, India*
A. MOR • *ICAR-National Institute of Animal Nutrition and Physiology, Bangalore, Karnataka, India*
BRUCE D. MURPHY • *Centre de Recherché en Reproduction et Fertilité, Faculté de Médicine Vé térinaire, Université de Montréal, St-Hyacinthe, QC, Canada*
J. B. NAGASHIMA • *Smithsonian Conservation Biology Institute, Front Royal, VA, USA*
S. NANDI • *ICAR-National Institute of Animal Nutrition and Physiology, Bangalore, Karnataka, India*
MARTHA SOFIA ORTEGA OBANDO • *Division of Animal Sciences, University of Missouri, Columbia, MO, USA*
MARÍA-TERESA PARAMIO • *Ciència Animal i dels Aliments, Universitat Autònoma de Barcelona, Barcelona, Spain*
RANDALL S. PRATHER • *Division of Animal Sciences, University of Missouri, Columbia, MO, USA; National Swine Resource and Research Center, University of Missouri, Columbia, MO, USA*
CATHY RAMSEY • *Oregon National Primate Research Center, Oregon Health and Science University, Beaverton, OR, USA*
I. J. REDDY • *ICAR-National Institute of Animal Nutrition and Physiology, Bangalore, Karnataka, India*
BETHANY K. REDEL • *Division of Animal Sciences, University of Missouri, Columbia, MO, USA; National Swine Resource and Research Center, University of Missouri, Columbia, MO, USA*
ROCÍO MELISSA RIVERA • *Division of Animal Sciences, University of Missouri, Columbia, MO, USA*
MARCELLO RUBESSA • *Department of Animal Sciences, University of Illinois, Urbana, IL, USA*
ISLAM M. SAADELDIN • *Department of Animal Production, College of Food and Agricultural Sciences, King Saud University, Riyadh, Kingdom of Saudi Arabia; Department of Physiology, Faculty of Veterinary Medicine, Zagazig University, Zagazig, Egypt*
TERRY SCHLENKER • *Colorado Center for Reproductive Medicine, Lone Tree, CO, USA*
JENNA KROPP SCHMIDT • *Wisconsin National Primate Research Center, University of Wisconsin-Madison, Madison, WI, USA*
LYNNE SELWOOD • *School of BioSciences, The University of Melbourne, Parkville, VIC, Australia*

Polani B. Seshagiri • *Department of Molecular Reproduction, Development and Genetics, Indian Institute of Science, Bangalore, India*

N. Songsasen • *Smithsonian Conservation Biology Institute, Front Royal, VA, USA*

Joanna Maria G. Souza-Fabjan • *Faculdade de Veterinária, Setor de Reprodução Animal, Universidade Federal Fluminense, Niteroi, RJ, Brazil*

Lee D. Spate • *Division of Animal Sciences, University of Missouri, Columbia, MO, USA; National Swine Resource and Research Center, University of Missouri, Columbia, MO, USA*

Jo Anne E. Stokes • *Equine Reproduction Laboratory, Department of Biomedical Sciences, Colorado State University, Fort Collins, CO, USA*

Ayman Abdel-Aziz Swelum • *Department of Animal Production, College of Food and Agricultural Sciences, King Saud University, Riyadh, Kingdom of Saudi Arabia; Department of Theriogenology, Faculty of Veterinary Medicine, Zagazig University, Zagazig, Egypt*

Ampika Thongphakdee • *Wildlife Reproductive Innovation Center, Research Department, Bureau of Conservation and Research, Zoological Park Organization under the Royal Patronage of H.M. the King, Bangkok, Thailand*

A. J. Travis • *Baker Institute for Animal Health, College of Veterinary Medicine, Ithaca, NY, USA; Atkinson Center for a Sustainable Future, Cornell University, Ithaca, NY, USA*

Paula Tríbulo • *Department of Animal Sciences, University of Florida, Gainesville, FL, USA*

Thi T. Truong • *School of BioSciences, University of Melbourne, Parkville, VIC, Australia*

Venkatappa Vani • *Department of Molecular Reproduction, Development and Genetics, Indian Institute of Science, Bangalore, India*

Chapter 1

Historical Note: Lessons Learned from Developing the First Semi-Defined Medium for Culture of One-Cell Rabbit Embryos to Expanding Blastocysts

Michael T. Kane

Abstract

In a review written in 1987 entitled "Pioneering mammalian embryo culture," John Biggers (The mammalian preimplantation embryo: regulation of growth and differentiation in vitro, Plenum Press, New York, 1987) summarized the history of early preimplantation embryo culture as follows "The culture of early mammalian embryos has been studied predominantly in two species: the rabbit and the mouse. Up to 1949, the rabbit was used almost exclusively, with media composed of ill-defined biological fluids. After this time, techniques for the culture of preimplantation mouse embryos rapidly developed using simple, defined media with relatively few components… In the 1960s, there was renewed interest in the culture of early rabbit embryos and a marked difference in the nutritional requirements of the two species was found." This chapter describes the history of the development of the semi-defined medium that stimulated this "renewed interest" referred to above and includes some of the initial difficulties in getting the work published. The phrase "semi-defined medium" is understood here to mean a defined medium (all of whose components are chemically known) supplemented with an undefined component such as albumin.

Key words History, Human, Rabbit, Preimplantation, Embryo, Blastocyst, Culture, Medium, Amino acids, Inositol

1 Development of the Medium

I came to Dr. R.H. Foote's laboratory in the Department of Animal Science at Cornell University in September 1965 from University College Dublin where I had completed an MAgSc degree with Dr. Ian Gordon in 1965. I enrolled for a PhD in the field of Physiology with minors in Biochemistry and Physical Biology and commenced research on rabbit preimplantation embryo culture. The Foote lab was well suited to that work because it had a large Dutch Belted rabbit colony. Another advantage was that I shared an office with Ralph Maurer, who was completing a MSc on repeated superovulation and embryo transfer in rabbits and went on to do a PhD on aging and embryonic mortality also in rabbits. Ralph

Jason R. Herrick (ed.), *Comparative Embryo Culture: Methods and Protocols*, Methods in Molecular Biology, vol. 2006, https://doi.org/10.1007/978-1-4939-9566-0_1, © Springer Science+Business Media, LLC, part of Springer Nature 2019

taught me the skills involved in recovering fertilized embryos from superovulated rabbits.

When we set out to culture rabbit embryos we were very much influenced by Brinster's 1963 work and visited his lab in Philadelphia to see his embryo culture setup. In hindsight, we were overly focused on his mouse embryo culture medium [1] and paid little attention to general work on tissue culture with complex media, which, as we were to find out later, was very relevant. Unfortunately, two- to four-cell rabbit embryos cultured in Brinster's medium stopped developing at the stage of poor looking morulae. A very significant factor in the markedly differing culture requirements between preimplantation mouse and rabbit embryos is that development of the mouse embryo from one-cell to periimplantation blastocyst is not accompanied by growth in protein or dry matter content, whereas rabbit embryo protein content increases by about 300-fold from one-cell to periimplantation blastocyst (reviewed in [2]).

Initial attempts to modify Brinster's medium for rabbit embryos involved altering the concentrations of bovine serum albumin (BSA) in the medium. The original formulation [1] contained 0.1% BSA. Increasing the level of BSA up to 3%, without other modifications to the medium, seemed to improve the quality of the morulae obtained, and, with 1.5% BSA, one morula started to blastulate. I named this medium Brinster's medium with 1.5% BSA, as BSM I (Basal Synthetic Medium I).

Around this time, Dr. Hideo Onuma joined the laboratory from Japan to work mainly on cattle embryo culture. He explored the use of serum for rabbit embryos—partly because the literature was not clear as to whether one could get blastocyst development in serum starting from the early cleavage stages. Working with Ralph Maurer, he quickly showed that one could culture two- to four-cell rabbit embryos to blastocysts in 100% bovine serum [3]. I picked up on this work and started adding various percentages of serum to the BSM I medium without BSA. I also started separating serum into dialyzable and nondialyzable fractions and adding back both fractions to BSM I with BSA. Adding back the dialyzable (low Mr) fraction supported development to the blastocyst stage, a significant advance over the development achieved in BSM I with BSA alone.

About this time, Krishnan and Daniel published a paper in *Science* on a uterine protein, they named blastokinin [4]; they claimed that this protein when used as a supplement to the complex culture medium, Ham's F10 [5], promoted development of late rabbit morulae to expanding blastocysts. This paper made quite an impression in embryo culture circles at the time because (a) it was published in *Science*, (b) for the first time it appeared that a specific uterine protein was promoting embryonic development, and (c) it had a catchy and interesting name for the uterine protein which

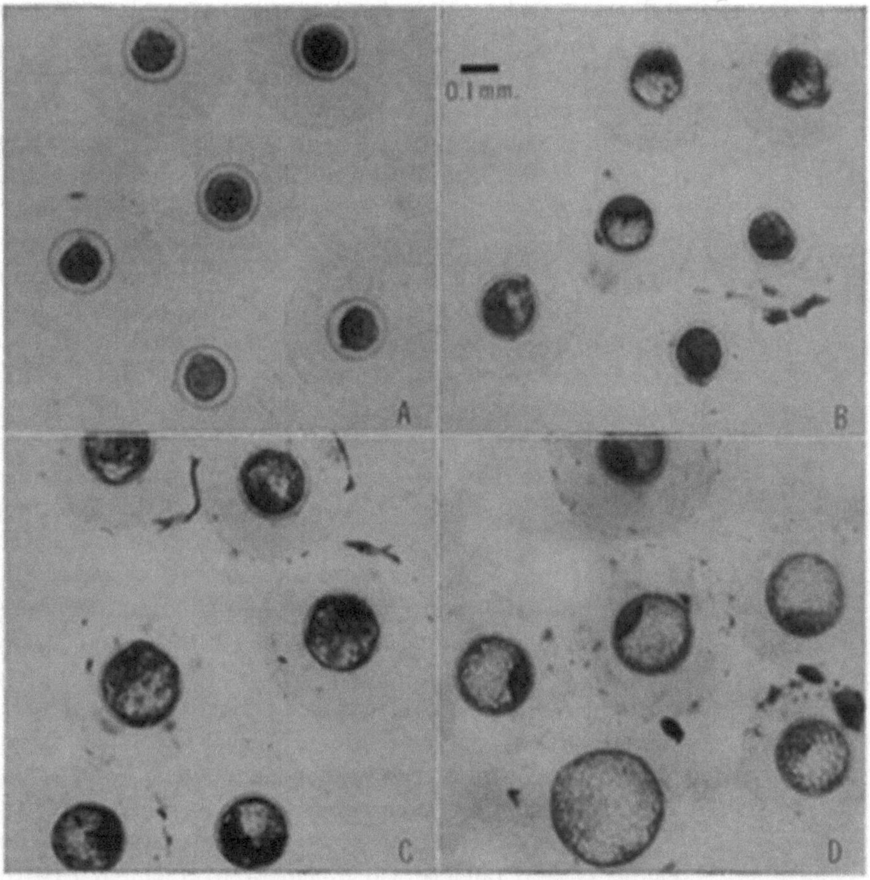

Fig. 1 Comparison of the effects of "blastokinin" and rabbit serum proteins on the development of rabbit morulae to blastocysts as compared with in vivo-grown blastocysts (as reported by Krishnan and Daniel in 1967 in *Science* 158:490–492 [4] and used with permission of *Science* here)

caught people's attention. However, if one carefully examined the data presented in the paper, there was little evidence that blastokinin was better than maternal serum proteins in stimulating rabbit blastocyst growth. When I took the paper to Bob Foote along with this comment, his immediate countercomment to me was to look at the figure on the last page and at the size of the blastocysts in the bottom, right panel of Fig. 2 of the paper (Fig. 1).

My response was that their Fig. 2 was very misleading especially to someone who just glanced quickly at the caption. The first part of the caption said, "Fig. 2. Photomicrograph of 3-day rabbit morulae cultured in vitro for 24 hours in Ham's F10." The eye of the casual reader might immediately go to the bottom right panel (D) and see the large expanded blastocyst in that panel and mistakenly think that the expansion is the effect of blastokinin. However, if one reads that part of the caption it says "normal 4-day blastocysts." These were NOT in vitro blastocysts grown under the influence of blastokinin but normal in vivo blastocysts grown in utero. The words "in vivo" should have been part of the caption of panel (D) to emphasize that fact. The proper comparison is between panel (B) "cultured in the presence of 0.3 mg of rabbit serum proteins per milliliter" and (C) "cultured in the presence of 0.2 mg of 'blastokinin' per milliliter." If one makes that proper comparison, there is clearly only a small difference between the effect of rabbit serum proteins and "blastokinin." The paper contains no numerical data on blastocyst diameters whatsoever to support the claims made for blastokinin. The publication of this paper in *Science* muddied the waters in relation to the function of this uterine protein, subsequently called uteroglobin [6] and it could be argued that its publication showed sloppy reviewing standards on the part of *Science* at the time. However, the paper itself and the use of F10 can be seen as a significant step on the way to BSM II, since it provided evidence that F10 supplemented by serum proteins could allow some degree of blastocyst development and growth from the late morula stage.

I immediately felt that if serum proteins could give this degree of blastocyst expansion in F10 starting with late morulae, why not try culturing two to four cell embryos in F10, the base medium from the Krishnan and Daniel paper, plus 1.5% BSA. The result was that a high proportion of these embryos developed to expanding blastocysts. The next step was to make a complex medium by taking BSM I (Brinster's 1963 medium plus 1.5% BSA) and then adding each of the four nutrient groups (amino acids, vitamins, trace elements, and nucleic acid precursors) present in F10 that were absent in BSM I. Development of embryos in this complex medium with all four nutrient groups from F10 present at the same concentrations as in F10 was then compared to development of embryos cultured in four new media that were prepared by individually omitting each nutrient group. The results of culturing two- to four-cell rabbit embryos in medium with 1.5% BSA were clear cut and very enlightening for rabbit embryo culture. Amino acids were essential for any degree of blastocyst formation, vitamins were necessary for significant blastocyst expansion, the nucleic acid precursor group appeared to inhibit blastocyst expansion, and there was no effect of the trace element group possibly due to the presence of these elements as contaminants of the major salt components, such as NaCl and $NaHCO_3$. BSM II was then formulated as the complex medium above minus the nucleic acid precursors and reported in my 1969 PhD thesis. It was rapidly shown that BSM II medium allowed culture of one-cell rabbit embryos to blastocysts [7].

2 Publication Difficulties

Bob Foote submitted these results to *Science*, probably in 1968, but the paper was rejected. As can be seen from his comment below on p41 of his account of his lifetime research (Foote, RH 1998 Artificial Insemination to Cloning: Tracing 50 Years of Research, https://ecommons.cornell.edu/handle/1813/3661) he was very unhappy with how this work was treated by the *Science* reviewers:

> "Kane undertook the task of determining the components in serum that promoted embryonic development (174, 176) and used this information to develop an excellent serum-free medium, but it contained BSA (172, 186). *The report by Kane and Foote (172) was the first to demonstrate that 2-cell rabbit embryos could be cultured to the blastocyst stage without adding serum or uterine fluids.* This paper was first sent to *Science*. Despite photographs and irrefutable statistical evidence, the paper was rejected because the reviewer, supported by the editor, said it was not possible to achieve this. This was a clear case of biased censorship. We resubmitted the paper elsewhere (172) after a cooling-off period, and it was accepted." (Note: Bob Foote's statement that the development of the medium was based on information from our work with serum is not correct; also I do not think that paper was illustrated by photographs).

I suspect that our paper may have been sent for review to J.C. Daniel, the senior author of the 1967 blastokinin paper in *Science*. However, I would be slow to agree that its rejection was necessarily due to biased reviewing. Our use of a high level of BSA, which even then was known (but not to us at the time) to be contaminated with variable amounts of substances (such as fatty acids, steroid hormones and citrate), may have given rise to serious questioning of our results. The variability of commercial batches of BSA showed up as a significant factor in a later work, particularly in relation to blastocyst expansion and cell proliferation [8].

The paper was finally published as "Kane, MT and RH Foote 1970 Culture of two-and four-cell rabbit embryos to the expanding blastocyst stage in synthetic media. Proceedings of the Society for Experimental Biology and Medicine **133**, 921–925" [9]. At the time, Foote remarked to me that a reviewer from this journal questioned the high level of BSA and that this was a further factor in the delay in publication. It was perhaps unfortunate that the paper was published in this journal rather than together with the three other (much less significant) papers from my PhD thesis, which were published in the newly established specialist journal, *Biology of Reproduction*.

The delay in publication, from when the paper was first written to when it appeared in press in 1970, was probably of the order of 2 years. In the meantime, a paper based on the medium was published in 1969 by Naglee et al. [10]. Naglee was a senior undergraduate at the time and did a project with Ralph Maurer. Ralph suggested that they should look at the effects of osmolarity on rabbit embryo culture using BSM II. I had no problem with that, but I took no part in the supervision of the project, except to emphasize that nucleic acid precursors should be omitted. Thus, the Naglee et al. paper was published in the year before the Proc. Soc. Exp. Biol. Med. paper but it cited the in press Proc. Soc. Exp. Biol. Med. paper for the composition of the medium.

There has been some confusion as to the name of the medium. While the names BSM I and BSM II were used in my 1969 PhD thesis, I never used either of those names in published papers and always referred to BSM II as a modified F10 medium with details of the modifications, primarily because I saw it as an evolving medium. However, the name BSM II has been used by other workers mainly in Germany in the laboratories of Henning Beier in Aachen and Berndt Fischer in Halle-Leipzig. In the work in my laboratory in Galway, this medium underwent various modifications with regard to energy substrates, vitamins, amino acids, and the role of BSA (reviewed in [2]). The medium that resulted from these modifications has been given the name RPECM—Rabbit Preimplantation Embryo Culture Medium [2].

3 Significance of this Work

The major significance of the development of the BSMII/F10 semi-defined medium was that it allowed for culture of embryos of a species, with a preimplantation growth pattern characterized by a marked increase in size and protein content, and culture requirements markedly different to the mouse embryo. The total absence of an early cleavage block in culture and the ability of the rabbit embryo to undergo three cleavage divisions without any exogenous energy source were also unique [11, 12].

The rabbit embryo also has an essential need for inositol for blastocyst growth, which is incorporated into the components of the phosphatidylinositol cycle [13]. This opened up an interesting area of investigation in preimplantation embryos, which ties in with the importance of inositol in postimplantation mouse embryo development and the significant numbers of neural-tube defects resulting from inositol deficiency [14].

Our discovery that citrate was the major contaminant of albumin responsible for the blastocyst promoting growth of some albumin samples [15] was not followed up by ourselves or others. In our case, this was partly because our embryo research was concentrating on the role of inositol but also partly because of a sense of disappointment that the active factor was not a more "interesting molecule," such as a novel peptide. This decision was unfortunate especially in view of recent information that citrate can have an important role in cell culture in serum-free media (reviewed in [2]).

One clear lesson that can be drawn from the manner in which the rabbit embryo medium was developed is the need to examine carefully the evidence underpinning what appear to be significant conclusions in a relevant paper. The volume of literature being published at the present time renders that advice more difficult to follow, but at the same time more necessary. Conclusions which authors may draw from their data may be totally invalid (as in the blastokinin paper), and this may be true even if the paper is published in a prestigious journal such as Science, Nature, or PNAS.

4 Benefits and Challenges of Working with Rabbit Embryos

Despite the renewed interest in the rabbit embryo in the 1960s that Biggers [16] referred to, the rabbit remains a little-studied, underutilized model species for studies of preimplantation embryo development. It is perhaps interesting to ask why this should be the case and also why it may be desirable to change this situation.

Probably the major reason the rabbit remains a little-studied, underutilized model species is the much higher cost per embryo and space requirements for rabbit housing associated with rabbit embryo culture as compared with mouse embryo culture. I became acutely aware of this issue when working with hamster embryos in Barry Bavister's lab in Wisconsin on a couple of summer visits in the 1980s. Because of the relatively low costs of housing hamsters, they were much more freely available than those of rabbits so that I could carry out culture experiments 3 days per week (as compared with at most 1 day per week with rabbit embryos in my lab in Galway), get 100–150 good embryos from four to five hamsters, and with each day's embryos representing one experimental block complete an experiment in a week and expect to achieve statistical significance. It should be mentioned here that anyone preparing to carry out rabbit embryo work should have an in-house rabbit colony probably using a small breed such as the Dutch Belted, or have a reliable supplier who can guarantee a supply of non-inbred females. Dutch Belted does give a good superovulatory response at 20 weeks of age. A factor which may compensate for the cost per embryo in some types of experiment is the amount of tissue per embryo in the two species: the mouse embryo across all preimplantation stages, including the blastocyst, has a protein content of about 20 ng whereas the rabbit embryo has a protein content for early, mid, late, and periimplantation blastocysts of about 440, 1550, 6850, and 50,000 ng [2]. Note that it is probable that if one is using superovulation to provide the embryos, blastocyst size and protein content may be smaller than the above figures and there may be some loss of embryos due to possible effects of crowding in the uterus [17].

Other factors that have tended to make the preimplantation rabbit embryo an under-studied model species include the following: (1) The seminal paper describing the essentiality of amino acids as a group for rabbit blastocyst formation, which was the first unambiguous statement of this fact for any mammalian species, was published in a somewhat obscure journal, popular enough at the time but later extinct, Proceedings of the Society for Experimental Biology; (2) interest in this paper was not helped by the fact that three other papers of much lesser significance to embryo culture, two of them studies of serum and serum extracts, were published by Kane and Foote at about the same time in Biology of Reproduction; (3) the use of two different names (BSM II and Modified F10) by two different labs using the medium and the failure to have a single distinctive name; (4) the complexity of the BSM II/Modified F10 medium (a medium not available commercially) and the resulting difficulty in preparing it discouraged its use. However, this medium has been used extensively by Professor Bernd Fischer's group in Halle-Leipzig to explore a variety of interesting reproductive biology questions (reviewed in [18]) and by my group at the University College, Galway (now National University of Ireland, Galway), which was then a small relatively unknown university and papers emanating from there tended to attract limited attention.

Although rarely used in this capacity, the rabbit embryo may well be a very useful model for studies of development to and growth of the blastocyst stage in human embryo culture, providing an interesting contrast to comparable studies with the murine embryo. The in vivo-grown mouse embryo at the blastocyst stage contains LESS protein and dry matter than it does at the one-cell stage and thus it is not surprising that one- and two-cell mouse embryos will develop in high proportions to blastocysts in simple media without amino acids, vitamins, and protein or growth factors. In contrast the in vivo-grown early rabbit blastocyst has about fourfold as much protein content and the periimplantation blastocyst about 300-fold as much protein content as the one-cell stage [2]. Thus it is not surprising that the rabbit embryo has an absolute requirement for a full range of amino acids in the culture medium for blastocyst formation. There was an earlier report stating that a group of ten amino acids were essential for blastocyst expansion; however, no data is provided to support the statement [19]. The rabbit is, to my knowledge, the only mammalian species known to have such an absolute requirement. There is also a need for water-soluble vitamins for even limited blastocyst expansion.

Much of the work on the effects of growth factors on embryos in culture has been carried out on mouse embryos. However, in view of the twin facts that mouse embryos even in vivo do not increase in protein content during the preimplantation stages and the one-cell mouse embryos in culture can readily grow to blastocysts in the absence of a wide range of nutrients, it would seem that the mouse model has serious limitations in studying the growth of the embryo at the blastocyst stage. Thus the rabbit embryo may be a better model. It is also possible that, since bovine blastocysts make considerable growth before implantation, such blastocysts produced from slaughterhouse material might also be a good model. Sheep and pig embryo culture may also have a role to play here.

A relevant question is this: "Do the complex culture requirements of the rabbit embryo make it a better model for culturing human embryos because these complex requirements may more closely resemble the needs of human embryos?" This depends on the answer to a further question: "Does the human embryo increase in cell number and undergo true growth, that is, increase in protein and dry matter content at or soon after blastocyst formation in vivo?" There are hardly any relevant data to answer that last question because of the ethical and other difficulties involved in obtaining such human data. However, very limited data from serial sections of early implanting human blastocysts [20] do not indicate a large increase in blastocyst cell numbers, at least nothing similar to the rabbit blastocyst, but some increase cannot be ruled out. It is unfortunate that no quantitative cell number data were obtained from those serial sections. It would be interesting, if those serial sections still survive to discover if such data could still be obtained from them.

Tentative indirect evidence that human blastocyst formation may involve true growth is that the early "simple" media used for murine embryos, which have been directly translated to human clinical use, have slowly become more and more complex, with many now resembling BSM II/Modified F10 as seen in amino acid analyses of common proprietary commercial media [21, 22]. I have the strong opinion that amino acid supplementation of culture media should consist of a full set of the 20 or so amino acids commonly found in proteins. While some amino acids have special functions (e.g., serine as a component of phosphatidylserine, glutamine as an effective energy substrate, and glycine as an osmolyte), the primary function of amino acids is as building blocks of proteins and here the distinction between essential and nonessential amino acids is somewhat dubious particularly in its generalization from its original use. To quote a review by Reeds [23] "from a functional perspective, all amino acids are essential." Had the human clinical world paid more attention to the rabbit literature, the use of a full range of amino acids might have occurred long before their introduction in the 1990s.

The analyses of commercial media by Morbeck et al. [21, 22] which were undertaken because the quantitative composition of human embryo culture media is not divulged by the manufacturers, unfortunately, did not include any data on water-soluble vitamin levels. The clinical world might still learn from the rabbit literature on vitamins showing that four water soluble vitamins, namely, myoinositol, pyridoxine, riboflavin, and niacinamide, were necessary for rabbit blastocyst expansion. The greatest effect was produced by inositol, with a marked stimulation of both cell proliferation and protein synthesis (reviewed in ref. [2]). Inositol was also found to be important in the culture of eight-cell hamster embryos to hatching blastocysts [24]. Another vitamin not one of the four, pantothenate, improved culture of one-cell hamster embryos to hatching blastocysts and also improved the yield of live fetuses recovered after transfer of cultured blastocysts [25]. In day 9 postimplantation rat embryos in culture, a group of vitamins, namely, pantothenate, riboflavin, inositol, folic acid, and niacinamide, was necessary for development, with inositol being one of the most important [26].

In view of inositol's ability in both rabbit and hamster to stimulate blastocyst growth and its known presence in oviduct and uterine fluid of the rabbit [27] and of the rat [28], inositol might well be a beneficial component of human embryo culture medium. As well as its role in the Ins(1,4,5)P_3 signaling system, it is a precursor of the phospholipid PtdIns which is a constituent of cell membranes and therefore necessary during embryonic cell proliferation even in the absence of true growth.

In a situation where due to ethical and other reasons, experiments on human embryos are constrained, it seems desirable that efforts should be made to collect human uterine fluid whenever it is ethically and practically possible to do so in order to obtain information on its composition, its amino acid and vitamin content, as well as the presence of peptide growth factors and other hormones. Information on the amino acid concentration for uterine fluid has recently been reported [29].

Because of the constraints on experiments on human embryo culture, the use of animal models is obviously essential. However, given the biological variation between species, it would seem blatantly obvious that there should not be a reliance on just one species, especially a species such as the mouse whose one-cell embryos develop in high proportions into viable blastocysts in simple media without amino acids, vitamins, protein, or growth factors! It would also seem obvious that a number of species should be used. I would suggest that in addition to the mouse two or three other species should be used—the rabbit, the hamster, and possibly as suggested earlier in vitro-produced embryos from the bovine or alternatively the porcine or ovine species. The case for adding a particular culture medium ingredient is strengthened when it has been shown to have a beneficial effect in more than one species.

Acknowledgments

Thanks are due to Professor H.J. Leese, Centre for Cardiovascular and Metabolic Research (CCMR), Hull York Medical School (HYMS), Kingston upon Hull, UK for very helpful criticism of the manuscript.

References

1. Brinster RL (1963) A method for in vitro cultivation of mouse ova from two-cell to blastocyst. Exp Cell Res 32:205–208
2. Kane MT (2018) Culture of preimplantation rabbit embryos. Methods Mol Biol in press 2006
3. Onuma H, Maurer RR, Foote RH (1968) In-vitro culture of rabbit ova from early cleavage stages to the blastocyst stage. J Reprod Fertil 16:491–493
4. Krishnan RS, Daniel JC Jr (1967) "Blastokinin": inducer and regulator of blastocyst development in the rabbit uterus. Science 158:490–492
5. Ham RG (1963) An improved nutrient solution for diploid Chinese hamster and human cell lines. Exp Cell Res 29:515–526
6. Beier HM (1968) Uteroglobin: a hormone-sensitive endometrial protein involved in blastocyst development. Biochim Biophys Acta 160:289–291
7. Kane MT, Foote RH (1971) Factors affecting blastocyst expansion of rabbit zygotes and young embryos in defined media. Biol Reprod 4:41–47
8. Kane MT (1983) Variability in different lots of commercial bovine serum albumin affects both cell multiplication and hatching of rabbit blastocysts in culture. J Reprod Fertil 69:555–558
9. Kane MT, Foote RH (1970) Culture of two-and four-cell rabbit embryos to the expanding blastocyst stage in synthetic media. Proc Soc Exp Biol Med 133:921–925
10. Naglee DL, Maurer RR, Foote RH (1969) Effect of osmolarity on in vitro development of rabbit embryos in a chemically defined medium. Exp Cell Res 58:331–333
11. Kane MT (1972) Energy substrates and culture of single-cell rabbit ova to blastocysts. Nature 238:468–469
12. Kane MT (1987) Minimal nutrient requirements for culture of one-cell rabbit embryos. Biol Reprod 37:775–778
13. Fahy MM, Kane MT (1993) Incorporation of [^{3}H]inositol into phosphoinositides and inositol phosphates by rabbit blastocysts. Mol Reprod Dev 34:391–395
14. Cockroft DL (1991) Vitamin deficiencies and neural-tube defects: human and animal studies. Hum Reprod 6:148–157
15. Gray CW, Morgan PM, Kane MT (1992) Purification of an embryotrophic factor from commercial bovine serum albumin and its identification as citrate. J Reprod Fertil 94:471–480
16. Biggers JD (1987) Pioneering mammalian embryo culture. In: Bavister BD (ed) The mammalian preimplantation embryo: regulation of growth and differentiation in vitro. Plenum Press, New York, pp 1–22
17. Adams CE (1962) Studies on prenatal mortality in the rabbit, *Oryctolagus cuniculus*: the effect of transferring varying numbers of eggs. J Endocrinol 24:471–490
18. Fischer B, Chavatte-Palmer P, Viebahn C, Navarrete Santos A, Duranthon V (2012) Rabbit as a reproductive model for human health. Reproduction 144:1–10. https://doi.org/10.1530/REP-12-0091
19. Daniel JC Jr, Krishnan RS (1967) Amino acid requirements for growth of the rabbit blastocyst in vitro. J Cell Physiol 70:155–160
20. Hertig AT, Rock J, Adams EC, Menkin MC (1959) Thirty-four fertilized human ova, good, bad and indifferent, recovered from 210 women of known fertility; a study of biologic wastage in early human pregnancy. Pediatrics 23:202–211
21. Morbeck DE, Krisher RL, Herrick JR, Baumann NA, Matern D, Moyer T (2014) Composition of commercial media used for human embryo culture. Fertil Steril 102:759–766.e759. https://doi.org/10.1016/j.fertnstert.2014.05.043
22. Morbeck DE, Baumann NA, Oglesbee D (2017) Composition of single-step media used for human embryo culture. Fertil Steril 107:1055–1060.e1051. https://doi.org/10.1016/j.fertnstert.2017.01.007

23. Reeds PJ (2000) Dispensable and indispensable amino acids for humans. J Nutr 130:1835S–1840S. https://doi.org/10.1093/jn/130.7.1835S
24. Kane MT, Bavister BD (1988) Vitamin requirements for development of eight-cell hamster embryos to hatching blastocysts in vitro. Biol Reprod 39:1137–1143
25. McKiernan SH, Bavister BD (2000) Culture of one-cell hamster embryos with water soluble vitamins: pantothenate stimulates blastocyst production. Hum Reprod 15:157–164. https://doi.org/10.1093/humrep/15.1.157
26. Cockroft DL (1988) Changes with gestational age in the nutritional requirements of postimplantation rat embryos in culture. Teratology 38:281–290. https://doi.org/10.1002/tera.1420380312
27. Gregoire AT, Gongsakdi D, Rakoff AE (1962) The presence of inositol in genital tract secretions of the female rabbit. Fertil Steril 13:432–435
28. Lewin LM, Yannai Y, Melmed S, Weiss M (1982) Myo-inositol in the reproductive tract of the female rat. Int J Biochem 14:147–150
29. Kermack AJ, Finn-Sell S, Cheong YC, Brook N, Eckert JJ, Macklon NS, Houghton FD (2015) Amino acid composition of human uterine fluid: association with age, lifestyle and gynaecological pathology. Hum Reprod (Oxford, England) 30:917–924. https://doi.org/10.1093/humrep/dev008

Chapter 2

Culture of the Mouse Preimplantation Embryo

David K. Gardner and Thi T. Truong

Abstract

The first culture media designed specifically to support development of the preimplantation mouse embryo were formulated over 50 years ago and were based on balanced salt solutions, containing the carbohydrates pyruvate, lactate, and glucose as the sole energy sources. Such media used a bicarbonate–carbon dioxide buffer system to maintain the desired pH, and were typically supplemented with serum albumin, but lacked free amino acids. In contrast to the complexity of a tissue culture medium, these original formulations of mouse embryo culture media were very simplistic. Over the intervening decades, as our understanding of the physiology and metabolism of the preimplantation embryo increased, together with a greater understanding of the environment within the female reproductive tract, culture media to support mouse embryo development in vitro have become more physiological and consequently more complex. A main addition to such media has been an array of amino acids. Although the media of today contain more components than their predecessors, their preparation remains relatively easy to accomplish, made feasible through the use of stock solutions, which also readily facilitates any changes to formulations to be made, an essential prerequisite for experimentation. As well as changes in media formulations, there have been exciting developments in incubator technology and design, such as the inclusion of time-lapse capability, redefining our ability to both culture and monitor embryo development in vitro.

Key words Amino acids, Blastocyst, Media, Metabolism, Physiology, Time-lapse

1 Introduction

John Hammond first showed it was possible to maintain the 8-cell mouse embryo in vitro for a few divisions, 70 years ago, using a simple salt solution, glucose, and egg white [1]. Wes Whitten went on to develop the first defined culture medium for the 8-cell embryo, consisting of Krebs–Ringer salts, glucose, and albumin [2]. These pioneering works were subsequently expanded extensively in the laboratories of John Biggers, David Whittingham, and Ralph Brinster [3, 4]. Further improvements to the formulations of embryo culture media were made possible through studies on the physiology and metabolism of the preimplantation embryo [5, 6], combined with a better understanding of the environment within the female reproductive tract [7, 8].

Jason R. Herrick (ed.), *Comparative Embryo Culture: Methods and Protocols*, Methods in Molecular Biology, vol. 2006, https://doi.org/10.1007/978-1-4939-9566-0_2, © Springer Science+Business Media, LLC, part of Springer Nature 2019

The preimplantation embryo is free floating within the lumen of the female reproductive tract and consequently is completely dependent upon the external milieu for the appropriate environment, nutrients, and signals. During this period of development, the embryo is responsive to its environment, and perturbations during this time have consequences for subsequent fetal development [9, 10]. Hence, great care must be taken to maintain a suitable and stable environment in vitro, including temperature and pH, both of which can dramatically affect embryo culture when outside of the physiological range.

The pronucleate oocyte and cleavage stage embryo are characterized by limited transcription and protein synthesis, and consequently energy demand and production are low. Up to the 8-cell stage, the embryo is dependent upon pyruvate, lactate, and aspartate for energy production, and it cannot utilize glucose as the sole energy substrate [3, 11]. Of note, it was held for many years that the mammalian zygote had an absolute requirement for pyruvate as an energy source to support the first cleavage division [3]. It has since been determined that the malate–aspartate shuttle is active in the mouse embryo, and hence lactate and aspartate can actually support the first cleavage division and embryo development in the complete absence of pyruvate [12]. However, it is not advocated that pyruvate be absent from an embryo culture medium, as it has an important role in regulating cytosolic redox and also serves as a powerful antioxidant.

As embryo development proceeds, there is a gradual activation of the embryonic genome [13], with the first transcripts being detected as early as the 2-cell stage, followed by a large activation of genes between the morula and blastocyst stages. As cell number increases exponentially, there are increased demands on biosynthetic pathways [14], and following compaction there is a significant increase in energy demand with the creation of the blastocoel, formed through the activity of Na^+/K^+ ATPases located on the basolateral membrane of the trophectoderm [15]. At this time, there is a large increase in demand for glucose, not only as an energy source but also for the synthesis of triacylglycerols and phospholipids, and as a precursor for complex sugars of mucopolysaccharides and glycoproteins [14]. Furthermore, glucose metabolism through the pentose phosphate pathway (PPP) is required to generate the ribose moieties required for nucleic acid synthesis and the NADPH required for the biosynthesis of lipids and other complex molecules. Rather than oxidizing all of the glucose consumed, the blastocyst converts around 50% to lactate in a process referred to as aerobic glycolysis, a trait blastocysts share with cancers [14]. Indeed, the lactate produced by the blastocyst is now considered to play a significant role during the various stages of implantation [16].

Before the 1990s, media for embryo culture contained carbohydrates but most lacked amino acids. Furthermore, only one formulation was typically used to support embryo development from the pronucleate oocyte to the blastocyst. Seminal studies in the laboratories of Bavister and Gardner on the role of amino acids in regulating embryo physiology and metabolism led to media formulations which included amino acids [17, 18]. A problem associated with conventional embryo culture in a static environment is the unintentional introduction of artifacts, an example of which is the production of ammonium from amino acids. Amino acids are labile at 37 °C and spontaneously release ammonium into the culture environment. For culture of somatic cells, it is well documented that ammonium production by amino acid breakdown results in compromised cell development [19]. Embryos, like somatic cells, have difficulty in dealing with an ammonium accumulation in the medium [20] and exhibit altered physiology, metabolism, gene expression, and compromised fetal development as a result of chronic exposure to ammonium [21, 22]. To add insult to injury, embryos themselves metabolize amino acids to produce ammonium. This issue has been addressed in three ways; renewing the medium every 48 h; the substitution of glutamine for a more stable dipeptide form such as alanyl- or glycyl-glutamine; and the titration of amino acids down to their lowest effective concentration.

Of great significance, within the lumen of the oviduct and uterus, carbohydrate and amino acids exist in gradients of concentration, which have been shown to affect embryonic development in vitro [23]. Hence the introduction of a biphasic or sequential media system based on the use of two media, in which embryos are moved from one medium to another after 48 h, was developed to mimic the maternal tract and to eliminate ammonium toxicity. Nutrient gradients are known to affect embryo metabolism in a stage-specific manner [24]. Metabolite concentrations in culture media have received recent reexamination, with the development of metaboloepigenetics, from which it is evident that metabolites and cofactors act as regulators of the epigenome [25, 26]; hence, nutrient concentration can affect the epigenetic status of the embryo [27, 28]. However, with the advent of time-lapse technologies, there has also been a resurgence of interest in the use of a single medium to support all stages of development, thereby minimizing the disturbance of the embryo in vitro, reducing both temperature and pH shifts. This situation reveals the paradox of life in vitro, whereby one must balance the sources of stress to which an embryo is exposed [29]. Consequently, when employing time-lapse systems we routinely choose to use sequential media, replenishing 80% the media on day 3 with the second phase medium.

As well as considering the culture media, it is essential to note at this point that success in the embryology laboratory is dependent upon several factors other than the formulations of the media per se. Hence, it is imperative that mammalian embryos are never exposed to any type of serum and their exposure to atmospheric oxygen (~20%) be kept to a minimum during collection and manipulation. All cultures should be performed at a reduced oxygen concentration (typically 5–7%) [23]. The addition of antioxidants can greatly increase embryo development and viability [30, 31]. Lights should always be dimmed and fluorescent lighting should be avoided in the embryology laboratory. Finally, it is paramount that quality control and assurance systems are in place to ensure all contact supplies, chemicals, and oils are prescreened for embryo toxicity before use [32]. Gardner and Lane have reviewed extensively embryo culture media formulations and their development [33].

2 Materials

2.1 Equipment and Supplies

1. Stereomicroscope with heated stage and magnification capacity of up to ×40.
2. Test tube warmers and heated stages (*see* **Note 1**).
3. Thermocouple to determine temperature within the drop of medium where the embryo is cultured.
4. Inverted microscope with magnification capacity of up to ×200.
5. Multigas incubator (to control CO_2 and O_2) with infrared (IR) sensor for CO_2, or modular chamber (in conjunction with a premixed gas cylinder, e.g. 6% CO_2, 5% O_2, 89% N_2), or a time-lapse incubation system.
6. Fyrite/Infrared CO_2 and O_2 analyzers to analyze the gas environment within the incubator.
7. Oil, paraffin (not mineral oil or blends).
8. Petri dishes (35 and 60 mm) or embryo-tested culture dishes.
9. Hyaluronidase (1000 IU/ml), preferably recombinant.
10. Borosilicate pipettes or glass capillaries, or gel loaders.
11. Pipetting system; mouth pipette, bulbs, syringes, etc.
12. Blunt 32-gauge needle, cut down to around 1 cm and 1 ml syringe.

2.2 Stock Solutions

When small amounts of media are required each week (e.g., 10–50 ml), or when one needs to alter the formulation for experimentation, it is practical to make media from stock solutions, as described in Table 1. All chemicals should be of the highest and/or tissue culture grade and prescreened using an appropriate bioassay [31] and tissue-culture grade water with a resistance of 18 MΩ

Table 1
Stock solutions used for preparation of embryo culture and handling media

	Mol weight	Weight (g)	Final conc. (mM)
Stock A			
Sodium chloride	58.44	5.2643	90.08
Potassium chloride	74.56	0.4108	5.5
Sodium phosphate (monobasic)	119.98	0.03	0.25
Magnesium sulfate heptahydrate	246.47	0.2465	1
Gentamycin		0.04	
×10, use for 3 months		in 100 ml	
Stock B			
Sodium bicarbonate	84.01	2.1003	25
Phenol red (*see* **Note 2**)		0.005	
×10, use for 1 month		in 100 ml	
Stock C			
Calcium chloride	147.02	0.2646	1.8
×100, use for 1 month		in 10 ml	
Stock D			
Glucose	180.16	0.0901	0.5
Sodium lactate (L-isomer only)	112.06	1.1766	10.5
Sodium pyruvate	110.04	0.0352	0.32
×10, use for 1 month		in 100 ml	
Stock E			
Glucose	180.16	0.5675	3.15
Sodium lactate (L-isomer only)	112.06	0.6578	5.87
Sodium pyruvate	110.04	0.011	0.1
×10, use for 1 month		in 100 ml	
Stock F			
EDTA disodium salt (Titriplex)	372.24	0.0037	0.01
×100, use for 1 month		in 10 ml	
Stock G—Cleavage stage amino acids			
Alanine	89.09	0.0089	0.1
Aspartate	133.1	0.0133	0.1
Asparagine-H_2O	150.14	0.015	0.1
Glutamate	169.1	0.0169	0.1
Glycine	75.07	0.00751	0.1
Proline	115.13	0.0115	0.1
Serine	105.09	0.0105	0.1
×100, use for 3 months (*see* **Note 3**)		in 10 ml	
Stock H			
Taurine	125.14	0.0125	0.1
×100, use for 3 months		in 10 ml	
Stock I			
Alanyl-glutamine	217.2	0.1086	0.5
×100, use for 3 months		in 10 ml	

(continued)

Table 1 (continued)

	Mol weight	Weight (g)	Final conc. (mM)
Stock J—Blastocyst amino acids			
Arginine–HCl	210.66	0.0632	0.3
Cystine	240.3	0.012	0.05
Histidine–HCl H_2O	191.62	0.0192	0.1
Isoleucine	131.17	0.0262	0.2
Leucine	131.17	0.0262	0.2
Lysine	146.19	0.0292	0.2
Methionine	149.21	0.0075	0.05
Phenylalanine	165.19	0.0165	0.1
Threonine	119.12	0.0238	0.2
Tryptophan	204.23	0.0051	0.025
Tyrosine	217.65	0.0218	0.1
Valine	117.15	0.0234	0.2
×100, use for 3 months (*see* **Note 4**)		in 10 ml	
Stock K—Vitamins			
Pantothenate	238.3	0.001	0.0042
Pyridoxal	203.62	0.001	0.0049
Riboflavin	367.37	0.0001	0.00027
Thiamine	337.27	0.001	0.00296
×100, use for 3 months		in 10 ml	
Stock L			
MOPS (*see* **Note 5**)	209.3	4.814	23
×10, use for 3 months		in 100 ml	

should be used. Stock solutions are prepared as ×10 or ×100 solutions by adding the amount of each component needed for 1 litre[1] of medium to 100 or 10 ml, respectively, to culture grade H_2O. All stock solutions should be filtered through a 0.2 μm filter immediately after preparation and stored at 4 °C in appropriate tissue culture ware. Always discard the first 2 ml as this contains potential harmful wash-off from the filter.

2.3 Embryo Culture and Handling Media

1. Water should first be added to the culture flask or tube using a sterile pipette.
2. Each component is then added using a displacement pipette (Table 2). Rinse each tip after the addition of each stock solution.
3. If more or fewer stocks are used than in Table 2, simply decrease or increase the amount of water added accordingly.
4. Upon preparation of the Cleavage medium, the unequilibrated pH should be 7.8 ± 0.05. Upon preparation of the Blastocyst medium, the unequilibrated pH should be 8.0 ± 0.05. A few μl of NaOH (2 M) might be required to attain this value. After equilibration with 6% CO_2, the pH of both Cleavage and Blastocyst media should be 7.30 ± 0.05.

Table 2
Preparation of example embryo culture media from stock solutions. Stocks can be modified and added according to the requirements of the culture or experiment

Stock	Cleavage stage medium (10 ml)	Blastocyst medium (10 ml)	Handling medium (10 ml)
H_2O	6.5	6.3	6.42
A	1.0	1.0	1.0
B	1.0	1.0	0.08
C	0.1	0.1	0.1
D	1.0	–	1.0
E	–	1.0	–
F	0.1	–	0.1
G	0.1	0.1	0.1
H	0.1	0.1	0.1
I	0.1	0.2	0.1
J	–	0.1	–
K	–	0.1	–
L	–	–	1.0

5. The osmolality of Cleavage and Handling media is 273 ± 5, while that of the Blastocyst medium is 270 ± 5.
6. Media are supplemented with serum albumin and hyaluronan. Albumin can be added as bovine serum albumin, human serum albumin, or recombinant human albumin. If the albumin preparation to be added is a solution, then the amount of water should be shorted to compensate for the amount of albumin solution. Albumin is typically added between 4 and 10 mg/ml. Hyaluronan is typically added as a solid and can be acquired in two forms; extracts from rooster comb, or as a fermented product from bacteria. The latter is endotoxin-free and should be considered the preferred source and is typically added between 0.125 and 0.5 mg/ml (*see* **Note 6**). Hyaluronan should be made as a stock solution and the amount of water in the medium shorted to account for volume of hyaluronan stock added.
7. Immediately after they are prepared, media should be filtered through a 0.2 μm filter and stored at 4 °C for 2-- 4 weeks. Discard the first 2 ml of medium that passes through the filter.

3 Methods

3.1 Superovulation of Mice

To increase the number of embryos available for an experiment and to reduce the number of animals that are required, it is common to induce multiple ovulations in mice by the administration of hormones (Table 3). Superovulation protocols also assist in the synchronization of embryo timing between females. Best results from superovulation protocols are achieved by using prepubertal females (3–5 weeks old). Two hormones are administered via the intraperitoneum. The first injection of pregnant mare's gonadotropin (PMSG) provides FSH that recruits follicles for ovulation. This is followed 48 h later by an injection of human chorionic gonadotropin (hCG), which mimics the LH surge and results in ovulation. The injection interval should be kept between 46 and 48 h. It is essential that the hCG injection is given prior to the release of the endogenous LH surge (especially important for animals >24 days old) which occurs around 15–20 h after the middle of the second dark cycle. Following hCG administration females are placed with males and allowed to remain overnight. Mating can be detected by the presence of a vaginal plug the following morning. Fertilized embryos can then be collected at discrete stages of development (see below).

Response to superovulation procedures will vary significantly between strains of mice. A dose of 2.5–5 IU per mouse at 3–4 weeks old is typical. As response to hormones vary, it is important to start at a lower dose and then gradually increase to 5 IU while assessing the number of embryos collected and any

Table 3
Timetable for hormone injections and embryo collections

PMSG injection	hCG injection	Zygote 9–10 am (day of plug)	2 Cell 9–10 am Day 2	8 Cell 6–8 am Day 3	Morula 3–4 pm Day 3	Blastocyst 8–9 am Day 4
Monday	Wednesday	Thursday	Friday	Saturday	Saturday	Sunday
Tuesday	Thursday	Friday	Saturday	Sunday	Sunday	Monday
Wednesday	Friday	Saturday	Sunday	Monday	Monday	Tuesday
Thursday	Saturday	Sunday	Monday	Tuesday	Tuesday	Wednesday
Friday	Sunday	Monday	Tuesday	Wednesday	Wednesday	Thursday
Saturday	Monday	Tuesday	Wednesday	Thursday	Thursday	Friday
Sunday	Tuesday	Wednesday	Thursday	Friday	Friday	Saturday

The light cycle in the mouse room is 12 h of daylight and 12 h of dark. Daylight is from 6:00 to 18:00. The room temperature is 21 °C ± 2 °C
Hormone injections are administered between 12 and 1 pm

fragmentation. The optimum dose will give the best-quality embryos, not necessarily the highest number of embryos.

Hormone solutions can be prepared by reconstituting lyophilized powder in sterile saline at a concentration of 50 IU/ml. This results in an injection of 0.1 ml per mouse to deliver 5 IU. Hormones can be stored for 2–3 weeks at −20 °C or for 2–3 months at −80 °C. There are significant differences in the activity of different batches of hormones; therefore, each new lot should be prescreened before general use by examining their effect on oocyte and embryo quality.

3.2 Dissection of Reproductive Tracts

1. Check for the presence of mating by visualization of a vaginal plug. Mated female mice can then be brought into the dissection area.
2. Instruments are washed, dried, and then sterilized by autoclaving.
3. A 35 mm Petri dish containing 2–3 ml of prewarmed handling medium with the lid on is placed on a calibrated warm stage adjacent to the dissection area (*see* **Note 1**). Do not leave the dish on the plate for extended periods of time to avoid evaporation.
4. The mouse is killed humanely by cervical dislocation and placed ventral side-up on absorbent paper.
5. Disinfect the abdomen with 80% ethanol.
6. Cut the peritoneum to expose the body cavity.
7. Move the coils of intestine to one side to expose the reproductive tract.
8. Hold the uterotubal junction with watchmaker forceps and strip the uterus of surrounding connective tissue.
9. For collection of embryos from the oviduct, make a cut between the oviduct and the ovary (avoid cutting the ovary). For collection of zygotes and 2-cell embryos, make a second cut below the uterotubal junction to isolate the oviduct from the uterus. For collection of 8-cell embryos make the second cut around one-third of the way down the uterine horn. Place the dissected tissue into the collection dish of handling medium.
10. For collection of blastocysts from the uterus, make a cut below the uterotubal junction of both uterine horns to isolate the oviduct from the uterus. Make a second cut through the cervix and place the connected uterine horns into the collection dish of handling media.

3.3 Collection of Zygotes

Before collection of the oviducts, warm 10 ml of handling medium and a 1 ml aliquot of hyaluronidase to 37 °C (*see* **Note 7**). Prepare pulled Pasteur pipettes with an internal diameter of 100–300 μm for manipulation of the embryos. A pipette that is just larger than the embryo is essential to ensure adequate washing and minimal transfer of medium between drops.

1. Place approximately 1 ml of handling medium into a new collection dish, either a 35 mm petri dish or organ well dish.
2. Transfer oviducts to the new collection dish, one at a time.
3. Stabilize the oviduct with fine forceps and locate the swollen ampulla.
4. Tear the ampulla of the oviduct to release the cumulus mass. Discard the oviduct and transfer a new oviduct to the collection dish. Repeat procedures until the cumulus masses from all oviducts are released. It is strongly recommended that a maximum of three mice be dissected at a time to ensure rapid collection and processing of the embryos.
5. Add 500 μl of warmed hyaluronidase solution to the dish to remove cumulus cells. Expose for 20–60 s.
6. During the exposure of the cumulus complex to hyaluronidase, set up three washing drops (~50 μl) of handling medium in a second collection dish (alternatively the lid of a collection dish can be used).
7. Collect the embryos using the pipette as soon as the cumulus cells begin to loosen (*see* **Note** 7).
8. Wash denuded embryos through three drops of handling medium.
9. Embryos are then ready to be placed into culture.

3.4 Collection of 2-Cell Embryos

1. Collect oviducts as above.
2. Attach a blunt 32-gauge needle to a 1 ml syringe loaded with prewarmed Handling medium.
3. Place approximately 1 ml of handling medium into a new collection dish, either a 35 mm petri dish or organ well dish.
4. Transfer oviducts to the new collection dish, one at a time.
5. Stabilize the oviduct with fine forceps and locate the infundibulum (opening of the oviduct by the ovary).
6. Insert the blunt 32-gauge needle and hold in position with fine forceps.
7. Gently flush the oviduct with ~0.25 ml of Handling medium. It is strongly recommended that a maximum of three mice be dissected at a time to ensure rapid collection and processing of the embryos.
8. Set up three washing drops (~50 μl) of Handling medium in a second collection dish (alternatively the lid of a collection dish can be used).

9. Wash 2-cell embryos through three drops of handling medium.
10. Embryos are then ready to be placed into culture.

3.5 Collection of 8-Cell Embryos

1. Collect oviducts with the top third of the uterus attached as above.
2. Attach a blunt 32-gauge needle to a 1 ml syringe loaded with prewarmed Handling medium.
3. Place approximately 1 ml of handling medium into a new collection dish, either a 35 mm petri dish or organ well dish.
4. Transfer oviducts connected to the top third of the uterus to the new collection dish, one at a time.
5. Stabilize the oviduct and uterus with fine forceps and locate the infundibulum.
6. Insert the blunt 32-gauge needle and hold in position with fine forceps.
7. Gently flush the oviduct with ~0.5 ml of Handling medium. It is strongly recommended that a maximum of three mice be dissected at a time to ensure rapid collection and processing of the embryos.
8. Set up three washing drops (~50 μl) of Handling medium in a second collection dish (alternatively the lid of a collection dish can be used).
9. Wash 8-cell embryos through three drops of handling medium.
10. Embryos are then ready to be placed into culture.

3.6 Collection of Blastocysts

1. Collect uteri as above.
2. Attach a blunt 32-gauge needle to a 1 ml syringe loaded with prewarmed Handling medium.
3. Place approximately 1 ml of handling medium into a new collection dish, either a 35 mm petri dish or organ well dish.
4. Transfer the uteri from one mouse to the new collection dish.
5. Stabilize the cervix with fine forceps.
6. Insert the blunt 32-gauge needle into one uterine horn through the cervix and hold in position with fine forceps.
7. Gently flush the uterus with ~0.5–1.0 ml of Handling medium. It is strongly recommended that a maximum of three mice be dissected at a time to ensure rapid collection and processing of the embryos.
8. Repeat **steps 6** and 7 for the contralateral uterine horn.
9. Set up three washing drops (~50 μl) of Handling medium in a second collection dish (alternatively the lid of a collection dish can be used).

10. Wash blastocysts through three drops of handling medium.
11. Embryos are then ready to be placed into culture.

Embryos are very sensitive to their environment. Therefore, it is important to minimize the time between extraction of the cumulus masses, cleavage-stage embryos, or blastocysts and their placement in culture. It is advisable to keep this time period to a maximum of 10 min. Therefore, it is preferable to do multiple smaller embryo collections rather than one large collection.

3.7 Embryo Culture in Microdrops

1. After 4 pm on the day before embryo culture, culture dishes for pronucleate oocytes should be prepared.
2. Rinse the pipette tip by taking up the required volume and then expelling into a discard dish. Rinsing the tip is a crucial step and should not be overlooked.
3. Using this prerinsed sterile tip, place the required number of 10 μl drops of culture medium into a petri dish. Immediately cover drops with 3.5 ml paraffin oil (7 ml for a 60 mm dish) to avoid evaporation. Always set up several wash drops in each dish. Optimally there should be one initial wash drop, followed by one wash drop for every culture drop. Prepare no more than 2–3 dishes at one time.
4. Once the initial 10 μl drops have been laid down, each drop is then increased to a final volume of 20 μl by adding an additional 10 μl of medium using a prerinsed tip. Use a new tip for each drop to avoid contaminating the culture media with oil.
5. Immediately place the dish in the incubator at 6% CO_2 and in reduced oxygen (5–7%). Gently remove the lid of the dish and set at an angle on the side of the plate. Dishes must equilibrate in the incubator with a semiopened lid for a minimum of 4 h (this is the minimal measured time for the media to reach correct pH under oil) and for a maximum of 18 h to prevent excessive ammonium production.

Day 1

1. Pronucleate oocytes are washed in handling medium and then further washed in the wash drops of cleavage medium. Washing entails picking up the embryos 2–3 times in a minimal volume and moving them around within the drop, in order to minimize medium carry over to the next drop. Embryos should be washed successively through the two wash drops in the culture dish before being transferred to the culture drops. The time that the culture dish is removed from the incubator should be less than 5 min.
2. Typically, ten embryos are cultured per 20 μl drop. If embryos are to be cultured individually, the droplet volume can be reduced to between 2 and 10 μl. Using volumes below 10 μl

is only safe if the covering paraffin oil has been screened using a 2 μl microdrop bioassay (as per ref. 31).

3. Dishes should be returned to the incubator and cultured for 48 h.

Day 2

1. After 4 pm on the day before embryo changeover into blastocyst medium, culture dishes should be prepared as described above.

Day 3

1. When embryos are changed over from Cleavage medium to Blastocyst medium, it is essential that embryos be thoroughly washed from one medium into another.
2. Embryos are removed from the drops of Cleavage medium and washed twice in wash drops of Blastocyst medium. Embryos should be washed successively in the two wash drops in the culture dish and then transferred to the new culture drop. Remember, the time that the culture dish is removed from the incubator needs to be less than 5 min.
3. Dishes should be returned to the incubator and cultured for a further 48 h.

3.8 Embryo Culture in a Time-Lapse System

There are numerous commercially available time-lapse systems available, each facilitating the imaging of embryos every 10–15 min. Further, such incubators provide a stable environment and are not as prone to variation in pH and temperature as a conventional box-type incubator. We employ sequential media for time-lapse culture, by removing 80% of the Cleavage stage medium and replacing with preequilibrated Blastocyst medium to the original final volume.

3.9 Embryo Scoring

Development of embryos should be monitored at a specific time on each successive day of development [34]. It is important to be consistent with the times of evaluation to increase the accuracy of assessing the embryonic stage reached under different culture conditions. For example, a blastocyst formed on the morning of day 4 is different to one that develops by 4 pm on the same day. Photomicrographs of culture mouse embryos throughout the preimplantation period are illustrated in Fig. 1.

3.10 Maintenance of Equipment

The maintenance of equipment is a key aspect to providing an environment that ensures consistent, high rates of embryo development. As discussed, all heated stages should be calibrated to ensure that the temperature within the culture drop is 37 °C. This will often mean (dependent on dish type) that the heated stages will be set 2–3 °C higher than the desired final temperature in the drops.

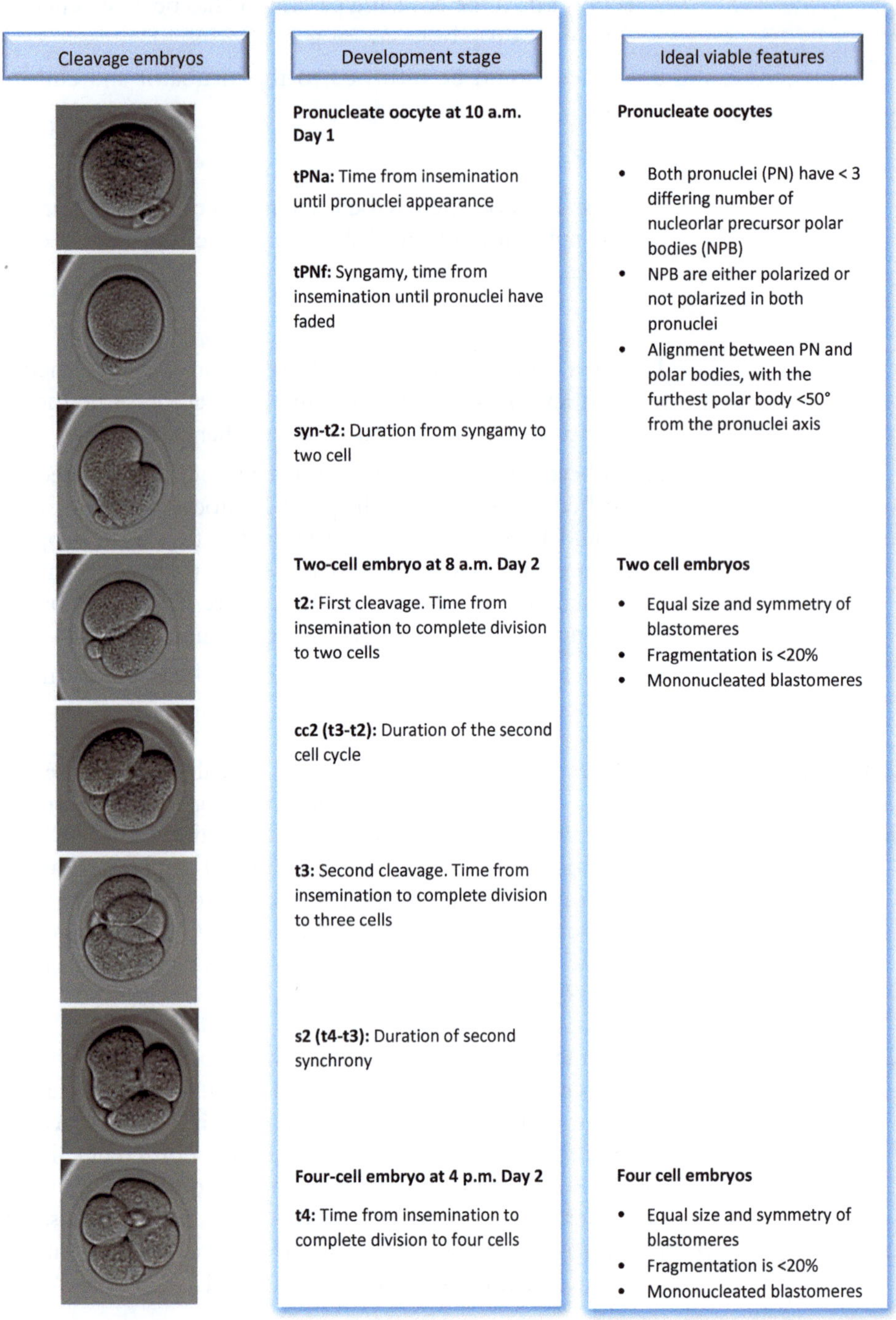

Fig. 1 Development and features of the mouse preimplantation embryo in culture. tPNa, time of pronuclei appearance; tPNf, the time of pronuclei fading; t2, the time from syngamy to cleavage to 2 cell; t3, t4, t5, t6, t7, t8, the time to cleavage to 3 cell, 4 cell, 5 cell, 6 cell, 7 cell, and 8 cell, respectively; tM, time of morula formation; tSB, time of initial appearance of blastocoel cavity; tB, time of blastocyst formation (i.e. blastocoel cavity is less than half the volume of the embryo); tEB, time to full blastocyst expansion; tHB, time of initial hatching event; tFHB, time to fully hatched. syn-t2 = duration from syngamy to two cell; cc2 = the duration of the second cell cycle; s2 = duration of second synchrony; cc3 = the duration of the third cell cycle; s3 = duration of third synchrony

Cleavage embryos	Development stage	Ideal viable features
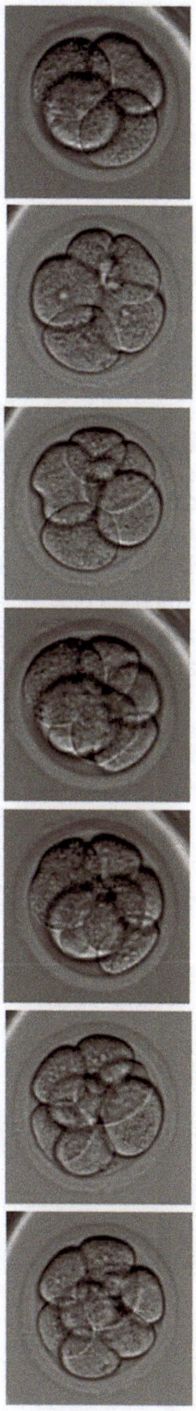	**cc3 (t5-t4):** Duration of third cell cycle	
	t5: Third cleavage. Time from insemination to complete division to five cells	**Five cell embryos** • Equal size and symmetry of blastomeres • Fragmentation is <20% • Mononucleated blastomeres
	t6: Time from insemination to complete division to six cells	
	t7: Fourth cleavage. Time from insemination to complete division to seven cells	
	Eight-cell embryo at 8 a.m. Day 3 **t8:** Time from insemination to complete division to eight cells	**Eight cell embryos** • Equal size and symmetry of blastomeres • Fragmentation is <20% • Mononucleated blastomeres

Fig. 1 (continued)

Compacting & cavitating embryos	Development stage	Ideal viable features
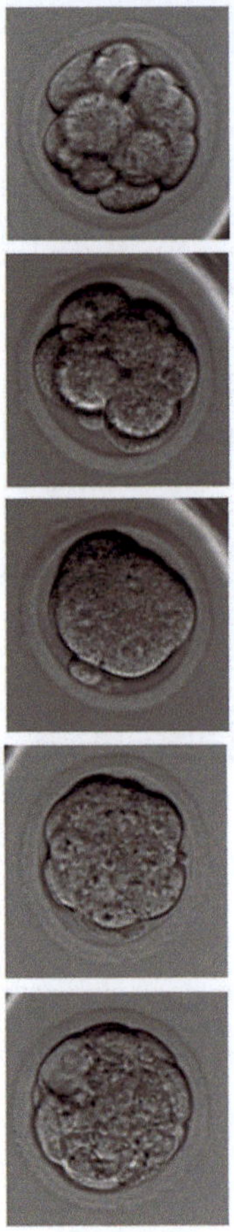	**s3 (t8-t5):** Duration of third synchrony	
	Beginnings of compaction at 8 a.m. Day 3 **t9:** Time from insemination to complete division to nine or more cells. Beginning of junction formation	
	Morula at 4 p.m. Day 3 **tM:** Time from insemination to formation of morula	**Morula embryos** • Merging of individual cells, with visible increase space between the embryo and zona pellucida and slight reduction in embryo volume • Lack of fragmentation
	Early blastocyst at 8 a.m. Day 4 **tSB:** Time from insemination to start of the appearance of the blastocoel	

Fig. 1 (continued)

Blastocyst embryos	Development stage	Ideal viable features
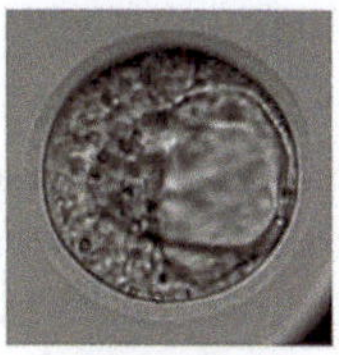	**tB:** Time from insemination to formation of blastocyst	
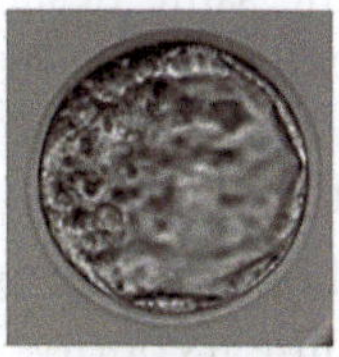	**Expanded blastocyst at 4 p.m. Day 4** **tEB:** Time from insemination to formation of expanded blastocyst. Blastocoel takes up half or more of the volume of the embryo	**Expanded blastocyst** • Blastocoel cavity completely fills the entire volume of the embryo. • Inner cell mass (ICM) is well formed with many cells tightly packed • Trophectoderm (TE) cells are numerous and form a cohesive epithelium • Zona pellucida thinning is apparent
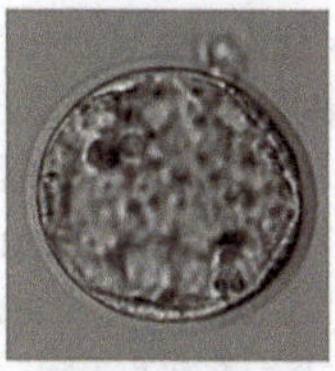	**Initiation of hatching at 4 p.m. Day 4** **tHB:** Time from insemination to start of hatching blastocyst.	
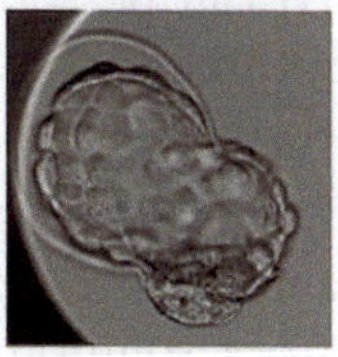	**Hatching blastocyst at 10 a.m. Day 5** Half of the blastocyst has emerged through the zona pellucida	
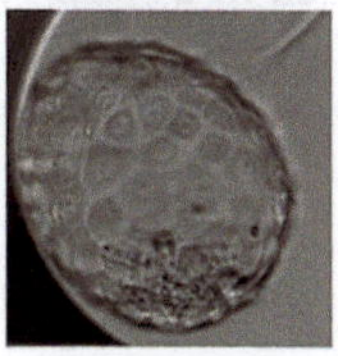	**Fully hatched blastocyst at 4 p.m. Day 5** **tFHB:** Time from insemination to fully hatched blastocyst	

Fig. 1 (continued)

It is critical to maintain the incubator environment because the embryo will spend much of its time in this environment. Incubator carbon dioxide and oxygen levels, as well as temperature, should be checked at least once a week. Levels should be checked using calibrated equipment, such as electronic meters with infrared (IR) sensors with a variance of only ±0.2%. It is essential that the first step to any equipment reading is to calibrate (or check) against a known carbon dioxide concentration (e.g., a certified gas cylinder) before taking the incubator reading. Similarly, temperature should be checked against a certified thermometer. All readings are best performed before an incubator is opened for the first time in a day, that is, taken from an undisturbed environment.

4 Notes

1. The temperature of all microscopes and heating stages needs to be calibrated to ensure that the temperature within the culture medium at the site of the embryos is actually 37 °C. This can be achieved using a thermocouple immersed on the medium on the dish. Note that not all dishes have the same design; for example, some have an air space on the bottom made by a lip, while others sit flush on the stage warmer. Hence the temperature of the stage needs to be calibrated for the type of dish used.
2. Phenol red is a useful pH indicator, although some people prefer to exclude it due to concerns regarding its purity. However, in general it is considered a practical component of media. To calibrate the color of the medium to its pH, Sorenson's phosphate buffer standards are prepared according to Table 4. Measure the pH with a pH meter, and adjust as required. Add solution A to lower the pH (make more acidic), add solution B to increase the pH (make more alkali). The pH standards should be filter-sterilized and can then be kept for up to 6 months. Please note that less phenol red can be added to the culture media if desired. In this case simply reduce the levels of phenol red in the Sorenson's buffers accordingly.
3. Stock G can be purchased in the form of Eagle's nonessential amino acids (MEM). Alternatively, the amino acids can be weighed out individually and made as a stock solution. Due to the small weights, making the ×100 stock in a total volume of 100 ml can be useful (i.e., 10× the weights).

Table 4
Preparation of color standards for pH of media

pH at 18 °C	Solution A[a] (ml)	Solution B[b] (ml)
6.6	62.7	37.3
6.8	50.8	49.2
7.0	39.2	60.8
7.2	28.5	71.5
7.4	19.6	80.4
7.6	13.2	86.8

[a]Solution A: 9.08 g KH_2PO_4 (0.067 M), 5 mg phenol red in 1 l of water
[b]Solution B: 9.46 g Na_2HPO_4 (0.067 M), 5 mg phenol red in 1 l of water

4. Stock J can be purchased in the form of Eagle's Essential amino acids (MEM). Alternatively, they can be weighed out individually and made as a stock solution. Due to the small weights, making the ×100 stock in a total volume of 100 ml can be useful (i.e., 10× the weights). In order to get the amino acids into solution, HCl (1 M) is typically required; for example, between 0.5 and 1.0 ml is titrated for 100 ml of amino acid stock.
5. Stock L is adjusted to pH of 7.3 ± 0.5 by using a concentrated stock solution of NaOH (2-5 M). Although the protocol described uses MOPS, an alternative buffer is HEPES. If HEPES is chosen over MOPS, then simply substitute based on molarity.
6. Hyaluronan is not currently considered a routine component of mammalian culture media. However, there is growing evidence that this macromolecule helps to maintain cellular ultrastructure and improve cryotolerance and viability (reviewed in ref. 23).
7. Exposure of embryos to hyaluronidase for too long, or at too high a concentration, is highly detrimental to the embryo. Therefore, ensure that the embryos are removed and washed as soon as the cumulus cells detach from the zona pellucida. This can be facilitated with gentle pipetting. Do not sit and watch, waiting for all the cumulus to simply fall off!

References

1. Hammond JM (1949) Recovery and culture of tubal mouse ova. Nature 163:28–29
2. Whitten WK (1956) Culture of tubal mouse ova. Nature 177:96–97
3. Biggers JD, Whittingham DG, Donahue RP (1967) The pattern of energy metabolism in the mouse oocyte and zygote. Proc Natl Acad Sci U S A 58:560–567
4. Brinster RL (1968) In vitro culture of mammalian embryos. J Anim Sci 27(Suppl 1):1–14
5. Leese HJ (1991) Metabolism of the preimplantation mammalian embryo. Oxf Rev Reprod Biol 13:35–72
6. Gardner DK (1998) Changes in requirements and utilization of nutrients during mammalian preimplantation embryo development and their significance in embryo culture. Theriogenology 49:83–102
7. Gardner DK, Lane M, Calderon I, Leeton J (1996) Environment of the preimplantation human embryo in vivo: metabolite analysis of oviduct and uterine fluids and metabolism of cumulus cells. Fertil Steril 65:349–353
8. Harris SE, Gopichandran N, Picton HM, Leese HJ, Orsi NM (2005) Nutrient concentrations in murine follicular fluid and the female reproductive tract. Theriogenology 64:992–1006
9. Lane M, Gardner DK (1994) Increase in postimplantation development of cultured mouse embryos by amino acids and induction of fetal retardation and exencephaly by ammonium ions. J Reprod Fertil 102:305–312
10. Gardner DK, Kelley RL (2017) Impact of the IVF laboratory environment on human preimplantation embryo phenotype. J Dev Orig Health Dis 8:418–435
11. Leese HJ (2015) History of oocyte and embryo metabolism. Reprod Fertil Dev 27:567–571
12. Lane M, Gardner DK (2005) Mitochondrial malate-aspartate shuttle regulates mouse embryo nutrient consumption. J Biol Chem 280:18361–18367
13. Hamatani T, Carter MG, Sharov AA, Ko MS (2004) Dynamics of global gene expression

changes during mouse preimplantation development. Dev Cell 6:117–131

14. Gardner DK, Harvey AJ (2015) Blastocyst metabolism. Reprod Fertil Dev 27 (4):638–654
15. Watson AJ, Natale DR, Barcroft LC (2004) Molecular regulation of blastocyst formation. Anim Reprod Sci 82–83:583–592
16. Gardner DK (2015) Lactate production by the mammalian blastocyst: manipulating the microenvironment for uterine implantation and invasion? BioEssays 37:364–371
17. Bavister BD, Arlotto T (1990) Influence of single amino acids on the development of hamster one-cell embryos in vitro. Mol Reprod Dev 25:45–51
18. Gardner DK, Lane M (1993) Amino acids and ammonium regulate mouse embryo development in culture. Biol Reprod 48:377–385
19. Schneider M, Marison IW, von Stockar U (1996) The importance of ammonia in mammalian cell culture. J Biotechnol 46:161–185
20. Wale PL, Gardner DK (2013) Oxygen affects the ability of mouse blastocysts to regulate ammonium. Biol Reprod 89:75
21. Gardner DK, Lane M (2005) Ex vivo early embryo development and effects on gene expression and imprinting. Reprod Fertil Dev 17:361
22. Lane M, Gardner DK (2005) Understanding cellular disruptions during early embryo development that perturb viability and fetal development. Reprod Fertil Dev 17:371–378
23. Gardner DK (2008) Dissection of culture media for embryos: the most important and less important components and characteristics. Reprod Fertil Dev 20:9–18
24. Lane M, Gardner DK (2000) Lactate regulates pyruvate uptake and metabolism in the preimplantation mouse embryo. Biol Reprod 62:16–22
25. Donohoe DR, Bultman SJ (2012) Metaboloepigenetics: interrelationships between energy metabolism and epigenetic control of gene expression. J Cell Physiol 227:3169–3177
26. Harvey AJ, Rathjen J, Gardner DK (2016) Metaboloepigenetic regulation of pluripotent stem cells. Stem Cells Int 2016:1816525
27. Tremblay R, Dufort I, Sirard MA (2017) Metabolic stress induces modifications in the epigenetic program of preimplantation bovine embryos. Mol Reprod Dev 85(2):117–127
28. Gaspar RC et al (2015) Oxygen tension affects histone remodeling of in vitro-produced embryos in a bovine model. Theriogenology 83:1408–1415
29. Wale PL, Gardner DK (2016) The effects of chemical and physical factors on mammalian embryo culture and their importance for the practice of assisted human reproduction. Hum Reprod Update 22:2–22
30. Truong TT, Soh YM, Gardner DK (2016) Antioxidants improve mouse preimplantation embryo development and viability. Hum Reprod 31:1445–1454
31. Truong T, Gardner DK (2017) Antioxidants improve IVF outcome and subsequent embryo development in the mouse. Hum Reprod 32:2404–2413
32. Gardner DK, Reed L, Linck D, Sheehan C, Lane M (2005) Quality control in human in vitro fertilization. Semin Reprod Med 23:319–324
33. Gardner DK, Lane M (2017) Embryo culture systems. In: Gardner DK, Simon C (eds) Handbook of in vitro fertilization, 4th edn. CRC Press, Boca Raton, pp 205–244
34. Gardner DK, Balaban B (2016) Assessment of human embryo development using morphological criteria in an era of time-lapse, algorithms and 'OMICS': is looking good still important? Mol Hum Reprod 22:704–718

Chapter 3

In Vitro Culture of Rat Preimplantation Embryos

Yuksel Agca

Abstract

The rat is one the most widely used laboratory animal species in many aspects of biomedical research, including the production of genetically engineered animal models to study human diseases and conditions. In addition to in vitro fertilization (IVF), the ability to grow IVF-derived or in vivo-collected zygotes to a desired preimplantation stage (zygote to blastocyst) entirely in vitro has a great importance for studies of developmental biology and genetic modification of laboratory rats for biomedical research. Although embryo biotechnologies are required to study or manipulate the genome effectively, such technologies for rat preimplantation embryos are not currently as successful as they are in the mouse. Here we provide a brief history of the development of rat in vitro culture systems and a step-by-step protocol to produce rat blastocyst stage embryos from zygotes under in vitro conditions from commonly used laboratory rat strains in biomedical research.

Key words Rat, In vitro preimplantation embryos, Culture, Osmolalities

1 Introduction

Preimplantation rat embryos, at a desired stage (from zygote to blastocyst), are often used in basic research, embryo development studies, transgenics, genome banking, and also during the establishment of embryonic stem cells [1–9]. Thus the ability to develop preimplantation rat embryos under in vitro conditions is crucial for biomedical research in terms of rapid advancement of embryo biotechnologies, such as transgenesis, cloning, and gene knockout technologies. While in vitro development of preimplantation embryos from 1-cell embryo to blastocyst has been very successful for most outbred/hybrid stocks and inbred mouse strains, to date, the success rates for rat embryos have been limited and vary among strains. Rat embryo culture systems such as R1ECM have been somewhat efficient in outbred Sprague-Dawley (SD) and Wistar rats, but they have not been widely validated for inbred strains including Lewis, Fisher 344, and Brown Norway. Similar to what has been previously observed for mouse embryos, one of the main reasons for this inefficiency was the failure of cultured rat zygotes to

Jason R. Herrick (ed.), *Comparative Embryo Culture: Methods and Protocols*, Methods in Molecular Biology, vol. 2006, https://doi.org/10.1007/978-1-4939-9566-0_3,

progress beyond 2- or 4-cells into later embryonic stages, such as morula or blastocyst [10, 11]. Initially the exact nature of this developmental block was not fully understood, but it has been largely attributed to detrimental effects of inorganic phosphate, glucose, and protein in the culture media [12, 13]. Chemically defined hamster embryo culture medium (HECM-1), which does not contain glucose or inorganic phosphate [14], was modified to develop rat embryos in vitro [15–17]. It has been reported that adjustment of inorganic phosphate concentration in the culture media based on the embryonic stage, significantly increased developmental potential for zygotes derived from outbred Wistar rats (80–90%), particularly for embryos collected from outbred stocks [18, 19]. Later, a two-phase chemically defined rat embryo culture system was developed that used potassium simplex optimized medium (KSOM), or human tubal fluid (HTF) for 18–22 h, followed by in vitro culture in mR1ECM medium [20]. This system allowed successful development of blastocyst stage embryos from both outbred (SD) and inbred (Wistar Furth, Lewis, F344, and PVG) strains of rat [20]. Recently, our group [21] and another report [22] showed that 2-cell arrested embryos had ultrastructural and molecular changes associated with oxidative stress and apoptosis and in vitro culture under low oxygen (5%) reduced the incidence of the 2-cell block in these rat embryos.

In addition to reduced (0.001–0.010 μM) inorganic phosphate, various other chemical and physical factors have been investigated to improve rat embryo culture systems. Although glucose neither inhibits nor promotes embryo development from the 1-cell to the morulae stage, the presence of 7.5–10 mM glucose had beneficial effects for progression of morula to blastocyst [17]. They concluded that glucose is an essential nutrient for development of preimplantation stage rat embryos. Both essential and nonessential amino acids have also been proven to be important for both in vitro embryo development and subsequent fetal development in the mouse [23, 24]. However, Miyoshi et al. [25] demonstrated that increase in osmolarity, altered with sorbitol, affects embryonic development in mHECM-1, with optimal blastocyst formation occurring when medium osmolality was 246 mOsm (63.8 mM NaCl and 30 mM sorbitol). However, higher osmolality (310 mOsm), achieved with increased concentrations of NaCl (110.0–130.0 mM), and the inclusion of bovine serum albumin (BSA) are necessary during IVF and during the early stages prior to pronuclear formation [26]. On the other hand, Han and Niwa [27] reported that while fetal bovine serum (FBS) has adverse effect on early cleavage-stage rat embryos, it enhances embryonic development from morula to blastocyst stage.

Most recently, Nakamura et al. [6] studied the efficacy of KSOM for in vitro development of rat embryos [23, 28, 29]. They investigated the effect of supplementing the amino acid derivative taurine, as well as glycine, glutamate, and alanine into KSOM (all at 1 mM) for rat embryo development, based on the findings suggesting that these compounds are abundant in mouse [30] and rat [6] oviductal fluid during the estrous cycle and the onset of pregnancy. Their findings suggested in the presence of three amino acids and taurine (1 mM) improved embryo development. They also determined that besides phosphate, phenol red had adverse effects on the development of rat zygotes and caused 2-cell block. Finally, embryonic development in medium containing amino acids and taurine was further improved, when the medium was changed daily, due to potential adverse effects of ammonium accumulation in the medium.

Here we describe a rat in vitro embryo culture protocol based on the above mentioned modifications to the mHECM-1 and R1ECM embryo culture systems, which is commonly used today and named as mR1ECM.

2 Materials

2.1 Equipment and Supplies

1. Physiologic saline solution.
2. Pregnant mare's serum gonadotropin (PMSG) dissolved in sterile saline solution and store at −80 °C freezer for up to 6 months.
3. Human chorionic gonadotropin (hCG) dissolved in sterile saline solution and store at −80 °C freezer for up to 6 months.
4. 1 mL syringes with attached 27-gauge needle.
5. Dissection tools (sharp scissors and micro dissection forceps).
6. Curved, serrated forceps.
7. Slide warmer.
8. Stereomicroscope.
9. 35 mm, sterile petri dishes, nontreated.
10. Filtered mineral or paraffin oil.
11. Hyaluronidase.
12. Heat pulled fine glass pipette with inner diameter of ~100 mM.

2.2 Rat Stock Solution

1. Fill an appropriate size Griffin Beaker with ~800 mL (~80% of the final volume) of embryo-tested, ultrapure culture-grade water (18.2 MΩ).
2. Add the components list in Table 1.
3. After all components are dissolved, transfer the contents to a 1 L volumetric flask and bring to desired volume by rinsing the Griffin Beaker with embryo-tested water and adding it to the

Table 1
The recipe of the stock solution used to prepare rat embryo culture media

Reagent	FW (g)	Final [mM]	g/L
NaCl	58.44	80	46.7520
KCl	74.55	3.2	2.3856
D-Glucose	180.2	7.5	13.5150
Penicillin G K Salt	372.2	100 μg/mL	0.7500
Streptomycin sulfate	728.69	50 μg/mL	0.5000
Sodium lactate[a] (60% syrup)	186.8	13.53	25.2740
$CaCl_2{\cdot}2H_2O$	147.02	2	2.9404
$MgCl_2{\cdot}6H_2O$	203.31	0.5	1.0166

[a]FW of 112.1 g/M. Given a 60% w:w solution = [112.1/(0.6)] = 186.8 g/mol

Table 2
The recipe of mR1ECM-HEPES (high NaCl): Rat 1-cell embryo holding medium

Reagent	FW (g)	Final [mM]	Amount per L
$NaHCO_3$	84.01	5	0.4200 g
HEPES	238.31	22	5.2428 g
Sodium pyruvate	110	0.5	0.0552 g
MEM NEAA 100×		0.5×	10 mL
MEM EAA 50×		0.5×	20 mL
GlutaMAX-1[a]		0.1	0.500 mL
Rat stock			100 mL
NaCl	58.44	80	1.7532 g

[a]In Vitrogen 35050-061

volumetric flask. Measure the osmolality, record and filter into an appropriately sized sterile bottle using a 0.22 μm filter.

4. Label the bottle with the date made and the expiration date.
5. Store at 4 °C for about 4 weeks ± 1 week.

2.3 mR1ECM-HEPES (High NaCl): Rat 1-Cell Embryo Holding Medium

1. Fill an appropriate size Griffin Beaker with ~500 mL (~50% of the final volume) of embryo-tested water.
2. Add the components in Table 2.
3. After all components are dissolved, check to ensure the pH of the solution is 7.4 ± 0.2 and adjust if needed. If any precipitates form in the solution, discard the solution and start over.

4. Transfer the contents to a 1 L volumetric flask and bring to desired volume by rinsing the Griffin beaker with embryo-tested water and adding it to the volumetric flask.
5. Check the osmolality of the solution and remake if it does not fall between 290 and 320 mOsm.
6. Sterilize the solution using a 0.22 μm filter unit.
7. Label the bottle with expiration date, pH, osmolality, initials, and batch number.
8. Store at 4 °C for up to 2 weeks.

2.3.1 Working Solution

1. Aliquot amount needed into a sterile tube using a sterile pipette (usually 50 mL).
2. Add 4 mg/mL fatty acid-free BSA. Place the tube in a 37 °C water bath to allow the BSA to dissolve.
3. Sterilize using 0.22 μm filter before use.

2.4 mR1ECM (High NaCl): Rat 1-Cell Embryo Culture Medium

1. Fill an appropriate sized Griffin Beaker with ~500 mL (~50% of the final volume) of embryo-tested water.
2. Add the following components in Table 3.
3. After all components have dissolved, transfer the contents to a 1 L volumetric flask and bring to desired volume by rinsing the Griffin beaker with embryo-tested water and adding it to the volumetric flask. If any precipitates form in the solution, discard the solution and start over.
4. Gently gas the solution with 5% CO_2, using a sterile Pasteur pipette for 15–30 min.
5. Check to ensure the pH of the solution is approximately 7.4 ± 0.2, and remake if the pH is outside this range.

Table 3
The recipe of mR1ECM (high NaCl): Rat 1-cell embryo culture medium

Reagent	FW (g)	Final [mM]	Amount per L
NaCl	58.44	80	1.7532 g
$NaHCO_3$	84.01	25	2.1003 g
Sodium pyruvate	110	0.5	0.0550 g
MEM NEAA 100×		0.5×	10 mL
MEM EAA 50×		0.5×	20 mL
GlutaMAX-1[a]		0.1	0.5 mL
Rat stock solution			100 mL

[a]In Vitrogen 35050-061

6. Check the osmolality of the solution and remake if the osmolality does not fall between 280 and 300 mOsm/kg.
7. Sterilize the solution using a 0.22 μm filter unit.
8. Label container with the current date, expiration date, pH, and osmolality.
9. Store at 4 °C for up to 2 weeks.

2.4.1 Working Solution

1. If media has been stored in the refrigerator, it must be gassed with 5% CO_2 for approximately 1 min prior to removing an aliquot.
2. Aliquot amount needed (usually 10 mL) using a sterile pipette and regas the bottle with 5% CO_2 before closing and additional storage.
3. Add 4 mg/mL fatty acid-free BSA.
4. Sterilize using a 0.2 μm syringe filter. It is important to discard the first few drops from the filter to eliminate the possible introduction of fibers from the syringe.
5. Make 30 μL drops in 35 mm culture dishes and cover with 3 mL embryo-tested mineral oil, which has preequilibrated in the incubator.
6. Immediately place dishes in an incubator preset to 5% CO_2, 5% O_2, and 37 °C and allow them to equilibrate overnight prior to use.

2.5 mR1ECM-HEPES (Low NaCl): Rat ≥ 2-Cell Embryo Holding Medium

1. Fill a Griffin beaker with ~500 mL (~50% of the final volume) of embryo-tested.
2. It is recommended to first dissolve PVA at 60 °C and then cool the media to room temperature for adding the rest of the components.
3. Add the components in Table 4.
4. After all components are completely dissolved, check the pH of the solution. The pH should be approximately 7.4 and should be adjusted if needed. If any precipitates form, discard the solution and start over.
5. Transfer the contents to a 1 L volumetric flask and bring to desired volume by rinsing the Griffin beaker with embryo-tested water and adding it to the volumetric flask.
6. Check the osmolality of the solution and remake if the osmolality does not fall between 240 and 255 mOsm/kg.
7. Sterilize the solution using a 0.22 μm filter unit.
8. Label container with date made, expiration date, pH, and osmolality.
9. Store at 4 °C for up to 2 weeks.

Table 4
The recipe of mR1ECM-HEPES (low NaCl): Rat ≥ 2-cell embryo holding medium

Reagent	FW (g)	Final [mM]	Amount per L
PVA		0.01%	0.100 g
$NaHCO_3$	84.01	5	0.4200 g
HEPES	238.31	22	5.2428 g
Sodium pyruvate	110	0.5	0.0552 g
MEM NEAA 100×		0.5×	10 mL
MEM EAA 50×		0.5×	200 mL
GlutaMAX-1[a]		0.1	0.500 mL
Rat stock solution			100 mL

[a]In Vitrogen 35050-061

Table 5
The recipe of mR1ECM (low NaCl): Rat ≥ 2-cell embryo culture medium

Reagent	FW (g)	Final [mM]	Amount per L
PVA		0.01%	0.10 g
$NaHCO_3$	84.01	25	3.004 g
Sodium pyruvate	110	0.5	0.055 g
MEM NEAA 100×		0.5×	10 mL
MEM EAA 50×		0.5×	20 mL
GlutaMAX-1[a]		0.1	0.5 mL
Rat stock solution			100 mL

[a]In Vitrogen 35050-061

2.5.1 Working Solution

1. Aliquot amount needed using a sterile pipette according to your needs.
2. Sterilize using a 0.22 μm syringe filter before use.

2.6 mR1ECM (Low NaCl): Rat ≥ 2-Cell Embryo Culture Medium

1. Fill an appropriate sized Griffin beaker with ~500 mL (~50% of the final volume) embryo-tested water.
2. It is recommended to first dissolve PVA at 60 °C and then cool the media to room temperature for adding the rest of the components.
3. Add the components in Table 5.
4. After all components have dissolved, transfer the contents to a 1 L volumetric flask and bring to desired volume by rinsing the

Griffin beaker with embryo water and adding it to the volumetric flask. If any precipitates form, discard the solution and start over.

5. Gently gas the solution with 5% CO_2, using a sterile Pasteur pipette, for 15–30 min.
6. Check to ensure the pH of the solution is 7.4 ± 0.2, and remake if the pH is outside this range.
7. Check the osmolality of the solution and remake if the osmolality does not fall between 235 and 255 mOsm/kg.
8. Sterilize the solution using a 0.22 μm filter unit.
9. Label the bottle with the date made, expiration date, pH and osmolality.
10. Store at 4 °C for up to 2 weeks.

2.6.1 Working Solution

1. Using a sterile Pasteur pipette, gas bottle with 5% CO_2 for approximately 1 min prior to removing an aliquot of the medium.
2. Aliquot amount needed using a sterile pipette and regas the top of the bottle with 5% CO_2 before closing and additional storage.
3. For 16-cell or morulae, make up a working solution with 10% fetal bovine serum (FBS).
4. Sterilize using a 0.22 μm syringe filter. It is important to discard the first few drops from the filter to eliminate the possible introduction of any fibers from the syringe.
5. Make the 30 μL drops in 35 mm the day before initiating and cover with 3 mL embryo-tested mineral oil, which has been preequilibrated in the incubator. Do not make more than five dishes at a time before adding mineral oil, since drops can evaporate in the flow hood changing osmolality and there can be a sharp rise in pH, both of which are deleterious to embryo development.
6. Immediately place dishes in an incubator preset to 5% CO_2, 5% O_2, and 37 °C and allow them to equilibrate overnight prior to use.

3 Methods

3.1 Rat Zygote Isolation

1. Intraperitoneal PMSG injection (20 IU per rat at 12:00 noon).
2. Intraperitoneal hCG injection (20 IU per rat at 12:00 noon) about 48 h post PMSG injection.

3. At approximately 9:00 am check for the presence of a vaginal plug in donor females that were paired with breeder males the night before.
4. Transfer females with vaginal plugs from the animal room and prepare the bench top with paper for dissection and get out the appropriate tools for dissection (e.g., petri dishes, 70% ethanol in spray bottle, curved serrated forceps, and a pair of fine and regular dissection scissors).
5. Euthanize rats using an institutional AUCAC approved method about 18–24 h post-hCG and subsequent mating.
6. Dissect out each oviduct/uterus using a standard aseptic dissection protocol and place the tracts on sterile filter paper to remove blood and fat.
7. Wash twice in 3 mL mR1ECM-HEPES (high NaCl) in 35 mm dishes to remove contaminants.
8. Transfer each oviduct into a 100 μL drop of mR1ECM-HEPES (high NaCl) containing hyaluronidase (1 mg/mL).
9. Using a fine forceps, release the clutches of zygotes by tearing the oviduct with a 27G needle under stereomicroscope.
10. Most of the zygotes will be partially free of cumulus cells, but some may require hyaluronidase treatment and repeated pipetting to completely remove the cumulus cells.
11. Place each dish containing the zygotes on a slide warmer set to 37 °C for about 5 min.
12. Once the presumptive zygotes are free of surrounding cumulus cells, transfer them (fine pulled glass pipette with ~100 mM inner diameter) to fresh mR1ECM-HEPES (high NaCl) and wash twice through 100 μL drops of mR1ECM-HEPES (high NaCl).
13. Check for the presence of a sperm tail within the cytoplasm and pronuclei under 20× objective (*see* Fig. 1a). It is important to note that pronuclei become more visible after 10:00 a.m.
14. The time of embryo collection depends upon the embryo stage that is needed for particular purpose. To collect 2-cell to blastocyst stage embryos from adult females, the infundibulum is

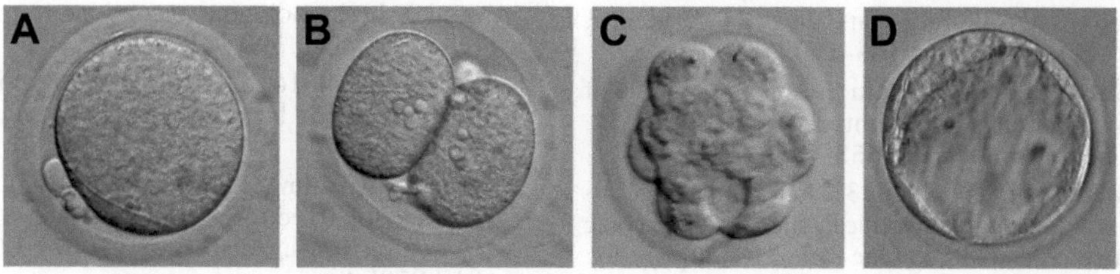

Fig. 1 Rat embryonic development in vitro

gently held using fine tweezers and each horn is flushed with about 100 μL mR1ECM-HEPES (high NaCl) to expel embryos. For example, one can obtain 2-cell embryos on the second, 8–16 cell on the third day, and morula and blastocyst stage embryos on the fourth and fifth day of post-hCG, respectively.

3.2 Rat Embryo Culture from Zygotes to Blastocyst

1. At approximately 10:00 am on the day of zygote collection (day 1), take out the petri dishes having mR1ECM (high NaCl) that were equilibrated overnight. Transfer the presumptive zygotes to drops of mR1ECM (high NaCl) (10–15 zygotes/drop) using a fine puled glass pipette.
2. The following day at approximately 9:00 am (day 2), move 2-cell embryos from mR1ECM (high NaCl) to mR1ECM (low NaCl). Wash embryos through three drops of the fresh medium and culture in groups of 10–15 embryos/drop.
3. In the morning of day 4, prepare dishes with fresh drops of mR1ECM (low NaCl) + 10% FBS and equilibrate. In the afternoon (3:00–5:00 pm), move morulae to drops of mR1ECM (low NaCl) + 10% FBS.
4. The morning of the following day (day 5), collect cultured blastocyst stage embryos.

Acknowledgments

Development of the described protocols has been supported by The National Institute of Health (# P40 OD011062) and the Rat Resources and Research Center (http://www.rrrc.us/).

References

1. Iannaccone PM, Galat V (2002) Production of transgenic rats. In: Pinkert C (ed) Transgenic animal technology. Academic, San Diego, CA, pp 235–250
2. Mullins LJ, Wilmut I, Mullins JJ (2004) Nuclear transfer in rodents. J Physiol 554:4–12
3. Agca Y, Critser JK (2005) Assisted reproductive technologies & genetic engineering in rats. In: Suckow MA, Weisbroth SH, Franklin CL (eds) The laboratory rats, 2nd edn. Academic, San Diego, CA, pp 165–190
4. Agca Y (2012) Genome resource banking of biomedically important laboratory animals. Theriogenology 78:1653–1665
5. Men H, Bryda EC (2013) Derivation of a germline competent transgenic Fischer 344 embryonic stem cell line. PLoS One 8:e56518. https://doi.org/10.1371/journal.pone.0056518
6. Nakamura K, Morimoto K, Shima K, Yoshimura Y, Kazuki Y, Suzuki O, Matsuda J, Ohbayashi T (2016) The effect of supplementation of amino acids and taurine to modified KSOM culture medium on rat embryo development. Theriogenology 86:2083–2090
7. Lan H, Li S, Guo Z, Men H, Wu Y, Li N, Bryda EC, Capecchi MR, Wu S (2016) Efficient generation of selection-gene-free rat knockout models by homologous recombination in ES cells. FEBS Lett 590:3416–3424
8. Miyoshi K (2016) Development of a culture medium for rat 1-cell embryos. J Mamm Ova Res 33:11–16
9. Meek S, Mashimo T, Burdon T (2017) From engineering to editing the rat genome. Mamm Genome. https://doi.org/10.1007/s00335-017-9705-8

10. Toyoda Y, Chang MC (1974) Fertilization of rat eggs *in vitro* by epididymal spermatozoa and the development of eggs following transfer. J Reprod Fertil 36:9–22
11. Whittingham DG (1975) Survival of rat embryos after freezing and thawing. J Reprod Fertil 43:575–578
12. Bavister BD (1995) Culture of pre implantation embryos. Facts and artifacts. Hum Reprod Update 1:91–148
13. Matsumoto H, Sugawara S (1998) Effect of phosphate on the second cleavage division of the rat embryo. Hum Reprod 13:398–402
14. Schini SA, Bavister BD (1988) Two-cell block to development of cultured hamster embryos is caused by phosphate and glucose. Biol Reprod 39:1183–1192
15. Kishi J, Noda Y, Narimoto K, Umaoka Y, Mori T (1991) Block to development in cultured rat 1-cell embryos is overcome using medium HECM-1. Hum Reprod 6:1445–1448
16. Ouhibi N, Warren J, English J, Sullivan N (1994) A culture system for the development of the preimplantation rat embryo. Theriogenology 41:535–551
17. Miyoshi K, Funahashi H, Okuda K, Niwa K (1994) Development of rat one-cell embryos in a chemically defined medium: effects of glucose, phosphate and osmolarity. J Reprod Fertil 100:21–26
18. Miyoshi K, Niwa K (1997) Stage-specific requirement of phosphate for development of rat 1-cell embryos in a chemically defined medium. Zygote 5:67–73
19. Nishikimi A, Uekawa N, Yamada M (2000) Involvement of glycolytic metabolism in developmental inhibition of rat two-cell embryos by phosphate. J Exp Zool 287:503–509
20. Zhou Y, Galat V, Garton R, Taborn G, Niwa K, Iannaccone P (2003) Two-phase chemically defined culture system for preimplantation rat embryos. Genesis 36:129–133
21. Agca C, Agca Y (2014) Molecular and ultrastructural changes of rat pre-implantation embryos during two-cell developmental arrest. J Assist Reprod Genet 31:767–780
22. Kito S, Kaneko Y, Yano H, Tateno S, Ohta Y (2008) Developmental responses of 2-cell embryos to oxygen tension and bovine serum albumin in Wistar rats. Exp Anim 57:123–128
23. Ho Y, Wigglesworth K, Eppig JJ, Schultz RM (1994) Preimplantation development of mouse embryos in KSOM: augmentation by amino acids and analysis of gene expression. Mol Reprod Dev 41:232–238
24. Biggers JD, McGinnis LK, Raffin M (2000) Amino acids and preimplantation development of the mouse in protein-free potassium simplex optimized medium. Biol Reprod 63:281–293
25. Miyoshi K, Abeydeera LR, Okuda K, Niwa K (1995) Effects of osmolarity and amino acids in a chemically defined medium on development of rat one-cell embryos. J Reprod Fertil 103:27–32
26. Oh SH, Miyoshi K, Funahashi H (1998) Rat oocytes fertilized in modified rat 1-cell embryo culture medium containing a high sodium chloride concentration and bovine serum albumin maintain developmental ability to the blastocyst stage. Biol Reprod 59:884–889
27. Han MS, Niwa K (2003) Effects of BSA and fetal bovine serum in culture medium on development of rat embryos. J Reprod Dev 49:235–242
28. Lawitts J, Biggers JD (1993) Culture of preimplantation embryos. Methods Enzymol 225:153–164
29. Erbach GT, Lawitts JA, Papaioannou VE, Biggers JD (1994) Differential growth of the mouse preimplantation embryo in chemically defined media. Biol Reprod 50:1027–1033
30. Harris SE, Gopichandran N, Picton HM, Leese HJ, Orsi NM (2005) Nutrient concentrations in murine follicular fluid and the female reproductive tract. Theriogenology 64:992–1006

Chapter 4

Enabling Hamster Embryo Culture System: Development of Preimplantation Embryos

Polani B. Seshagiri and Venkatappa Vani

Abstract

Development of preimplantation embryos, from fertilization to hatched-blastocyst stage, has been a challenging task, regardless of the mammalian species being studied. While the mouse model has been versatile for studying in vitro development of early embryos, other rodent species are important to gain insights into comparative early embryogenesis. The golden hamster (*Mesocricetus auratus*) offers unique advantages to study cellular and molecular regulation of gamete maturation, fertilization and preimplantation development, including the phenomenon of blastocyst hatching. Achieving in vitro fertilization and first cleavage division is relatively easy; however, subsequent development past the two-/four-cell stage had been difficult in hamsters. Pioneering research, carried out over three decades has markedly enabled successful in vitro development of one-cell embryos to blastocysts. This article provides a comprehensive perspective (historical and current) on the embryo culture systems and details an optimized culture protocols to achieve normal and viable development of preimplantation embryos in the golden hamster.

Key words One-cell embryo, Developmental block, Energy substrates, Hamster embryo culture medium (HECM), Hatching, Viability, Vitamins

1 Introduction

Infertility is one of the major reproductive health challenges that affect millions of human couples desirous of having their progeny [1]. Assisted reproductive technology (ART) has enabled infertile couples to have offspring; however, challenges exist to improve on the efficiency and outcome of ART in humans [2, 3], as well as in nonhuman primates [4–6]. There are even greater challenges to achieve efficient development of preimplantation (non)manipulated embryos in many species such as the mouse [7–9], hamster [10–13], rat [14], and rabbit [15]. Regardless of the species, extensive research is dedicated to improving embryo culture technology because of the need to (1) augment fertility in economically profitable species [16] (2) generate genetically modified species for basic research [17] and (3) harness full potential of stem cell technologies

Jason R. Herrick (ed.), *Comparative Embryo Culture: Methods and Protocols*, Methods in Molecular Biology, vol. 2006, https://doi.org/10.1007/978-1-4939-9566-0_4, © Springer Science+Business Media, LLC, part of Springer Nature 2019

that are dependent on the manipulation of early embryogenesis and allow generation of genetically modified organisms [18, 19]. Invariably, application of these embryo-based technologies is governed by our (in)ability to culture biologically normal, viable preimplantation embryos that are capable of producing live births [6, 9, 13, 16, 20–22].

1.1 Comparative Preimplantation Embryology

Among model organisms studied, the mouse has been the mainstay model for a number of studies involving early embryogenesis [7, 23–26]. Comparatively, other animal models, such as rat [27, 28], hamster [20, 21, 29–38], and rabbit [39] have not caught up with the advancement of embryo culture technologies. Nevertheless, nonmouse species offer excellent opportunities to perform comparative early embryogenesis studies. Since, specific chapters in this book deal with embryo cultures in other species, this article will focus only on the golden hamster (*Mesocricetus auratus*). Here, we present historical perspective, current state of knowledge, and the experimental methodologies available on the hamster culture system that enable efficient development of one-cell embryos through to blastocysts.

1.2 Hamster Embryo Development (Historical Status)

About a half-century ago, it was demonstrated, for the first time, that freshly recovered ova are easily fertilized in vitro and their first cleavage division achieved in the golden hamster [40, 41]. Similar work has followed [12, 42]. However, in the past, obtaining development beyond two-/four-cell stages, and particularly to the blastocyst stage, has been extremely difficult due to a developmental block (i.e., the "two-cell block") exhibited by hamster embryos [10, 11, 41, 43]. This had greatly hampered the use of hamster model for studying preimplantation development. Development of hamster embryos during the entire preimplantation period is governed by environmental variables and the culture media components to which embryos are exposed during culture [13, 20, 44, 45]. These in vitro developmental hurdles were overcome by pioneering embryo culture experiments, carried out primarily by Professor Barry Bavister and his colleagues [10, 13, 20–22, 29–34, 46–50], as well as by others [37, 38, 51–54], in the last three decades.

1.3 Hamster Embryo Development (Current Status)

Both by empirical and evidence-based approaches, ten different formulations of hamster embryo culture medium (HECM-1 to -10; *see* Table 1) have been developed [21, 30, 34–36, 49, 55, 56]. They vary in terms of the composition of energy substrates, amino acids, vitamins and bovine serum albumin (BSA) included in individual media (Table 1). In addition, gas tensions (10% CO_2 and 10% O_2) used in the incubators for culturing embryos was optimized to achieve the best embryo development [44, 57]. Various formulations of HECMs were substantially refined and embryo

Table 1
Composition of various formulations of hamster embryo culture medium (HECM)

	HECM formulation											
Medium components	**HECM-1**	**HECM-2**	**HECM-2m**	**HECM-2h**	**HECM-3**	**HECM-4**	**HECM-5**	**HECM-6**	**HECM-7**	**HECM-8**	**HECM-9**	**HECM-10**
PVA (mg/ml)	1.0	1.0	0.1	0.1	0.1	0.1	0.1	0.1	0.1	0.1	0.1	0.1
NaCl (mM)	98.0	116.5	116.5	116.5	113.8	103.3	113.8	113.8	113.8	113.8	113.8	113.8
KCl (mM)	3.2	3.2	3.2	3.2	3.0	3.0	3.0	3.0	3.0	3.0	3.0	3.0
$CaCl_2{\cdot}2H_2O$ (mM)	2.0	2.0	2.0	2.0	1.9	1.9	1.9	1.9	1.9	1.9	2.0	1.0
$MgCl_2{\cdot}6H_2O$ (mM)	0.5	0.5	0.5	0.5	0.5	0.5	0.5	0.5	0.5	0.5	0.5	2.0
$NaHCO_3$ (mM)	25.0	25.0	25.0	25.0	25.0	25.0	25.0	25.0	25.0	25.0	25.0	25.0
1 M HCl (μl)	–	–	–		140	–	140	140	140	140	140	140
Na-lactate (mM)	10.0	10.0	10.0	10.0	4.5	4.5	4.5	4.5	4.5	4.5	4.5	4.5
Pyruvate (mM)	0.5	0.25	0.2	0.2	–	–	–	–	–	–		
BSA (mg/ml)	–	–	–	3	–	–	–	–	–	–	–	–
Amino acids[a]	20	4	4	4	1	20	3	10	10	10	10	10
Vitamins[b]				3							1	1
Osmolality (mOsmols)	290	290	280–295	280–295	275	290	275	275	275	275	275	275
Reference[c]	21	30	35	36	49	55	34	34	34	34	60	56

Abbreviations: *HECM* hamster embryo culture medium, *PVA* polyvinyl alcohol, *BSA* bovine serum albumin (fraction-V)

[a]The Number in parenthesis indicates the number of amino acids included for the indicated medium formulations. For details refer to [34]; Shown here are amino acids indicated by their one-letter code. The 20 amino acids in HECM-1 and -4 are A, R, N, D, C, E, Q, G, H, I, L, K, M, F, P, S, T, W, V, and Tu; the four amino acids in HECM-2, -2m, -2h are Q, I, M, and F; the amino acid in HECM-3 is Q; The three amino acids (in HECM-5) include Q, G, Tu; the ten amino acids in HECM-6 and -10 are N, D, C, E, Q, G, H, K, P, and S

[b]HECM-2h is supplemented with three vitamins: choline chloride (5 μM), inositol (3 μM), and pantothenate (3 μM). HECM-9/-10 is supplemented with pantothenate (3 μM)

[c]Details of media composition shown in the table are derived from the respective cited publication shown for each medium in the column

Table 2
Culture media suitable for development of hamster preimplantation embryos

Stage of development	Culture medium	% Development	References
1-cell to 2-cell	TLP-PVA	45	[41]
2-cell to 4-cell	TLP-PVA HECM-1/-2	53 93	[20] [32]
2-cell to 8-cell	TLP-PVA HECM-1/-2	75 82.5/67.9	[21] [32]
8-cell to blastocyst	TALP-PVA HECM-1/-2 HECM-2m	90 89–91 100	[29, 30] [32] [35]
8-cell to hatched blastocyst	TALP TALP-PVA HECM-2h	38 37 100	[46] [47, 48] [36]
Further improvisation:			
1-cell to blastocyst	HECM-3 HECM-5 HECM-6 HECM-9	18 41 59 75–78	[34] [34] [34] [50]
2-cell to blastocyst	HECM-9	90	[56]

Abbreviations: *HECM* hamster embryo culture medium; *PVA* polyvinyl alcohol; *TLP* Tyrode's solution with lactate and pyruvate; *TALP* Tyrode's solution with albumin, lactate, and pyruvate

culture technology and SOPs developed (*see* Subheadings 2 and 3). They are highly reproducible, simple, and efficient to achieve maximal and normal development of all stages of preimplantation embryos, that is, beginning from one- or two- or eight-cell stage to blastocysts [29–31, 34, 35, 49, 50, 58–60], including hatching of blastocysts [37, 38, 46, 54, 61, 62]. Interestingly, unlike other mammalian species, hamster preimplantation embryos, beginning from the one-cell stage, can be grown in chemically defined, protein-free media (Table 2) and resulting embryos are biologically viable [34, 49, 60].

1.4 Hamster Embryo Development (Future Prospects)

The currently available embryo culture technology provides excellent opportunities to understand the molecular regulators involved in preimplantation development [34, 37, 38, 59, 63], including blastocyst hatching [38, 47, 54, 61–65] and implantation [53]. Moreover, because hamster blastocyst hatching is associated with inflation–deflation cycles and because perihatching blastocysts exhibit trophectodermal projections (TEPs; [37, 63, 64]), the hamster model offers excellent opportunities to investigate dynamic cellular, biochemical, and molecular events occurring during perihatching development [38, 54, 63, 65]. Interestingly, because hamster embryos can be cultured in protein-free, chemically defined media [34, 50, 56], fine details on embryonic secretome during the prehatching and perihatching development can be investigated using modern mass spectrometric analyses [66, 67]. These exciting research outcomes have immense implications in our understanding of hatching and implantation failures and for developing strategies to mitigate them in mammals, including humans [2, 9, 16, 63, 68].

The following sections provide comprehensive information for the interested researcher on the experimental protocols to recover and culture all stages of preimplantation embryos in the golden hamster. We have avoided description of experimental systems for prefertilization studies. For this, readers are encouraged to refer to published articles on hamster sperm capacitation [69], oocyte maturation, and fertilization [12, 41, 70].

2 Materials

2.1 Animal Resource

1. Mature cycling, female golden hamsters (age: 2–4 months old).
2. Proven fertile male golden hamsters (age: 4–10 months old).
3. Polypropylene sterilized cages with suitable bedding material.
4. Animal housing with 14:10 light–dark photoperiod.

2.2 Culture Media

1. Embryo culture grade water (sensitivity: 18 MΩ cm at 25 °C) and chemical reagents (*see* **Note 1**).
2. Autoclaved glass vessels or sterile disposables (150 ml bottles, 500 ml beakers, 100 ml measuring cylinders, and 100 or 250 ml volumetric flasks) (*see* **Note 2**).
3. Tyrode's Lactate—polyvinyl alcohol medium (TL-PVA): PVA (0.1 mg/ml), Phenol Red (0.01 mg/ml), NaCl (114 mM), KCl (3.16 mM), Glucose (5 mM), NaH_2PO_4 (0.35 mM), $NaHCO_3$ (25 mM), $MgCl_2{\cdot}6H_2O$ (0.5 mM), $CaCl_2{\cdot}2H_2O$ (2 mM), and sodium lactate (10 mM). When necessary, TALP-PVA medium can be supplemented with 0.2 mM sodium pyruvate and 3 mg/ml BSA (*see* **Notes 3** and **4**).
4. TL-HEPES medium: This is similar to TL-PVA except that it should contain only 2 mM $NaHCO_3$ and supplemented with 9.2 mM HEPES. Adjust the pH of the medium to 7.4. Use this medium (maintained at air-atmosphere) for processing dissected oviducts and uterine horns.
5. Na-Pyruvate Stock: 11 mg/ml dissolved in saline (0.9% NaCl in culture-grade water).
6. Hamster Embryo Culture Medium (HECM): *See* formulations in Table 1 and prepare according to **Notes 3** and **4**.

7. Equilibrated Oil: Take equal volumes (100 ml each) of silicone oil and autoclaved culture-grade water, shake them vigorously in an autoclaved 250 ml-separating funnel, and keep on a stand for phase separation (funnel covered with aluminum foil). Repeat this process three times at 2 h intervals. After the phases have separated, discard the water. Combine equal volumes of "washed" silicone oil and 0.9% saline, shake as described above, allow for phase separation, and discard the saline. Filter-sterilize (0.22 μM) the oil and store it at RT in a sterile glass bottle with an equal volume of filter-sterilized 0.9% saline [11].

2.3 Hormone

1. Pregnant mare's serum gonadotropin (PMSG): Reconstitute in normal saline (0.9% NaCl in culture-grade H_2O) to a final concentration of 1000 IU/100 μl and store at −20 °C. On the day of hormone injection, suitably dilute (with normal saline) the hormone into Eppendorf tubes to have about 25 IU injection per hamster.

2.4 Surgical Equipment

1. Culture-washed and autoclaved four forceps (two blunt + two pointed), two watchmaker's forceps, two scissors (one big + one small), two iris-scissors, 100 ml beaker, and two spatulas.
2. Flushing stainless-steel needle (30 G) with blunted and fire-polished bevel.
3. Embryo transfer microscopic stand with stereo-zoom binocular head and fiber optic incident light source units.
4. Ketamine hydrochloride anesthetic.
5. Hair clippers.
6. Betadine or absolute alcohol.
7. Sterile surgical drapes.
8. 37 °C heating block to hold culture tubes.
9. Suture.

2.5 Molecular Biology and Immunolocalization Reagents

1. Dynabeads mRNA DIRECT kit (Ambion by Life Technologies).
2. DynaI Magnetic Particle Concentrator.
3. RT-PCR kit.
4. Paraformaldehyde (PFA).
5. Phosphate buffered saline.
6. Triton X-100.
7. Desired primary antibody.
8. Anti-IgG, fluorescence-tagged secondary antibody or Cy-3 or Cy5 or suitable Alexa-conjugated secondary antibodies.
9. DAPI and propidium iodide.

10. RNAse.
11. Slides and coverslips.
12. Antifade solution (*see* **Note 5**).

3 Methods

3.1 Preparation of Embryo Culture Dishes

1. Dispense HECM into culture dishes: 2 ml HECM in a 35 mm dish; 6 ml HECM in a 60 mm dishes, and 4 ml of equilibrated silicone oil in a 35 mm dish.
2. Make 4–5 drops of 50–100 μl of HECM in 35- or 60-mm culture dishes and overlay with 3 or 6 ml of equilibrated silicone oil, respectively.
3. Equilibrate all dishes in the CO_2 incubator at 37 °C for 2 h.

3.2 Superovulation

Hamsters are capable of providing large numbers of preimplantation embryos (40–60 per donor), following PMSG-superovulation protocol [11, 59, 71]. Choose animals on day 1 of their estrous cycle (day of postestrus discharge, PED) [72]. In the morning (before 10 AM), give intraperitoneal (ip) injections of PMSG in the lower abdominal region (*see* **Note 5**). Three days (post-PMSG injection) later, introduce a stud male into the cage (male–female ratio at 2:1) late evening (*see* **Note 5**). Observe and record lordosis behavior, a sign of mating behavior of the female. The following morning, with the help of a toothpick, take a sample of the vaginal postestrus discharge and microscopically examine for the presence of sperm (indication of insemination). Recover embryos from excised reproductive tracts of mated females [11, 35–37, 55] (*see* **Note 5**).

3.3 Embryo Recovery

Anesthetize the animal with ketamine (25 mg/kg body weight) and/or perform cervical dislocation. When the animal is completely unconscious, sterilize the ventral lower-abdominal area with ethanol and cut open the skin and abdominal wall with sterile sets of forceps and scissors. With another fresh set of surgical tools, exteriorize the reproductive tracts, excise them and place them into a 35-mm culture dish containing 2 ml of TL-HEPES medium. Transfer the tissue into another fresh 35-mm culture dish containing 2 ml of TL-HEPES medium [10, 59]. Carry this to the embryo recovery station in the culture room (*see* **Note 6**).

Neatly dissected oviducts or uterine horns are placed in 35-mm culture dishes containing equilibrated TL-PVA medium to proceed with the recovery of 1-, 2-, 4-cell embryos (from oviducts) or eight-cell embryos, morulae, or early blastocysts (from uterine horns). Recovery of specific preimplantation embryonic stages is time-dependent [10, 54], and it is as follows: 1-cell, 10 h after egg activation (PEA, [42]); 2-cell, 20–24 h PEA [10, 20]; 4-cell, 48 h

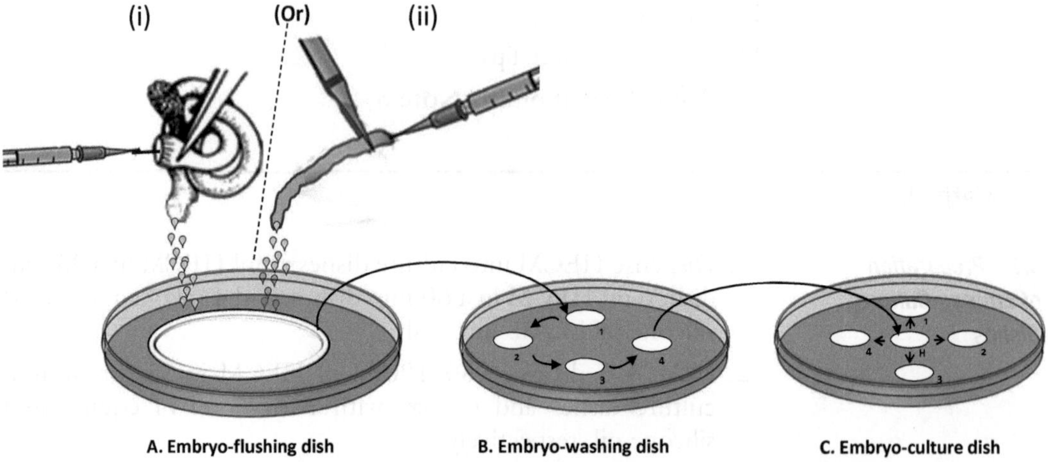

Fig. 1 A schematic diagram showing the embryo recovery procedure in the golden hamster. Either oviducts (i) or uterine horns (ii) are flushed with the HECM to recover 1-, 2-, 4-cell stage embryos (oviducts) or 8-cell embryos, morale, blastocyst (uterine horns). The embryo culture dish will be incubated and assessed for embryo development. The schematic diagram of the oviduct (i) is a modified version from "Manipulating the mouse embryo: a laboratory manual by B Hogan et al. (1986)"

PEA [45]; 8-cell, 56–58 h PEA [29, 57, 59]; morula, 60 h PEA [46]; and early blastocyst, 70 h PEA [48]. A schematic diagram showing the flushing and embryo recovery procedure is shown in Fig. 1. A 1 ml syringe is filled with equilibrated flushing medium (HECM) and fitted with the flushing needle. Then, the needle is gently inserted into the ampulla of the oviduct or into the uterine horn. The medium is gently pushed into the lumen and flushates are collected into a 60 mm organ culture dish containing equilibrated silicone oil. Dishes are placed onto stage warmer for about 2–3 min and then observed under the stereo-zoom microscope to locate embryos. All embryos are gently picked using the mouth-pipette fitted with Pasteur pipettes (Fig. 1) (*see* **Note 6**).

3.4 Embryo Culture System

To achieve maximal in vitro development of preimplantation embryos, ideal environmental conditions and choice of a particular HECM formulation is critical (*see* **Note 7**). Table 2 shows various HECM formulations that are suitable to achieve development of early stage (one- or two-cell) embryos to blastocysts. Recovered embryos can be allocated in groups of 5–10 embryos to 50–100 μl drops of culture medium under mineral (silicone) oil overlay (Fig. 1). Culture dishes containing embryos should be incubated at 37.5 °C in a humidified atmosphere of 5% or 10% (optimal) CO_2 in air [11, 59]. For achieving improved development of one- or two-cell embryos to blastocyst [34, 50, 56], cultures should also be carried out with 5% or 10% (optimal) O_2 along with either of the above two CO_2 concentrations [49, 57]. This requires a specialized trigas incubator. Embryos can be cultured for a desired (finite)

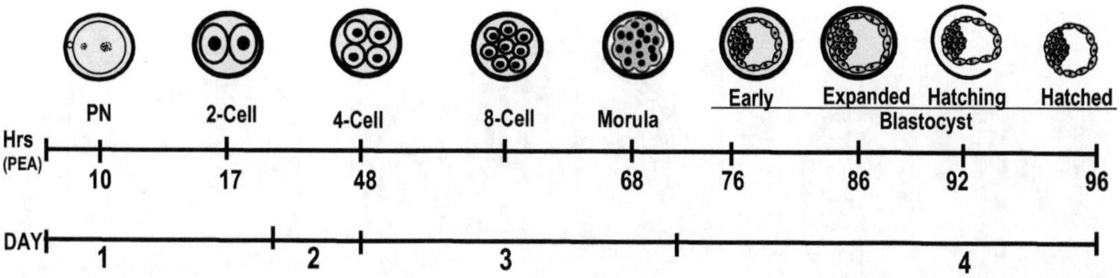

Fig. 2 A schematic diagram showing the time frame for hamster preimplantation embryo development in vitro. *Abbreviations*: *PEA* post-egg activation, *PN* pro-nucleate-stage zygote. Times shown in hours (Hrs) and day are not drawn to the scale. Time taken for embryo development in vitro is generally slower than the time-course observed in vivo; the time delay is ~4 h (early-cleavage stages) to about 36–68 h (peri-hatching stages). Data from McKiernan and Bavister [59]

period (e.g., 24–72 h) depending on the stage of embryos being cultured and the developmental end-points being targeted or assessed (Fig. 2; see section below).

3.5 Experimental Design

Developmental competence of embryos is dependent on embryo donors and also, males used for mating. Because between-animal variations are often observed, it is always desirable to plan block-design experiments with adequate replication [11, 38, 54]. It is important to make sure that all recovered embryos form a single donor are randomly distributed among all treatments and control drops in a culture dish (*see* Fig. 1). Depending on the experimental design, number of embryos per drop can range from 5 to 20. This design will facilitate analysis of multiple factors in multifactorial experiments [11]. Moreover, such an approach is useful when one or more regulatory factors (or, dose-effects) are tested on embryo development.

3.6 Developmental Endpoints

It is necessary to assess robust developmental endpoints at specified time points during the culture period. The most commonly used endpoint is the proportion of embryos developing to the two-, four-, or eight-cell noncompact stage when one-cell embryos are cultured. Similarly, formation of morulae and blastocysts could be scored after 76, 50, or 24 h after one-, two-, or eight-cell embryos are placed in culture [29, 30, 32, 34, 35, 47–50, 56]. But, scoring blastocysts quality can become subjective since early, mid, late, and expanded blastocyst stages can all be observed. Furthermore, hatching and fully hatched stages can be observed (Fig. 3). These assessment criteria require some level of expertise by the investigator [36]. Representative images of cultured blastocysts (zona-intact and azonal blastocysts) are shown in Fig. 3.

An important criterion to judge the quality of developed blastocysts is to determine the total number of cells and/or the number of trophectoderm and inner cell mass cells in the blastocyst

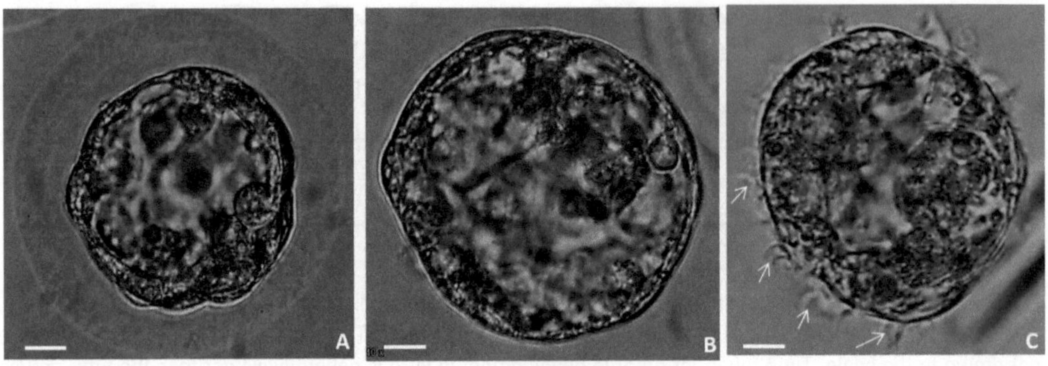

Fig. 3 Representative photomicrographic images of cultured hamster blastocysts. Shown are zona-intact expanded blastocyst (**a**), azonal hatched blastocyst (**b**) and hatched blastocyst with trophectodermal projections: TEPs (**c**). Arrows in panel **c** indicate TEPs on the periphery of blastocyst. Bar = 10 μm

[22, 36–38]. These numbers can be quantitated by DNA-binding fluorescence dyes using immunofluorescence microscopy. Typically, expanded or hatched blastocysts contain a total of about 30–40 cells, with about 7:3 ratio of the TE:ICM cells [22, 36–38].

Blastocyst attachment and trophoblast (TB) outgrowth provide an easy in vitro model to assess the attachment and proliferation potential of hatched blastocysts that can be quantitatively measured by morphometric analysis and videomicrography [36, 37]. Hatched blastocysts are placed in either FBS-coated or extracellular matrix protein-coated culture dishes and cultured in BSA-containing HECM-2h for a period of 96–120 h. During this period, TBs proliferate and exhibit outgrowth. The perimeter and area of these outgrowths can be measured [36, 37].

To obtain real-time developmental kinetics, continuous time-lapse videomicrography is an extremely versatile technique [64]. However, this requires a sophisticated stage-top, gas-controlled, mini-chamber, coupled with a video-imaging system. These monitoring mechanisms provide excellent details on cellular and other developmental dynamic events. Of particular interest in hamsters are the appearances of TEPs in perihatching blastocysts (Fig. 3c; [37, 38, 64]).

Morphological assessments of embryonic development can also be combined with measurements of a variety of marker(s) of embryonic development and function [66]. These can be examined by RT-PCR and/or immunofluorescence imaging methods ([37, 38, 54]; *see* **Notes 8** and **9**). More sophisticated embryomics approaches requiring high-end mass-spectrometry equipment with bioinformatics evaluation capabilities, can also be performed [66, 67]. The ultimate endpoint of biological viability of cultured embryos would be their transfer to pseudopregnant recipient females and monitor progression of pregnancy, fetal development, and the number of live pups born. Detailed methodologies are described elsewhere [10, 11, 55, 60].

4 Notes

1. Obtain only embryo-culture tested or embryo-quality, analytical-grade culture media components sourced from reputed vendors. Caution must be exercised with regard to proper storage of all components as per the vendors' specifications. Medium components such as salts, energy substrates, amino acids and BSA are to be stored in small sterile culture tubes (4-, 15-, 50-ml) and placed in desiccators either at RT or at 4 °C in the refrigerators, depending on the storage specifications of reagents. Repeat opening and closing of bulk reagents bottles lead to its deterioration in quality. Stock solutions in small aliquots (e.g., amino acids) must be kept frozen at −20 °C.
2. All embryo culture-related materials, glassware and other (non-plastic) cell culture-ware items, spatulas, and surgical tools used must be sterilized (autoclaved) prior to their use. Washed glassware and other items must be rinsed at least twice with cell culture-grade water, dried, and then sterilized. They may be coming into contact either directly with embryos or indirectly through solutions or medium being used. Nonautoclavable items can be sterilized by ethylene oxide gas exposure.
3. It is important to diligently follow the culture medium preparation protocol and other SOPs. To prepare 250 ml of the desired medium on the day before the experiment, take 150 ml of culture-grade water in a 500 ml glass beaker with stir bar and place onto a magnetic stirrer. Slowly sprinkle PVA into water with gentle stirring. To this, add the required medium components into the solution. Check the pH and osmolality of the medium (pH: 7.8; osmolality: 280–290 mOsmols/kg water). Make up the volume to 250 ml with culture-grade water in a volumetric flask. Sterilize the medium by passing through a 0.2 μm sterile, disposable filter units. Prepared medium, after filter sterilization, must be stored in autoclaved glass (100-ml) bottles covered with aluminum foil at 4 °C in the refrigerator for up to 3 weeks. Any stored medium showing particulate matter or sediments must be discarded. While antibiotics such as penicillin G and gentamycin are added normally, their absence in the culture medium is desirable for culturing at least up to 72 h if sterile techniques are followed while preparing the culture medium and its use during embryo culture [73].
4. Potentially labile and unstable medium components such as pyruvate, amino acids (e.g., glutamine), and vitamins must be added to embryo culture medium on the day of the experiment (Table 1). Prior to setting up the embryo culture drops, the freshly prepared medium must be filter-sterilized and collected into sterile culture tubes (15- or 50-ml).

5. The successful induction of superovulation in hamsters depends on several variables such as the concentrations of PMSG being injected; age, weight of females and the time of hormone injection. The best age for superovulation is usually 4–6 weeks. Healthy hamsters which exhibit three consecutive 4-day estrous cycles must be used as embryo donors. Animal health is normally ascertained by the body weight, adequate body fat, proper musculature, and smooth texture of the fur. Animals will be injected with the PMSG on the day of the postestrus discharge: PED (day 1). The time of hormone administration relative to the light–dark cycle of the animal room affects both the number of ovulated oocytes and their postfertilization development. Ensure that the diurnal cycle for the animal room is set to be 14 h light–10 h dark. It is important to administer PMSG in animal weight-dependent dosage manner; hormone range: 10–25 IU for about 90–140 g weight [71]. After PMSG injection, females should be mated to fertile males 3 days later. Not more than two females are placed in individual cage. The mating behavior of males varies. They should not be used on consecutive days, and it is important to keep a record of their mating success. Males which fail to mate should be identified and eliminated for mating purpose; new proven-fertile stud males should be regularly recruited.
6. While recovering embryos, donor animals should be handled and their euthanasia and dissection should be carried out outside the dedicated embryo culture room. Only sterilized surgical tools must be used, one set for external body dissection and another set for dissecting and excising the reproductive tracts. Tissues must be placed in sterile TL-HEPES medium in 35 mm petri dishes, give a couple of washes by transferring them into fresh TL-HEPES medium [11]. Only washed reproductive tracts suspended in sterile TLP-HEPES medium in culture dishes must be brought to the culture room. Prior to flushing the reproductive tracts to recover embryos, sterilize the culture hood with UV for 5–10 min before setting up culture experiments. As shown in Fig. 1, separate culture dishes should be set up for recovery, wash and culture of embryos. A separate set of sterilized surgical instruments (watchmaker's force-ups and stainless flushing needles) and freshly drawn sterile Pasteur pipettes must be kept ready adjacent to the stereo-zoom microscope. These items can be kept protected in 4 or 15 ml culture tubes in order to avoid contamination. Cleaned mouth-pipette fitted with Pasteur pipettes should be used to pick up and transfer embryos from recover to wash to culture drops. While sucking and releasing embryos, care must be taken to avoid introducing air bubbles into the culture drops. The exact timing for developmental stage of embryos being recovered

would be dependent on light–dark schedules (14:10, day-–night) and the geographical location of the animal housing facility. It may be necessary to ascertain the above timings of in vivo development of embryos in specific geographical areas where animal research facility is located.

7. While a number of experimental systems could be similar to those used with other model organisms (e.g., the mouse), it is important to adhere to specific requirements (right culture medium and environmental conditions) for culturing hamster embryos in order to achieve maximal and potentially viable preimplantation development. List of HECM formulations suitable for achieving maximal development of hamster preimplantation embryos are described in Table 2. Besides, experimental systems for embryo culture vary depending on the research investigation being pursued. For development of one- and two-cell embryos to blastocysts HECM-1 or -2 could be used [34]; it is also desirable to use HECM-9: HECM-6 supplemented with 3 μM pantothenate [50, 60]. This medium is reported to be superior to obtain blastocysts which show high rates of postimplantation (fetal) development [60]. Using HECM-9, under optimized conditions of 37.5 °C, 10% CO_2, 5–10% O_2, it is possible to achieve over 26% blastocyst development from two-cell embryos [44]. For achieving variable levels of CO_2 and O_2, it is necessary to have a specialized trigas incubator which can process regulated injections of CO_2 and O_2 and N_2. For a good embryo culture practice, it is desirable to verify (1) the set chamber temperature using a thermometer, (2) the set levels of CO_2 and O_2 using an external measuring device (Fyrite Apparatus), and (3) levels of humidifying water tray in the incubator. Measured values may be recorded twice daily. To achieve hatched blastocysts, from the eight-cell stage, supplementation of vitamins and BSA is required, as in TALP-PVA [47, 48] or HECM-2h [36] (*see* Table 2).
8. Embryonic mRNA expression: To isolate mRNA use Dynabeads mRNA DIRECT kit. At least 5–15 embryos are required [37, 38, 54]. They are lysed using the Tris lysis buffer. The lysate is loaded onto 20 μl of Dynabeads oligo(dT)25 beads. Loaded beads are washed with buffer-I then with buffer-II. Using DynaI Magnetic Particle Concentrator, the poly-A mRNA is adsorbed and then specifically eluted in 20 μl distilled water, after heating at 65 °C for 90 s. The enriched mRNA is subjected to RT-PCR analysis using commercial kit [74].
9. Embryonic immunofluorescence imaging: Fix embryo(s) with PFA (4%)-PBS (1 h) followed by washing in PBS-PVA (0.1%); permeabilize embryos with 0.2% Triton X-100 (30 min), followed by blocking in 0.3% BSA (1 h). After rinsing in

PBS-PVA, incubate embryos (overnight at 4 °C) with the primary antibody, against specific embryonic protein being examined; wash three times in PBS-PVA and incubate embryos with anti-IgG (fluorescence-tagged) secondary antibody at RT for 1 h. Counter stain nuclear DNA with DAPI (1 μg/ml) or PI (10 μg/ml), followed by washings of embryos in PBS-PVA. Mount embryos on coverslips with an antifade solution. Examine embryos for immunofluorescence using a confocal microscope: imaging system [37, 38].

Acknowledgments

P.B.S. would like to dedicate this chapter to Prof. Barry D. Bavister who is the pioneer of hamster embryo culture technology. P.B.S. is immensely indebted to Prof. Bavister for his invaluable inputs during P.B.S.'s initial research phase on hamster preimplantation embryogenesis and for his generous encouragement of P.B.S.'s research career. We thank Ms. Madhulika Pathak and M.S. Pamdavathy for help during the preparation of the manuscript. We gratefully acknowledge the funding support received from the Indian Institute of Science and the Department of Science and Technology, Government of India, New Delhi. High-end confocal imaging facility provided by the Department of Biotechnology (Government of India, New Delhi) is gratefully acknowledged.

References

1. Polis CB, Cox CM, Tunçalp Ö, McLain AC, Thoma ME (2017) Estimating infertility prevalence in low-to-middle-income countries: an application of a current duration approach to Demographic and Health Survey data. Hum Reprod 32:1064–1074
2. Shahbazi MN, Jedrusik A, Vuoristo S, Recher G, Hupalowska A, Bolton V, Fogarty NNM, Campbell A, Devito L, Ilic D, Khalaf Y, Niakan KK, Fishel S, Zernicka-Goetz M (2016) Self-organization of the human embryo in the absence of maternal tissues. Nat Cell Biol 18:700–708
3. Kushnir VA, Barad DH, Albertini DF, Darmon SK, Gleicher N (2017) Systematic review of worldwide trends in assisted reproductive technology 2004–2013. Reprod Biol Endocrinol 15:1–9
4. Bavister BD, Boatman DE, Collins K, Dierschke DJ, Eisele SG (1984) Birth of rhesus monkey infant after in vitro fertilization and nonsurgical embryo transfer. Proc Natl Acad Sci U S A 81:2218–2222
5. Wolf DP (2004) Assisted reproductive technologies in rhesus macaques. Reprod Biol Endocrinol 2:37
6. Simerly C, Schatten G (2012) Utility of animal models for human embryo culture: nonhuman primates. Methods Mol Biol 912:39–58
7. Brinster RL (1963) A method for in vitro cultivation of mouse ova from two-cell to blastocyst. Exp Cell Res 32:205–207
8. Chatot CK, Ziomek CA, Bavister BD, Lewis JL, Torres I (1989) An improved culture medium supports development of randombred 1-cell mouse embryos in vitro. J Reprod Fertil 86:361–368
9. Rossanth J (2016) Making the mouse blastocyst: past, present, and future. Curr Top Dev Biol 117:275–288
10. Bavister BD, Leibfried ML, Lieberman G (1983) Development of preimplantation

embryos of the golden hamster in a defined culture medium. Biol Reprod 28:235–247

11. Bavister BD (1987) Studies on the developmental blocks in cultured hamster embryos. In: Bavister BD (ed) The mammalian preimplantation embryos: regulation of growth and differentiation in vitro. Plenum, New York, pp 219–249
12. Bavister BD (1989) A consistently successful procedure for in vitro fertilization of Golden hamster eggs. Gamete Res 23:139–158
13. Bavister BD (1995) Culture of preimplantation embryos: facts and artifacts. Hum Reprod Update 1:91–148
14. Kishi J, Noda Y, Narimoto K, Umaoka Y, Mori T (1991) Block to development in cultured rat 1-cell embryos is overcome using medium HECM-1. Hum Reprod 6:1445–1448
15. Kane MT (1987) Minimal nutrient requirement for culture of one-cell rabbit embryos. Biol Reprod 37:775–778
16. Tibary A (2017) Grand challenge animal reproduction-theriogenology: from the bench to application to animal production and reproductive medicine. Front Vet Sci 4:1–6
17. Loi P, Toschi P, Zacchini F, Ptak G, Scapolo PA, Capra E, Stella A, Marsan PA, Williams JL (2016) Synergies between assisted reproduction technologies and functional genomics. Genet Sel Evol 48:53
18. Horii T, Hatada I (2016) Production of genome-edited pluripotent stem cells and mice by CRISPR/Cas. Endocr J 63:213–219
19. Kim EJ, Kang KH, Ju JH (2017) CRISPR-Cas9: a promising tool for gene editing on induced pluripotent stem cells. Korean J Intern Med 32:42–61
20. Schini SA, Bavister BD (1988) Development of golden hamster embryos through the "two-cell block" in chemically defined medium. J Exp Zool 245:111–115
21. Schini SA, Bavister BD (1988) Two-cell block to development of cultured hamster embryos is caused by phosphate and glucose. Biol Reprod 39:1183–1192
22. Seshagiri PB, Bavister BD (1990) Assessment of hamster blastocysts derived from eight-cell embryos cultured in hamster embryo culture medium-2 (HECM-2): cell numbers and viability following embryo transfer. J In Vitro Fert Embryo Transf 7:229–235
23. Biggers JD (1987) Pioneering mammalian embryo culture. In: Bavister BD (ed) The mammalian preimplantation embryos: regulation of growth and differentiation in vitro. Plenum, New York, pp 1–22
24. Lane M, Gardner DK (2007) Embryo culture medium: which is the best? Best Pract Res Clin Obstet Gynaecol 21:83–100
25. Kang JH, Hwang JS, Yoon JA, Jun JH, Lim HJ, Yoon TK, Song H (2011) Activation of peroxisome proliferators activated receptor δ (PPARδ) promotes blastocyst hatching in mice. Mol Hum Reprod 17:653–660
26. Brusentsev EY, Igonina TN, Amstislavsky SY (2014) Traditional and modern approaches to culture of preimplantation mammalian embryos *in vitro*. Russ J Dev Biol 45:53–65
27. Casanova EA, Okoniewski MJ, Cinelli P (2012) Cross-species genome wide expression analysis during pluripotent cell determination in mouse and rat preimplantation embryos. PLoS One 7:e47107
28. Nakamura K, Morimoto K, Shima K, Yoshimura Y, Kazuki Y, Suzuki O, Matsuda J, Ohbayashi T (2016) The effect of supplementation of amino acids and taurine to modified KSOM culture medium on rat embryo development. Theriogenology 86:2083–2090
29. Seshagiri PB, Bavister BD (1989) Glucose inhibits development of hamster 8-cell embryos *in vitro*. Biol Reprod 40:599–606
30. Seshagiri PB, Bavister BD (1989) Phosphate is required for glucose inhibition of development of hamster 8-cell embryos in vitro. Biol Reprod 40:607–614
31. Seshagiri PB, Bavister BD (1991) Glucose and phosphate inhibit respiration and oxidative metabolism in cultured hamster eight-cell embryos: evidence for the "Crabtree effect". Mol Reprod Dev 30:105–111
32. Seshagiri PB, Bavister BD (1991) Relative developmental abilities of hamster 2- and 8-cell embryos cultured in hamster embryo culture medium-1 and -2. J Exp Zool 257:51–57
33. Seshagiri PB, McKenzie DI, Bavister BD, Williamson JL, Aiken JM (1992) Golden hamster embryonic genome activation occurs at the two-cell stage: correlation with major developmental changes. Mol Reprod Dev 32:229–235
34. McKiernan SH, Clayton MK, Bavister BD (1995) Analysis of stimulatory and inhibitory amino acids for development of hamster one-cell embryos *in vitro*. Mol Reprod Dev 42:188–199
35. Ain R, Seshagiri PB (1997) Succinate and malate improve development of hamster eight-cell embryos in vitro: confirmation of viability of embryo transfer. Mol Reprod Dev 47:440–447
36. Mishra A, Seshagiri PB (1998) Successful development in vitro of hamster 8-cell embryos

to 'zona-escaped' and attached blastocysts: assessment of quality and trophoblast outgrowth. Reprod Fertil Dev 10:413–420

37. Sireesha GV, Mason RW, Hassanein M, Tonack S, Navarrete Santos A, Fischer B, Seshagiri PB (2008) Role of cathepsins in blastocyst hatching in the golden hamster. Mol Hum Reprod 14:337–346
38. Sen Roy S, Seshagiri PB (2013) Expression and function of cyclooxygenase-2 is necessary for hamster blastocyst hatching. Mol Hum Reprod 19:838–851
39. Fischer B, Chavatte-Palmer P, Viebahn C, Santos AN, Duranthon V (2012) Rabbit as a reproductive model for human health. Reproduction 144:1–10
40. Yanagimachi R, Chang MC (1963) Fertilization of hamster eggs in vitro. Nature 200:281–282
41. Yanagimachi R, Chang MC (1964) In vitro fertilization of golden hamster ova. J Exp Zool 156:361–376
42. Ji W, Bavister BD (2000) Development of zona-free hamster ova to blastocysts in vitro. Theriogenology 54:827–834
43. Wittingham DG, Bavister BD (1974) Development of hamster eggs fertilized in vitro and in vivo. J Reprod Fertil 38:489–492
44. McKiernan SH, Bavister BD (1990) Environmental variables influencing in vitro development of hamster 2-cell embryos to the blastocyst stage. Biol Reprod 43:404–413
45. Monis H, Bavister BD (1990) Analysis of the inhibitory effect of inorganic phosphate on development of four-cell hamster embryos in vitro. J Exp Zool 256:75–83
46. Kane MT, Carney EW, Bavister BD (1986) Vitamins and amino acids stimulate hamster blastocysts to hatch *in vitro*. J Exp Zool 239:429–432
47. Kane MT, Bavister BD (1988) Protein-free culture medium containing polyvinyl alchohol, vitamins and amino acids supports development of eight-cell hamster embryos to hatching blastocyst. J Exp Zool 247:183–187
48. Kane MT, Bavister BD (1988) Vitamin requirements for development of eight-cell hamster embryos to hatching blastocysts *in vitro*. Biol Reprod 39:1137–1143
49. McKiernan SH, Bavister BD, Tasca RJ (1991) Energy substrate requirements for in-vitro development of hamster 1- and 2-cell embryos to the blastocyst stage. Hum Reprod 6:64–75
50. Ludwig TE, Lane M, Bavister BD (2001) Differential effect of hexoses on hamster embryo development in culture. Biol Reprod 64:1366–1374
51. Paria BC, Reese J, Das SK, Dey SK (2002) Deciphering the cross-talk of implantation: advances and challenges. Science 296:2185–2188
52. Seshagiri PB, Mishra A, Ramesh G, Rao RP (2002) Regulation of peri-attachment embryo development in the golden hamster: role of growth factors. J Reprod Immunol 53:203–213
53. Reese H, Wang H, Ding T, Paria BC (2008) The hamster as a model for embryo implantation: insights into a multifaceted process. Semin Cell Dev Biol 19:194–203
54. Sen Roy S, Seshagiri PB (2016) The NF-κB signaling system is required for blastocyst hatching in the golden hamster: mediated by the expression of hatching-promoting cathepsins. J Reprod Health Med 2:74–82
55. McKiernan SH, Bavister BD (1993) What is the most appropriate endpoint for evaluating preimplantation embryonic development in vitro? In: Bavister BD (ed) Preimplantation embryo development. Springer, New York, p 320
56. Ludwig TE, Squirrell JM, Palmenberg AC, Bavister BD (2001) Relationship between development, metabolism, and mitochondrial organization in 2-cell hamster embryos in the presence of low levels of phosphate. Biol Reprod 65:1648–1654
57. Carney EW, Bavister BD (1987) Regulation of hamster embryo development in vitro by carbon dioxide. Biol Reprod 36:1155–1163
58. Bavister BD, McKiernan SH (1993) Regulation of hamster embryo development. In vitro by amino acids. In: Bavister BD (ed) Preimplantation embryo development. Springer, New York, pp 57–72
59. McKiernan SH, Bavister BD (1994) Timing of development is a critical parameter for predicting successful embryogenesis. Hum Reprod 9:2123–2129
60. McKiernan SH, Bavister BD (2000) Culture of one-cell hamster embryos with water soluble vitamins: pantothenate stimulates blastocyst production. Hum Reprod 15:57–164
61. Mishra A, Seshagiri PB (2000) Evidence for the involvement of a species-specific embryonic protease in zona escape of hamster blastocysts. Mol Hum Reprod 6:1005–1012
62. Mishra A, Seshagiri PB (2000) Heparin binding-epidermal growth factor improves blastocyst hatching and trophoblast outgrowth in the golden hamster. Reprod Biomed Online 1:87–95
63. Seshagiri PB, Sen Roy S, Sireesha G, Rao RP (2009) Cellular and molecular regulation of

mammalian blastocyst hatching. J Reprod Immunol 83:79–84

64. Gonzales DS, Boatman DE, Bavister BD (1996) Kinematics of trophectoderm projections and locomotion in the peri-implantation hamster blastocyst. Dev Dyn 205:435–444
65. Seshagiri PB, Vani V, Madhulika P (2016) Cytokines and blastocyst hatching. Am J Reprod Immunol 75:208–117
66. Katz-Jaffe MG, Schoolcraft WB, Gardner D (2006) Analysis of protein expression (secretome) by human and mouse preimplantation embryos. Fertil Steril 86:678–685
67. Herrick JR, Silva E, Krisher RL (2016) Using new analytical tools to produce better embryos *in vitro*. Anim Reprod 13:182–190
68. Wang H, Dey SK (2006) Roadmap to embryo implantation: clues from mouse models. Nat Rev Genet 7:185–199
69. Libersky EA, Boatman DE (1995) Effects of progesterone on in vitro sperm capacitation and egg penetration in the golden hamster. Biol Reprod 53:483–487
70. Gwatkin RB, Haidri AA (1973) Requirements for the maturation of hamster oocytes *in vitro*. Exp Cell Res 76:1–7
71. Lee ST, Kim TM, Cho MY, Moon SY, Han JY, Lim JM (2005) Development of a hamster superovulation program and adverse effects of gonadotropins on microfilament formation during oocyte development. Fertil Steril 1:1264–1274
72. Orsini MW (1964) The external vaginal phenomenon characterizing the stages of the estrous cycle, pregnancy, pseudopregnancy, lactation and the anestrous hamster. Proc Anim Care Panel 11:193–206
73. Zhou H, Mckiernan SH, Ji W, Bavister BD (2000) Effect of antibiotics on development *in vitro* of hamster pronucleate ova. Theriogenology 54:999–1006
74. Livak KJ, Schmittgen TD (2001) Analysis of relative gene expression data using real time quantitative PCR and the $2^{-\delta\delta Ct}$ method. Methods 25:402–408

Chapter 5

Culture of Preimplantation Rabbit Embryos

Michael T. Kane

Abstract

It is surprising that so little attention is currently given to in vitro culture of preimplantation rabbit embryos, even though the rabbit is the only laboratory animal in which there is very considerable embryo growth before implantation, resulting in a 300-fold increase in protein content of embryonic cells during the preimplantation period and the formation of more than a 100,000 cells in the blastocyst. This growth pattern explains why blastocyst formation in vitro has an absolute requirement for amino acids, and vitamins, particularly inositol, are esssential for blastocyst growth. A semi-defined medium supplemented with 1.5% BSA (variously known as BSM II or modified F10) was developed at Cornell University at the end of the 1960s and allowed the systematic investigation of the requirements for development of 1-cell rabbit embryos to blastocysts. However, the requirements for in vitro blastocyst growth comparable to in vivo growth still remain an unsolved problem. Citrate, often found as a contaminant in serum albumin, may have an essential role in rabbit blastocyst growth, which would fit in with its role in the development of serum-free media for culture of various types of mammalian cells.

A comprehensive account of the methodology is given to enable a researcher with experience culturing embryos of a different species to work on the rabbit embryo. This account covers medium preparation, hormonal stimulation of superovulation, natural breeding/artificial insemination, and collection of embryos of different stages from 1-cell to blastocyst either after euthanasia or under anesthesia. Peculiarities of the rabbit embryo such as the presence of the mucoprotein coat and its effects on behavior of cultured and transferred embryos are described. Suggestions are made for future avenues of research.

Key words Rabbit, Preimplantation, Embryo, Blastocyst, BSA, Citrate, Culture, Medium, Amino acids, Vitamins

1 Introduction and Background

1.1 Early History

Development of the methodology for culture of preimplantation embryos has been essential for the advancement of assisted reproduction technologies such as IVF and for the use of transgenic methods to study genetic aspects of mammalian development and physiology. While the mouse embryo is currently the preeminent model, almost all early culture work was carried out on the rabbit embryo, possibly because of the ease of timing of egg stage and recovery in an induced ovulator. This work dates back at least as far as Brachet who in Belgium in 1912 cultured rabbit blastocysts in

Jason R. Herrick (ed.), *Comparative Embryo Culture: Methods and Protocols*, Methods in Molecular Biology, vol. 2006, https://doi.org/10.1007/978-1-4939-9566-0_5,

plasma [1]. Early work was also carried out on one-cell and cleavage-stage embryos using serum or plasma [2]. However, the discovery of Whitten in 1957 [3] that 2-cell mouse embryos could be cultured in a simple salt solution supplemented with bovine albumin led to a revolution in the study of preimplantation mammalian embryos. This work was taken up by Biggers, Brinster, McLaren, and their collaborators and extended to the 1-cell mouse embryo and to the demonstration that cultured mouse blastocysts transferred to recipient mothers led to viable pregnancies [4].

Most of the work of developing a defined or semidefined culture medium for rabbit embryos was carried out by R.H. Foote's group [5, 6] at Cornell University (1967–1971; 1990–1997) and my group mainly in Galway in Ireland (1970–1994).

1.2 Preimplantation Embryo Growth Patterns and Culture Requirements

A major reason for the popularity of the mouse embryo model was that culture from the 1-cell stage to the blastocyst stage required only a very simple medium: a Krebs–Ringer bicarbonate salt solution supplemented with pyruvate, lactate, and 0.1% bovine serum albumin (BSA). These extraordinarily limited requirements are explained by the growth pattern of the preimplantation mouse embryo: the mouse blastocyst implants about 4.5 days postovulation with about 100 cells [7], and a significantly lower content of protein [8] and dry matter [9] than at the 1-cell stage. Thus, the mouse embryo, even in vivo, does not undergo true growth while developing from 1-cell to blastocyst. The reported mean protein contents [8] for different stages were as follows (embryo stage, ng protein ± SEM): 1-cell, 27.8 ± 0.51; 2-cell, 26.1 ± 0.41; 8-cell, 23.4 ± 0.32; morula, 20.6 ± 0.74, early blastocyst, 23.9 ± 0.63; and late blastocyst, 20.1 ± 0.74 (the decrease at the late blastocyst stage is due the loss of the zona pellucida during hatching.). From these data it is very understandable why amino acids, vitamins, etc. are not required in the medium for culture of 1-cell mouse embryos to blastocysts.

The growth pattern of the preimplantation rabbit embryo is in striking contrast: the blastocyst starts to implant about 7 days postovulation [10] at which stage it has between 99,000 cells at 150 h [11] and 255,000 at 168 h [12]. The mean protein content [13] (exclusive of embryonic coverings—zona pellucida, mucin coat, neozona, and gloiolemma) was as follows (embryo stage, hours post-hCG, μg protein ± SEM): 1-cell, 18–23 h, 0.16 ± 0.01; early morula, 42–48 h, 0.16 ± 0.02; late morula, 70–72 h, 0.22 ± 0.01; early blastocyst, 94–96 h, 0.44 ± 0.05; mid blastocyst, 117–125 h, 1.55 ± 0.15; late blastocyst, 142–144 h, 6.85 ± 0.79; and periimplantation blastocyst, 166–170 h, 50.38 ± 1.51. Thus there is about a 300-fold increase in the protein

Table 1
Effect of omission of nutrient groups from culture medium on development of 2- and 4-cell rabbit embryos to blastocysts

	% Embryos in different stages after culture for 4 days			
	Preblastocysts	Early blastocysts	Blastocysts	Expanding blastocysts
Complex medium (CM)	18	23	18	42
CM – amino acids	100	0	0	0
CM – vitamins	46	42	7	5
CM – trace elements	21	26	14	39
CM – nucleic acid precursors	19	5	12	63

Data extracted from [5, 14]. Number of embryos per treatment was 57. An embryo with the blastocoel just visible was classified as an early blastocyst, an embryo with the blastocoel at least half the size of the embryo as a blastocyst, and an embryo that was clearly starting to expand as an expanding blastocyst. The complex medium used was composed from a modification of Brinster's mouse medium [15] plus the amino acids (a full complement of 20 amino acids), vitamins (biotin, calcium pantothenate, choline chloride, myoinositol, niacinamide, lipoic acid, pyridoxine–HCl, riboflavin, thiamine–HCl, folic acid, vitamin B12), trace elements ($FeSO_4$, $CuSO_4$, $ZnSO_4$) and nucleic acid precursors (hypoxanthine, thymidine) at the same concentrations as Hams F10 medium plus 1.5% BSA [16]

content of the embryonic cells from 1-cell to the periimplantation blastocyst.

This kind of growth pattern is consistent with Table 1 showing the effects of omitting in turn one of four nutrient groups (amino acids, water-soluble vitamins, trace elements, and nucleic acid precursors) from a complex medium on the development of 2–4-cell rabbit embryos.

In Kane's PhD thesis [14], the optimal medium derived from these results, which was similar to the complex medium in Table 1 with the omission of the nucleic acid precursors, was named Basal Synthetic Medium II (BSM II); this name was also used by Henning Beier's group in Aachen and Berndt Fischer's group in Halle-Leipzig (e.g. [17, 18]). However, in published papers by Kane and coworkers, the medium was always referred to as a modified F10 medium.

There was a total absence of blastocyst formation in the absence of amino acids and very limited blastocyst expansion in the absence of the water-soluble vitamins [5]. The omission of nucleic acid precursors from the medium seemed to improve blastocyst expansion, suggesting that either hypoxanthine and/or thymidine were inhibitory to blastocyst expansion.

The reason for the absolute necessity for amino acids for blastocyst formation is obvious in view of the 2.75-fold increase in protein content in vivo at the early blastocyst stage as compared with the 1-cell stage [13]. Examination of which amino acids were most important indicated methionine, serine, and threonine in that order [5], but most amino acids tested appeared to provide some benefit and given the huge increase in protein content during blastocyst growth (and the significant increase at the late morula stage), it seems obvious that all 20 amino acids should be included in any medium used for culture of morula or blastocyst stages.

The absence of water-soluble vitamins reduced blastocyst formation but its most obvious effect was in reducing blastocyst expansion. Examination of the effects of the 11 vitamins resulted in large, statistically significant decreases in blastocyst expansion when inositol, pyridoxine, riboflavin, and niacinamide were individually omitted from medium for culture of morulae to blastocysts, with the effect of inositol being the most serious [19]. The optimal level of inositol lay in the range 15–75 μM [20, 21] as compared with the 3 μM found in F10 medium. It was later demonstrated that inositol was transported into the cleavage stages via a sodium-dependent saturable active transport system whereas uptake at the blastocyst stages was nonsaturable and possibly occurred by a channel mechanism [22]. Inositol taken up by blastocysts was incorporated into the phosphoinositides and inositol phosphates of the phosphatidylinositol signaling system [23]. The omission of vitamin B12 significantly increased blastocyst expansion, suggesting that B12 at the concentration found in F10 is toxic to rabbit blastocysts.

The inhibitory effect of the nucleic acid precursors on the formation of expanding blastocysts was due to a toxic effect of thymidine [24] at the level in F10 (0.727 μg/ml).

The maximum degree of blastocyst expansion shown in this medium was about 400–600 μm [5, 24], much smaller than the final blastocyst diameters of about 5 mm seen in vivo at around the time of implantation [10].

1.3 Refining the Rabbit Embryo Culture Medium and Culture Conditions

A number of other discoveries emerged from the availability of this modified F10/BSM II medium.

1. Exogenous energy requirements for early cleavage: 1-cell embryos in medium with polyvinyl alcohol (PVA), but without an exogenous fixed nitrogen source or energy substrate, cleaved to a mean of 10.4 cells per embryo [25]. Addition of glucose, pyruvate, a group of 20 amino acids from Ham's F-10 medium, or BSA gave a statistically significant doubling of the cleavage rate.
2. There was no early cleavage block in rabbit embryos such as the 2-cell block in the mouse; 1-cell rabbit embryos readily developed to blastocysts in vitro [24, 26].
3. The optimal osmolarity for culture from 2–4 cell to blastocyst was found to be 270 mOsm/l. Blastocyst formation occurred over the entire range of osmolarities studied from 230 to

340 mOsm/l, with higher osmolarities being more deleterious than lower osmolarities [27].

4. With a 5% CO_2 gas phase, the optimal pH for culture of 1-cell embryos, based on the proportion of blastocysts formed, was about 7.3, corresponding to a bicarbonate concentration of about 22 mM, but variations in development between pH 7.2 and 7.7 was small [28]. Blastocyst expansion was greatest at pH 7.55, corresponding to a bicarbonate concentration of about 40 mM. It is possible that blastocyst formation and expansion may have different optimal pH requirements. This would tie in with the finding by Cross that bicarbonate was actively transported into the blastocoel of the rabbit blastocyst [29].
5. Also unlike in the mouse, 1-cell rabbit embryos can develop to morulae in a HEPES-buffered medium without bicarbonate or carbon dioxide and then, on transfer to a HCO_3/CO_2 buffered medium, develop to blastocysts. Exogenous CO_2 only becomes essential at the blastocyst stage [30].
6. A level of 10% CO_2 in the gas phase is superior to 5% in the modified F10/BSM II medium [31] and in Medium RD (RPMI 1640 and Dulbecco's MEM, 1:1 v/v) [32]. 10% CO_2 should decrease the pH slightly but not significantly change the range of pH between 7.2 and 7.7 mentioned above at which embryo development is optimal.
7. A level of 5% oxygen in the gas phase is superior to the conventional 20% level with modified F10/BSM II medium [33]. It is particularly significant that this work involved culture of late morulae to blastocysts for up to 48 h at a time when significant blastocyst growth occurs and one might have expected demand for oxygen to be high. It also confirmed earlier work culturing 1-cell embryos to blastocysts in Medium RD [34].
8. In vitro fertilization of ovulated oocytes followed by culture for 72 h in the modified F10/BSM II medium to morulae and early blastocysts [35], followed by transfer to recipient does, resulted in birth of live young.

The medium Rabbit Preimplantation Embryo Culture Medium (RPECM) described in Subheading 2.5 is based on this modified F10/BSM II medium with some minor changes.

1.4 The BSA Problem: Role of Fatty Acids and Citrate

After the initial discovery of the importance of amino acids and water-soluble vitamins, the main problem that emerged in relation to rabbit embryo culture was not culture to the blastocyst stage but culture and growth of blastocysts to a size, cell number, and functional capacity attained by day 6 in vivo blastocysts. The obvious problem was that the degree of blastocyst expansion attained in vitro (400–600 μm) was much smaller than the final blastocyst diameter of about 5 mm found in vivo around the time of implantation [36]. This problem still remains.

The comparatively high level of 1.5% BSA used initially in the medium was chosen because, in preliminary experiments culturing 2–4-cell rabbit embryos and adding levels of BSA from 0.1% to 7% to a simple medium without amino acids, vitamins or trace elements, it was found that levels of 1.5% and 3% seemed to produce the most normal looking morulae [14]. In an early paper [24] this medium was described as "defined." However, given the presence of 1.5% BSA, this statement is obviously wrong because of "the extraordinary ligand binding properties" of albumin [37]: (a) A major function of albumin in blood is to transport small molecules especially, but not restricted to, lipid-soluble molecules (e.g., fatty acids [37] and steroid hormones [38]); (b) it accounts for most of the antioxidant capacity of serum [37]; and (c) it binds heavy metals [39].

The innate variability in commercial samples of BSA was reflected in a corresponding variability in the ability of different lots of BSA to stimulate blastocyst growth. Culture of early morulae for 4 days using two different lots of BSA (same supplier and catalogue number) resulted in blastocyst cell counts (mean ± S.E.M., $n = 99–101$) of 742 ± 46 and 345 ± 25 [40].

Examination of the role of albumin-bound fatty acids showed that long chain fatty acids, such as myristic, palmitic, linoleic, and oleic acid, could fulfill the role of energy substrate and substitute for pyruvate in development of 1-cell embryos to morulae in a simple medium without amino acids or vitamins, but supplemented with 0.1% charcoal-treated, defatted BSA [41]. Also while palmitic and oleic acids were capable of stimulating blastocyst formation in the absence of pyruvate when added to a complex medium with amino acids and vitamins, they did not stimulate blastocyst growth [42]. Extraction of a batch of BSA with high blastotrophic ability with 5% formic acid followed by filtration through a membrane with a cutoff M_r of 10,000 and freeze-drying of the filtrate resulted in material with an apparent $M_r \leq 10{,}000$ that stimulated cell proliferation and blastocyst expansion [43]. This is illustrated in Table 2 and Fig. 1.

Later work showed that the major low molecular weight factor responsible for the blastotrophic effect of BSA was citrate [44]. Citrate concentrations in commercial samples of BSA have been reported to range from 10 to 40 μmol/g of BSA [45]. This means that by adding serum or albumin in cell culture, one may be adding unknown amounts of citrate. Very limited experiments that involved adding citrate to the culture medium for 6 day blastocysts suggested that citrate stimulated both blastocyst expansion and radiolabelled iron uptake (Morgan and Kane, unpublished data). It must be noted here that there was little or no evidence that citrate stimulated development to the blastocyst stage, only

Table 2
Effect of low molecular weight (LMW) extract of BSA on rabbit blastocyst cell count and expansion after 4-day culture from the morula stage

		Basic medium + LMW extract (mg/ml)		Basic medium + unextracted BSA (mg/ml)	
	Basic medium	0.04	0.2	1	5
No. of embryos	51	50	48	56	57
Embryo cell count (mean ± sem)	99 ± 7	210 ± 29	807 ± 65	316 ± 55	1128 ± 22
Embryo diameter (μm, mean ± sem)	155 ± 7	191 ± 9	349 ± 13	214 ± 12	457 ± 27

The basic medium to which additions were made was the modified F10 medium supplemented with 0.5% charcoal-treated defatted BSA [43]

that it stimulated growth at the blastocyst stage; there is tentative evidence from work separating albumin with blastotrophic activity into high and low molecular weight fractions that the blastotrophic factor may inhibit development of 1-cell rabbit embryos to the blastocyst stage [42].

1.5 Work by the Foote Group to Develop a Protein and Macromolecule-Free Medium

In the last 25 years, with the exception of R.H. Foote's lab in Cornell, very little work has been carried out to develop improved media for rabbit embryo culture. Most groups have been content to use existing media to study particular aspects of rabbit embryo development. The work of the Foote group concentrated mainly on developing the use of a protein- and macromolecule-free medium. Using Medium RD, they cultured 1-cell rabbit embryos to blastocysts in protein- and macromolecule-free medium (except for any macromolecule carried over in the washing procedure). Transfer of some blastocysts to recipients after 65 h in this medium resulted in day 15 fetuses [6]. This work does not conflict with work on BSA and its associated contaminants because, while Medium RD allowed for blastocyst formation without the addition of BSA, the cell counts (119 ± 12) reported are a long way (Table 2) from the cell counts (1128 ± 122) for unextracted BSA at 5 mg/ml or the counts (807 ± 65) for the low molecular weight BSA extract at 0.2 mg/ml (Table 2). This was, however, a very significant step in showing that a protein-free, macromolecule-free medium could be used to culture 1-cell embryos to blastocysts. As an aside, it should be noted that it is almost impossible to obtain a totally protein- or macromolecule-free environment for culture of preimplantation embryos of any species. In our experience and that of other workers, if one washes embryos thoroughly in a macromolecule-free solution, using three or four washes, the embryos become impossible to handle—they stick to handling pipettes, culture dish surfaces, and air bubbles.

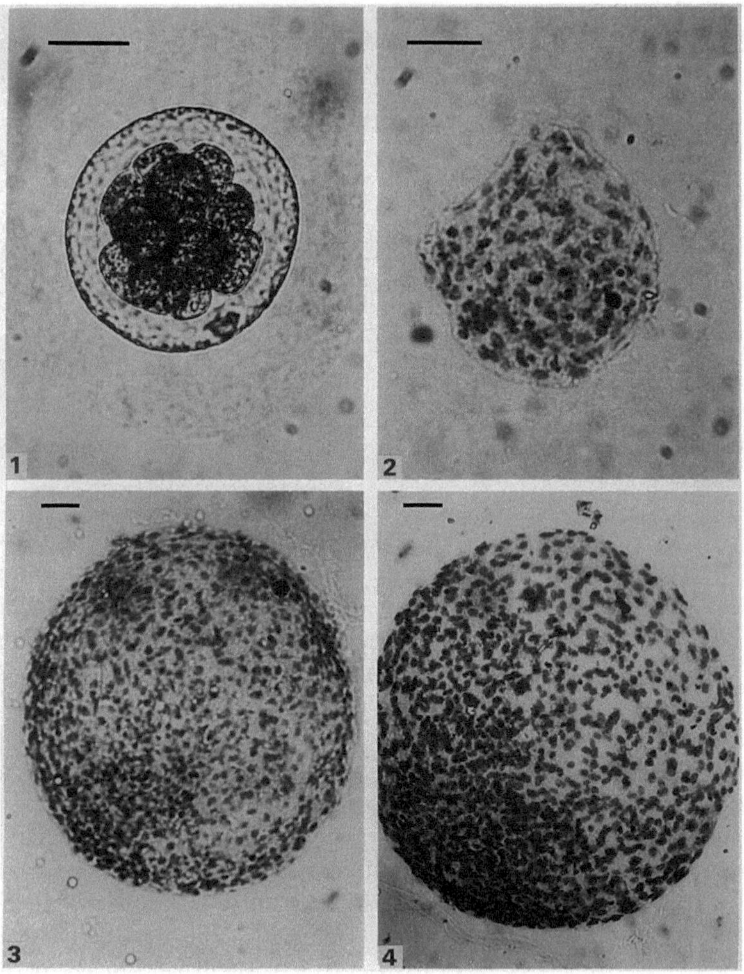

Fig. 1 Photomicrographs of (1) an unstained rabbit morula before culture, and stained blastocysts after culture for 4 days: (2) blastocyst after culture in basic medium without any additive, (3) blastocyst after culture in basic medium plus 0.2 mg/ml low molecular weight extract and (4) blastocyst after culture in basic medium plus 5 mg/ml unextracted BSA (bar = 50 μm). Black dots represent stained nuclei

1.6 Preimplantation Embryonic Coverings in the Rabbit

Preimplantation rabbit embryos have a number of embryonic coats or coverings, some of which can affect embryo development in culture. According to Denker [36], these coverings are as follows: (a) the zona pellucida, produced by the oocyte in the ovary [46]; (b) the mucoprotein layer or mucin coat, derived from oviduct secretions [47]; (c) the neozona, formed under the mucin coat in the uterus by trophoblast secretions; this replaces the zona pellucida which has been completely dissolved away at the blastocyst stage around 4.5 days postcoitum and (d) the gloiolemma derived from uterine secretions. Thus the late rabbit blastocyst has three coats: neozona, mucin coat, and gloiolemma.

Mouse embryos undergo zona shedding or hatching both in vivo [48] and in vitro [15]. Rabbit embryos do not hatch in vivo—they maintain their covering right up to the time of implantation [36]. However, in some circumstances rabbit blastocysts can hatch in vitro. Rabbit embryos cultured from the 1-cell stage with only very thin mucin coats (0–5 μm) hatch or shed the zona at the blastocyst stage in a manner similar to mouse embryos [49]. Embryos with thick mucin coats (59–112 μm) shed the zona much less readily and zona shedding requires considerable growth associated with high cell counts [40]. Embryos collected 20 or 25 h postcoitum and then cultured to blastocysts did not implant on transfer to the uteri of recipients [50]. However, if such embryos were first transferred to the oviduct of recipient does for 24 h to add mucin, and then retransferred to the uteri of synchronized does, live young were produced.

1.7 Use of Embryo Culture Media to Study Development of Rabbit Embryos

Researchers looking for ideas on how rabbit embryo culture may be used to study problems of embryonic development might usefully consult the work of Bernd Fischer's group at Halle in Germany. This group, in almost 20 papers, using the original Modified F10/BSM II medium, has made extensive use of rabbit embryo culture as a tool to explore questions of embryo development such as gene expression [51], mechanisms of insulin action [52], and effects of maternal diabetes [53].

Other workers have used standard culture media, mostly TCM 199 but also DMEM and Ham's F10. Media were generally supplemented with fetal bovine serum but occasionally with rabbit serum or BSA.

1.8 Suggestions for Future Research

In the study of the embryos of laboratory species, a unique advantage attached to the rabbit embryo is the fact that the rabbit blastocyst undergoes a very large degree of growth while it is still a "free floating" entity before implantation. This gives a special interest to the use of rabbit blastocysts cultured in vitro to study the control of early embryonic growth. However, the major problem at this time in rabbit embryo culture is the inability to achieve in vitro blastocyst growth to a level equivalent to in vivo growth. I would suggest that two media might be useful in this endeavor: the RPECM described in the next section, which is based on the original modified F10/BSM II medium, and Medium RD (RPMI-1640 and Dulbecco's MEM, 1:1 v/v).

The procedure for preparing RPECM involves the use of frozen stock solutions for the various nutrient groups together with a very limited number of dry components, mainly NaCl and $NaHCO_3$, thus making it readily possible to study any particular group of components separately. Because 1–2 cell rabbit embryos readily form blastocysts in a medium with KSOM salts supplemented with PVA and only Eagle's essential and nonessential amino acids [54], it would be interesting to explore the effect of substituting the salts in the present RPECM with KSOM salts.

The fact that inositol is essential for rabbit blastocyst growth and that the elements of the phosphatidylinositol signaling system, phosphoinositides and inositol phosphates [23], are present in rabbit blastocysts raise the question of the role of the signaling system and what are its normal stimulators in blastocysts. This could prove a fertile field for research.

The role of citrate in blastocyst growth and cell proliferation is of considerable interest because citrate is currently being used in serum-free media (e.g., with hybridoma [55] and other cells [56]) mainly as a chelator to aid iron uptake. A number of issues arise in relation to studying the effects of citrate on cultured rabbit embryos. One issue is the need to examine its effects over all stages from morulae to late blastocysts. Because Medium RD supports blastocyst development in the absence of protein, it might be a useful medium to study the effects of citrate on rabbit blastocysts, possibly using ferric citrate. Ferric citrate has been recommended as a supplement for CHO cells in serum-free medium [57]. There is evidence that ferric citrate binds to a specific citrate binding site on the cell membrane, which facilitates internalization of both citrate and iron [58].

The interaction of albumin, citrate and iron is complicated: albumin binds citrate [59] and citrate binds iron by chelation [55]. There is evidence that citrate and iron bind to albumin as a citrate-iron complex [59]; thus albumin may have a role in getting citrate and iron into the blastocyst. These facts are important in investigating the role of citrate in blastocyst growth and in how albumin, citrate, and iron might be added to culture medium.

An intriguing possibility for the role of citrate in rabbit blastocyst growth and cell proliferation arises from the suggestion [60] that a type of Warburg effect may be operating during embryo cell proliferation, particularly at the blastocyst stage. Oxidative phosphorylation and glycolysis may be operating independently of each other with glucose and glycolysis supplying building blocks for DNA, protein and lipid synthesis, including glycoproteins and glycolipids for cell junction formation. In such a situation, citrate may fulfill a major role in replenishing TCA cycle intermediates.

An important question is whether a macromolecule should be used and, if yes, what macromolecule should be used—PVA or albumin; ordinary albumin or defatted albumin; with or without the addition of citrate to the culture medium. Choosing the batch of defatted BSA to use may require considerable care because charcoal-treated, defatted albumin may be toxic to cells due to contaminants from the charcoal treatment or by an action whereby such albumin strips low molecular weight metabolites from cells due to its high affinity for such molecules. The use of PVA might substitute for albumin and should be tried, but it is unlikely to have the same relationship with citrate as albumin. Another approach might be to use recombinant albumin, but the problems due to albumin's ligand binding properties are not removed by using recombinant albumin. Perhaps an initial useful approach might be to take ordinary Fraction V or crystallized BSA with a known measured citrate content, supplement this with added citrate so as to give different known concentrations of citrate in the medium when using the BSA at levels of 0.1% and 0.5%, and study the effects of this on expansion and cell proliferation of cultured early blastocysts.

Another potential area of research arises because the trophoblast of the rabbit blastocyst is a type of epithelium that transports nutrients and other substances (apparently including citrate). Thus, it could function as a model for studies on metabolite transport and electrophysiology, such as in the studies of Cross [29].

2 Materials

2.1 Equipment

1. Water purification system to produce ultrapure water with not less than 18 mΩ resistivity. The outlet water should be sterilized by a filter unit with 0.22 μm diameter pores. This quality of water should be used for making up all media and for washing and rinsing glassware.
2. Inverted microscope.
3. Small animal clippers (Oster Model A500).

2.2 Rabbits

1. Breeds: New Zealand White or Dutch Belted rabbits. Dutch Belted rabbits can be superovulated at 20 weeks of age; New Zealand White rabbits a few weeks older.
2. Hormones for superovulation: Folltropin-V (pFSH preparation with reduced LH), equine chorionic gonadotropin (eCG) also known as pregnant mare's serum gonadotropin (PMSG), and human chorionic gonadotropin (hCG).
3. Rabbit Artificial vagina (Ref: 007220, IMV Technologies) for collecting semen.

4. Drugs for rabbit anesthesia: Hypnorm™ (0.20 mg/ml fentanyl base and 10 mg/ml fluanisone) and diazepam (oil solution 5 mg/ml).
5. Tegaderm sterile transparent dressings size 4 × 4¾ in.

2.3 Culture Dishes and Syringe Filter Units

1. Sterile 60 mm × 15 mm tissue-culture-treated (TC) petri dishes for embryo culture.
2. 4-well, 12-well, and 24-well dishes (not TC treated) for embryo culture.
3. Sterile 35 mm × 10 mm petri dishes for embryo collection.
4. Syringe filter units for sterilizing small volumes of media with 0.22 μm pores and 13 mm diameter.

2.4 Chemicals

Use analytical grade reagents throughout.

1. Paraffin oil, paraffin liquid, liquid paraffin, mineral oil for use as an oil overlay, type meeting specifications of the US, European, or British Pharmacopoeias (USP, Ph. Eur., or BP) which should help obviate potential problems of oil toxicity.
2. Polyvinyl alcohol (PVA), cold water soluble.

2.5 Rabbit Preimplantation Embryo Culture Medium (RPECM)

RPECM (*see* Table 3 for detailed composition) is composed of a simple Krebs-Ringer type salt solution (derived from Brinster's Medium; [15]) combined with the amino acids, and trace elements of Ham's F10 medium [16] with some modifications [61]. Inositol is at 45 μM (15× the F10 concentration) and serine at 0.3 mM (3× F10). Nucleic acid precursors (thymidine, hypoxanthine) are omitted, as is vitamin B12. The medium is normally supplemented with 0.5% Fraction V or crystallized BSA. Charcoal-treated BSA (0.1% or 0.5%) may be used where one is adding in fatty acids or citrate. Polyvinyl alcohol (0.1%) may be used in certain situations. No antibiotics or phenol red are used in the medium. The medium is made up from dry components (NaCl, $NaHCO_3$, BSA, or PVA), and frozen stock solutions which have been stored at −20 °C, with 50 ml medium being made up at one time, usually sufficient for one day's experiment (Table 4). The advantage of this arrangement is that any particular set of components of the medium can be studied as desired without altering other components (*see* **Note 1**).

Stock solutions for preparation of RPECM are as follows:

1. STGP (salts, trace elements, glucose, and pyruvate at 50/15× concentration in final medium) Stock Solution (Table 5).
2. Amino Acids (50/9× concentration in final medium) Stock Solution (AA, Table 6).
3. Vitamins (50× concentration in final medium) Stock Solution (V, Tables 7 and 8).

Table 3
Composition of Rabbit Preimplantation Embryo Culture Medium (RPECM)

	Component	*Concentration (mM)*	*Concentration (g/l)*
Salts	NaCl	108.00	6.311
	KCl	4.78	0.356
	$CaCl_2 \cdot 2H_2O$	1.71	0.251
	$MgSO_4 \cdot 7H_2O$	1.19	0.294
	KH_2PO_4	1.19	0.162
	$NaHCO_3$	25.07	2.106
Energy substrates	Glucose	1	0.180
	Na Pyruvate	0.5	0.055
	Component	*Concentration (μM)*	*Concentration (mg/l)*
Trace elements	$FeSO_4 \cdot 7H_2O$	3	0.834
	$CuSO_4 \cdot 5H_2O$	0.01	0.0025
	$ZnSO_4 \cdot 7H_2O$	0.1	0.0288
Amino acids	L-Arginine–HCl	1000	211
	L-Histidine–HCl	100	21.0
	L-Lysine–HCl	100	29.3
	L-Tryptophan	3	0.6
	L-Methionine	30	4.48
	L-Phenylalanine	30	4.96
	L-Tyrosine	10	1.81
	L-Alanine	100	8.91
	Glycine	100	7.51
	L-Serine	300	31.5
	L-Threonine	30	3.57
	L-Aspartic acid	100	13.3
	L-Glutamic acid	100	14.7
	L-Asparagine H_2O	100	15.0
	L-Glutamine	1000	146.2
	L-Isoleucine	20	2.6
	L-Leucine	100	13.1
	L-Proline	100	11.5
	L-Valine	30	3.5
	L-Cysteine	200	24.2
Vitamins	Ca-pantothenate	3	0.715
	Choline chloride	5	0.698
	Myoinositol	45	8.109
	Niacinamide	5	0.611
	Pyridoxine–HCl	1	0.206
	Riboflavin	1	0.376
	Thiamine–HCl	3	1.012
	Biotin	0.1	0.024
	Lipoic acid	1	0.206
	Folic acid	3	1.32

To allow blastocyst development from the 1-cell stage, add 0.5% Fraction V or crystallized BSA

Table 4
Preparation of RPECM

Add the following components to about 20 ml ultrapure water in 50 ml graduated cylinder (*see* Notes 2 and 3)		
Dry salts - allow to dissolve	NaCl $NaHCO_3$	0.3156 g 0.1053 g
Pipet in the following stock solutions	15 ml STGP 9 ml AA 1 ml V	
Make up to	50 ml with H_2O	
Finally add in and dissolve macromolecules	According to the experiment either Fraction V or defatted BSA or polyvinyl alcohol (PVA)	

Medium is made up fresh for an experiment on one day. All solutions are made up with ultrapure water. If desired the medium can be made up first not to 50 ml but to 40 ml and then divided into ten aliquots of 4 ml (assumes ≤10 treatments). To each of these a different compound (or concentration of a compound) might be added to give a particular treatment and then each aliquot made up to 5 ml with water

Table 5
Stock solution STGP

Add the following dry components to 900 ml ultrapure water in a 1L graduated cylinder. Allow to dissolve one at a time (*see* Note 3). Then add trace elements in Solution T. Make up to 1L. Because STGP is diluted at 15 ml in 50 ml of final medium, the concentrations in STGP are at 50/15 the concentrations in RPECM

Component	Concentration (g or ml/l)	Solution T components	Concentration (g/l)
KCl	1.187	$FeSO_4 \cdot 7H_2O$	0.8340
$CaCl_2 \cdot 2H_2O$	0.837	$CuSO_4 \cdot 7H_2O$	0.0025
$MgSO_4 \cdot 7H_2O$	0.980	$ZnSO_4 \cdot 7H_2O$	0.0288
KH_2PO_4	0.540		
Sodium pyruvate	0.183		
Glucose	0.600		
Solution T—trace elements (made up as on right)	3.33 ml		

Note: Solution STGP is stored frozen at −20 °C in aliquots of about 20 ml

Table 6
Stock solution AA (amino acids)

Solution	Component	Stock Soln (g/100 ml)
Soln A_1	L-Arginine HCl	2.11
	L-Histidine HCl	0.21
	L-Lysine HCl	0.29
	L-Tryptophan	0.006
Soln A_2	L-Methionine	0.0448
	L-Phenylalanine	0.0496
	L-Tyrosine	0.0181
Soln A_3	L-Alanine	0.0891
	Glycine	0.0751
	L-Serine	0.3150
	L-Threonine	0.0357
Soln A_4	L-Aspartic Acid	0.1330
	L-Glutamic Acid	0.147
	L-Asparagine H_2O	0.150
	L-Glutamine	1.462
Soln A_5	L-Isoleucine	0.026
	L-Leucine	0.131
	L-Proline	0.115
	L-Valine	0.035
Soln A_6	L-Cysteine	0.242

The procedure for preparation of the stock solution is as follows: Solutions A_1, A_2, A_3, A_5, A_6 are weighed out into slightly less than 100 ml ultrapure water in a 100 ml graduated cylinder and allowed to dissolve (*see* **Note 3**); then as each is completed, combined in a 1000 ml graduated cylinder. Because of the low level of tryptophan in solution A_1, it may be more accurate to make a preliminary more concentrated solution and then add some of this to solution A_1. Solution A_4 needs some care in dissolving the amino acids. The aspartic and glutamic acids are weighed out first into about 100 ml water and dissolved by adding a few drops of approx. 1 N NaOH and by vigorous shaking. The process requires some little time. The glutamine and asparagine are then added together with another few drops of 1 N NaOH and dissolved with vigorous shaking. The solution A_4 is then added to the others in the 1000 ml graduated cylinder and the cylinder made up to the 600 ml mark with H_2O. After mixing the solution it is then diluted 1:2 with water. This may be done most simply by discarding half the solution and leaving a volume of 300 ml in the cylinder. This volume is then made up to 900 ml with water, mixed and stored in small aliquots of about 20 ml at −20 °C. It should be noted that the increased dilution of amino acids as compared with F10 is compensated for by a proportionate increase in the volume of amino acid solution added

Table 7
Stock solution V (vitamins)

Solution	*Component*	*Weigh out (g)*	
Solution A	Folic acid	0.0132	200 ml Solution V is made up as below using ultrapure water. Because Solution V is diluted at 1 ml in 50 ml of RPECM, the concentrations in V are at 50× the concentrations in RPECM 1. Weigh out folic acid into 100 ml water in a 100 ml graduated cylinder. Add a few drops of 1 N NaOH to dissolve folic acid (Solution A). 2. Weigh out all the constituents of Solution B except Biotin and lipoic acid into 400 ml water in a 500 ml graduated cylinder. 3. Weigh out Biotin into 100 ml water in a graduated cylinder. Add 50 ml to 400 ml of Solution B above. 4. Weigh out lipoic acid, dissolve in a few drops of 1 N NaOH. Make up to 50 ml and add 10 ml of this solution to 450 ml Solution B. Make up this solution to 500 ml. Add 100 ml of this solution to 100 ml of Solution A to make 200 ml of Solution V.
Solution B	Ca Pantothenate	0.0357	
	Choline chloride	0.0349	
	Myoinositol	0.4054	
	Niacinamide	0.0305	
	Pyridoxine–HCl	0.0103	
	Riboflavin	0.0188	
	Thiamine–HCl	0.0506	
	Biotin	0.0024	
	Lipoic acid	0.0515	

Solution V is stored in 2–4 ml frozen aliquots

Table 8
Preparation of flushing medium (Solution FM)

Component	*Concentration (g/l)*	*mM*	
NaCl	7.338	125.56	Add the components on the left to about 980 ml ultrapure water in a 1000 ml graduated cylinder. Neutralize solution to pH 7.4 with a 1 N NaOH solution (usually about 6–7 ml/l is required). Carry out neutralization and pH measurement at incubator temperature (usually 39 °C) so to give correct pH for use at that temperature. Add sufficient water to bring the volume up to the 1000 ml mark. Mix thoroughly. If embryos are to be stored in FM for any length of time, sodium pyruvate should be added at the rate of 0.055 g/l or 0.0055 g/100 ml. Store aliquots of 50 or 100 ml frozen at −20 °C. PVA or BSA may be added with other components or just before use. BSA may take some time to dissolve and should be allowed time to do so without shaking. While no antibiotics are used in the culture medium, relatively high levels are used in the Flushing Medium
KCl	0.356	4.78	
$CaCl_2{\cdot}2H_2O$	0.251	1.71	
KH_2PO_4	0.162	1.19	
Mg_2PO_4	0.294	1.19	
HEPES	2.383	10.00	
BSA or PVA	1.000		
Penicillin G	500,000 IU/l		
Streptomycin SO_4	500 mg/l		

3 Methods

3.1 Provision of Embryos

3.1.1 Superovulation of Rabbits

Rabbits are constant estrous animals when on a 12L:12D light cycle and induced ovulators. They can be induced to ovulate at any time either by mating or by injection of luteinizing hormone (LH), gonadotropin-releasing hormone (GnRH), or hCG. Two types of superovulation treatment may be used to increase the number of recovered embryos (~20 to 50 per female):

1. FSH Treatment: Folltropin-V twice daily (a.m. and p.m.) for 3 days via a subcutaneous (sc) injection (0.3 mg/injection for rabbits 20 weeks old and 0.4 or 0.5 mg for adult rabbits or those injected for second time). hCG (100–150 IU) injected intravenously (iv) on the morning of fourth day (*see* **Note 4**).
2. eCG (PMSG): One sc injection of eCG (100–150 IU) followed by iv injection of hCG (100–150 IU) 72 h later, at the same time as mating or artificial insemination. This treatment is not usually as effective as the FSH treatment.

3.1.2 Breeding of Rabbits

Natural Breeding

1. Does should be caged individually for 18 days prior to mating to avoid the possibility of pseudopregnancy.
2. Does should always be brought to the buck's cage for mating. Some does may not accept natural service immediately, and they may have to be brought to the buck a second time.
3. If copulation is completed, ovulation will usually occur, but injection with hCG ensures ovulation.

Artificial Insemination

1. Semen collection: Collect semen with the artificial vagina (AV) as follows (*see* **Note 5**): Prepare the AV by filling the space between liner and casing with water at 50–55 °C (temperature of the water in the AV should be about 42 °C at the time of semen collection). Lubricate the AV with KY or petroleum jelly and attach the collection tube. Bring the teaser doe to the buck's cage, holding her by the scruff of the neck and allow the buck to mount her, but collect the semen with the AV by interposing it between the buck's penis and the doe as he mounts her. Ejaculation is seen to be complete when the buck falls sideways off the doe's back. The yield of semen can be maximized by the use of up to three "false mounts" in which the male is brushed off the teaser doe's back before he can ejaculate. The collected semen should be milky white; if it is yellow, it suggests the presence of urine and should be discarded. Presence of urine may indicate the temperature of the AV was too hot. The semen should be examined under a microscope at 100× for wave motion. Semen from two bucks may be combined, and 0.1 ml of undiluted semen is used for insemination.

2. Insemination: The inseminator should sit down on a chair with a lab coat draped loosely over the knees and between the legs. Invert the doe with the head down between the legs facing the inseminator and the hind limbs supported by the knees. The inseminating catheter consists of a fire-polished glass tubing 15 cm long, with a 4–5 mm outer diameter. About 2 cm from one end a 15–20° bend is made, so that the catheter can be directed over the pelvic arch. Attach a 1 ml syringe to the catheter using 0.3 cm diameter rubber tubing and aspirate 0.1 ml of semen into the catheter. Spread the vulva slightly, gently insert the catheter, and rotate the catheter to clear the pelvic arch. Gently continue insertion of the catheter until the anterior end of the vagina is reached. Withdraw the pipette a few millimeters, expel the semen, and completely withdraw the pipette. Inject hCG into the marginal ear vein at about the same time as insemination.

3.1.3 Embryo Collection

1. Embryos may be collected from rabbits either after humane euthanasia or under anesthesia. Animal care facilities in one's institution should always be consulted for both procedures since regulations differ not just between countries but also between institutions within countries. Possible methods of humane euthanasia [62] include cervical dislocation after stunning by a trained operator or injection of sodium pentobarbital (200 mg/kg) into the marginal ear vein.
2. Anesthesia of rabbits for embryo collection in our laboratory was carried out using Hypnorm and diazepam [63]. Hypnorm was given im (0.5 ml/kg) and diazepam iv (5 mg/kg in a 5 mg/ml oil solution). The fentanyl in the Hypnorm functions as an analgesic and the fluanisone as a tranquilizer, and the diazepam acts as a tranquilizer and muscle relaxant. The analgesic effects of Hypnorm are short lived (20 min), but the effects of the diazepam last for many hours. In the hands of experienced operators, 20 min is amply sufficient for surgery, embryo collection, and suturing the rabbit afterward. It should be noted that regulations apply to the use of Hypnorm in many countries because fentanyl is an opioid (*see* **Note 6**). Anesthetize the rabbit about 10 min before operating. Prepare the rabbit for surgery during that waiting time.
3. The next part of the procedure is the same whether done under anesthesia or after euthanasia. Clip the abdomen of the rabbit with a fine clippers and swab the abdominal area with 70% alcohol. Cover the abdominal surface with a 4 × 4¾ in sterile Tegaderm dressing and place surgical cloths all round and attach with towel forceps. The purpose of the dressing is to maintain the sterility of the skin adjacent to the incision.
4. Make a mid-ventral incision with a scalpel through the dressing to access the uterus and oviduct. Embryos from 1-cell to early

morulae (i.e., from about 12 to 60 h after hCG injection) may be found in the oviduct, day 3 embryos may be in the oviduct or uterus or both, and later stages are in the uterus. Detailed information on the timing of developmental stages in the rabbit (and other species) is present in the review by Blandau [64].

5. Flushing the oviducts: Insert a polyethylene catheter (PE 160) into the fimbrial end of the oviduct and hold it in place with the fingers. Then take a 5 ml syringe filled with sterile flushing medium (Table 8, prewarmed in the incubator to 39 °C) and attach it to a sterile filter and a sterile 18 gauge needle with a blunt tip (sandpapered to remove sharp edges). Insert the blunt needle into the tip of one uterine horn (the rabbit has two uteri and two cervices) near the uterotubal junction, thread the needle through the uterotubal junction into the first few cm of the oviduct, and clamp it tightly in place with either the fingers or a hemostat. In the case of surgically operated animals a hemostat may damage the oviduct. Gently force flushing medium (2.5 ml) out of the syringe through the filter and needle, through the oviduct, and out through the flushing cannula into a 35 × 10 mm plastic culture dish. Note that the polyethylene catheter and the blunt 18 G needle are stored in 70% alcohol and washed out with sterile medium before use.
6. Flushing the uterine horns: Uterine embryos (blastocysts) may be recovered in a number of ways. One way (to be used only with euthanized animals) is to cut out the uterine horns and flush them with medium from the cervical end to the uterotubal junction. A second method that can be used with both anesthetized and euthanized animals is as follows: Make a slit (which can later be closed with one or two sutures) in the wall of the vagina, insert a glass (no sharp or rough edges) or plastic cannula through the slit, and gently thread it through one cervix into the appropriate uterine horn. Then insert a sterile, blunt 18 G needle into the uterus near the uterotubal junction and force 4–5 ml flushing medium through the uterus. The fluid plus blastocysts comes out through the cannula and is collected in a suitable plastic tube. Push any remaining fluid out through the cannula by gently squeezing the fluid from the uterotubal junction end of the uterus to the cervical end with the fingers.

3.1.4 Post-embryo Collection Procedures

For terminal surgeries, the animal should be killed while it is still under anesthesia according to approved methods. If the donor rabbit is allowed to recover, close incision with two sets of sutures: at the subcutaneous layer with one continuous stitch using a synthetic absorbable material such as polydioxanone [65], and at the skin with interrupted stitches using a nylon thread. Monitor recovery from anesthesia and surgery carefully in the first 2 h after the operation (while it may still be kept in the operation room) and over the following couple of days after it is returned to the animal house. Check particularly that the skin stitches are all intact as rabbits tend to pick at the stitches. Institutional veterinary advice should be consulted on the advisability of using antibiotics after surgery.

3.2 Culture Procedures

3.2.1 Preparation of Culture Dishes

Since rabbit embryos can vary in size from less the 0.2 mm to over 5 mm, various types of dish/well culture systems may be used. The principal points to be considered are type and size of culture dish, the use of an oil overlay, and how these may suit various sizes and stages of embryo. There is also room for researcher preferences particularly in relation to cleavage stage embryos and whether an oil overlay is used with multiwell dishes (*see* **Note** 7). It is very important that culture dishes and multiwell plates should be prepared with medium and oil (if used) the night before to allow for gas phase equilibration in the incubator.

1. Tissue culture dishes with a paraffin oil (paraffin liquid) overlay.

   Cleavage stage embryos can be cultured in drops of medium under paraffin liquid (paraffin oil, US terminology) in 60 × 15 mm plastic TC-treated petri dishes. Label the bottom of the petri dishes with the drop number(s) using an oil-based marking pen. If there is more than one drop, appropriate divisions should also be marked on the bottom of the dish. Using more than one drop per dish has the disadvantage that they may run together under the paraffin oil. Pipet 10 ml of paraffin into each dish. Place 0.5 ml drops of medium under the paraffin using a 25 g needle attached to a sterile, disposable 1 ml tuberculin syringe and syringe filter. The tip of the needle is placed under the paraffin and a droplet of 0.5 ml medium is forced through the filter and under the oil (*see* **Note 8**).

2. Multiwell dishes w or w/o paraffin overlay.

   Four-well dishes (non-TC treated) are particularly suitable for blastocyst stages, but can also be used for cleavage stages. Embryos are cultured in 0.5 or 1 ml volume of medium with or without a thin (1 mm) overlay of paraffin (*see* **Note** 7). Smaller medium drops (0.1 or 0.2 ml) may be used if one is only working with cleavage stages and not blastocysts.

3.2.2 Preparation of Embryos

1. All embryo handling operations after collection should be carried out in a Class 2 Biological Safety Cabinet using a high quality binocular dissecting microscope with a diascopic stand.
2. Wash embryos free of oviductal components and debris by successive transfers to culture dishes (e.g., size 35 × 10 mm) containing approximately 2 ml of sterile flushing medium. Pick up the embryos in as small a fluid volume as possible with a suitable pipette and then transfer them to the culture dish (*see* **Note 9**). Make three transfers of embryos from washing dish to

washing dish, and after each transfer either gently rock the dish to spread the oviductal components or move embryos about the dish with the pipette.

Avoid cold shock of the embryos as far as possible. Also avoid temperatures above 40 °C as these can readily damage embryos.

3.2.3 Placement of Embryos in Culture

The actual placing of embryos in the treatment drop should be done by taking up the full complement of embryos for that treatment in as small a volume of fluid as possible and gently expelling the embryos into the culture medium.

Statistical considerations should inform how one places embryos in culture. If embryos are collected from a number of does on one particular day, pooled, and then divided as equally as possible and allocated randomly between the treatments, all of which have been set up together at one time and placed in culture in dishes, all of which have been set up together at one time and in media which have a common origin, that day's experiment constitutes a single block of data. If the experiment is repeated over a number of days in a similar fashion then one has a randomized complete block design (RCB).

An alternative approach, if one has enough embryos from each superovulated doe, is to randomly allocate an equal number of embryos from each doe to each treatment. In that case, the medium drops associated with each doe constitute one block of an experiment. However, it is desirable that the number of embryos per drop should lie between about four and ten, at least for cleavage stages, as there is evidence that increasing numbers per drop up to about ten increases blastocyst formation [61]. Allocating embryos in this fashion enables valid experimental comparisons to be made and facilitates proper statistical analysis of the data. The situations where particular treatments are omitted from particular days, or particular does, can make experimental conclusions very dubious.

3.2.4 Temperature and Gas Phase in the Incubator

1. Temperature. The body temperature of the rabbit is about 39 °C [66] and therefore rabbit embryos should probably be cultured at 39 °C. Since high temperatures will rapidly kill embryos and temperature control in incubators can be defective, it is useful to check the incubator temperature regularly with a certified, calibrated thermometer.
2. Gas Phase. Incubator gas phase should probably have 5% O_2 (but *see* **Note 10**) and 10% CO_2 [31], even though 5% CO_2 is more commonly used. One can use the Bacharach Fyrite gas analyzer to check both O_2 and CO_2 concentrations. One can also use the Henderson–Hasselbalch equation to check the pH for various CO_2 and HCO_3 concentrations, which will differ slightly from the actual pH of the medium due to the buffering action of other medium components such as amino acids. In

checking medium pH with a pH meter, one should always check at the temperature at which the medium will be used and the meter should also be calibrated with standards at that same temperature. A useful reference explaining issues of oxygen, carbon dioxide, and water vapor partial pressures in incubators and the effects of altitude above sea level is [67].

3.3 Examination of Embryos

3.3.1 Routine Examination

1. Generally embryos should be examined at 24 h intervals, or more frequently, depending on the experiment, using an inverted compound microscope. As far as possible avoid shocking the embryos on a cold microscope stage. One-cell embryos in culture cleave to the early morula stage (round ball of cells containing 16 blastomeres) after about 24 h of culture, a late morula at about 48 h, and blastocysts at 72–84 h [61]. The beginnings of blastocoel formation are first visible as a series of cracks in the morula usually arising from one spot, which then becomes a clear or light area, the blastocoel cavity. The blastocoel enlarges to leave a thin rim of cells, the trophoblast and a thicker clump of smaller cells to one side, the inner cell mass. Blastocysts cultured from 1-cell embryos routinely expand and then escape or "hatch" through a hole in the zona and leave the zona behind. These "hatched" blastocysts may continue expansion, depending on the medium, up to a diameter of 500–700 μm.
2. Without using special methods for examination of the embryos such as embryo staining, it is not possible to be certain that cleavage stages of embryos (i.e., preblastocyst stages) contain normal, nucleated blastomeres. Fragmentation of blastomeres occurs readily in poor culture conditions. Nevertheless, the processes of blastocyst formation, hatching, and expansion can be followed without using special methods.
3. In examining the embryos it may be necessary to adjust the microscope's iris diaphragm of the condenser somewhat to gain contrast, which is normally very poor in unstained tissue. With 3D structures such as embryos, this technique can give better contrast than a phase contrast microscope. Haloing from out-of-focus structures with phase contrast microscopy causes problems, although apodized optical systems can prevent this effect.

3.3.2 Measuring Blastocyst Diameter

Blastocyst diameters can readily be measured using an eyepiece graticule. An eyepiece graticule is a glass disc fitted into the eyepiece of the microscope; the disc is marked with a fine scale from 0 to 100. The eyepiece graticule must be calibrated for each objective lens with a stage micrometer, which is a glass slide engraved with an accurate scale. A hemacytometer grid is a readily available scale that can be used. Eyepiece graticules can be fitted to existing eyepieces or eyepieces purchased with fitted graticules. See https://www.microscopyu.com/tutorials/reticlecalibration for a tutorial on calibrating an eyepiece. If the blastocyst is not spherical but oval, then one should measure both the long and short axis and use the mean of both measurements.

3.3.3 Staining of Embryos

Hoechst 33342 Method

1. Fix embryos in 1% formalin overnight.
2. Prestain them for 3 min with 0.1% trypan blue in 2.3% sodium citrate dihydrate to quench the fluorescence of sperm heads attached to the mucin coat.
3. Stain them for 3 min with the DNA-specific fluorochrome, Hoechst 33342 (1 mg/ml water).
4. Count nuclei at 500–800 magnification under epifluorescence illumination using an eyepiece micrometer grid to facilitate counting (for details of this method, *see* refs. 6, 68).

Grayson Method

1. Place embryos in a small drop of medium on a slide precoated with Mayer's albumin and allow air-dry overnight (*see* **Note 11**).
2. Fix the air-dried slides in ethanol–acetic acid (2:1, v/v) for 45 min, stain in aceto-orcein–HCl for 30 min (*see* **Note 11**).
3. Dehydrate successively in 50%, 60%, 80%, and 100% ethanol (5 min intervals), clear in xylene using the sequence of 50% xylene–50% ethanol (5 min) followed by two changes to 100% xylene (each 5 min). Mount in mounting medium under a coverslip.
4. Count the number of cells per blastocyst under high power with the aid of an eyepiece micrometer grid (for details of this method, *see* refs. 40, 69).
5. The Grayson method works with stages from oocyte to blastocyst, but there may be some difficulty with cleavage stages of the rabbit embryo due to the presence of the mucin coat if these are grown in vivo; mucin may be removed with acidic PBS and pronase [70].

4 Notes

1. When one is adding new media components or increasing, decreasing or removing existing components, one may have to adjust the NaCl concentration to compensate for osmolarity changes. This is not necessary if the change is less than 1 mOsm and does not really become important until the change is of the

order of 5–10 mOsm or more. The basic rule is that osmolarity (and therefore the adjustment made in NaCl) is based on the number of dissolved particles (e.g., 1 molecule of glucose or inositol = 1 particle, 1 molecule of NaCl = 2 particles, and 1 molecule of Na_2SO_4 = 3 particles). Thus, a 1 mM change in NaCl can compensate for a 2 mM change in inositol or glucose and a 0.66 mM change in Na_2SO_4.

2. When making up all media and stock solutions, always add $NaHCO_3$ last or near to last (just prior to the addition of BSA, when used), after all other components in the medium are dissolved. The reason for this is that the concentrations of calcium and phosphate ions are very close to their solubility product (above which precipitation as a calcium phosphate salt may occur) and the addition of $NaHCO_3$ makes the pH more alkaline and thus may cause calcium phosphate precipitation. Also add the $NaHCO_3$ to a near full volume of the medium, as precipitation is also promoted by its addition in the presence of high concentrations of calcium and phosphate ions in reduced volumes.
3. To fully dissolve all components, cover the top of the graduated cylinder with a double layer of Parafilm and invert the cylinder a number of times. For some solutions, vigorous shaking, as indicated in table caption, is necessary to dissolve chemicals.
4. Superovulation procedures using FSH in rabbits are based on work with Armour porcine FSH preparations [71, 72]. There was, however, an error in Kennelly and Foote's paper [71] with respect to the dose of FSH, which is corrected in the Varian, Maurer, and Foote's paper [72]. The error was that the Kennelly and Foote's paper [71] gave the dose of FSH as "per day" when it was meant to be "per injection."
5. For a description of the Walton type of commercially available AV *see* ref. 73. However, another simpler type of AV [74] has distinct advantages in that (a) a number can be readily fabricated, (b) semen does not normally come into contact with the liner of the AV and (c) it can easily be washed and sterilized without disassembly. This has the added advantage that a different AV can be used for each buck, thus reducing the possibility of the spread of disease. The only downside is that they are not available commercially and have to be fabricated in house. The use of this type of AV has been recently reviewed [75].
6. An alternative to the Hypnorm diazepam combination for rabbit anesthesia is the Hypnorm midazolam combination [76]. Midazolam is a water-soluble benzodiazepine unlike diazepam which is poorly soluble in water. It should be noted that in the USA fentanyl because of its opioid nature is a

Schedule II controlled drug subject to regulation under the Controlled Substance Act.

7. TC-treated dishes have a hydrophilic surface which allows cells to adhere to the surface and proliferate. It also allows culture medium drops under a layer of oil (such as the layer formed when 10 ml oil is pipetted into a 60 × 15 mm dish) to adhere to the dish surface and remain in place. This system is probably most useful for culturing cleavage stage embryos from 1-cell to morula in small volumes of medium ($\leq$0.5 ml). It also means that when oocytes surrounded by follicular cells, are cultured in similar conditions, the follicular cells can adhere to the dish surface and proliferate.

   On the other hand non TC-treated dishes have a hydrophobic surface which does not allow cells or aqueous drops to adhere to the surface. This works very well in multiwell dishes, such as 4-well plates, where the surface of the well is covered with medium. This system is very useful for blastocyst culture, but it can also be used with cleavage stage embryos. The system can be used with or without a paraffin overlay. The advantages of using paraffin both in the culture dish system and multiwell system are that it helps to reduce evaporation from the medium and can also act as a barrier against contamination by microorganisms. The disadvantages are that it adds another unknown factor and occasionally some batches have proved to be toxic to embryos. Using oil which meets the USP, Ph. Eur., or BP specifications should reduce this risk. Embryo-tested oil can be obtained from some sources. Oil can also be washed with an excess of medium if problems of toxicity arise, but this may not always be compatible with maintaining sterility.

8. We had a practice of sterilizing medium *immediately before the point of use*. When embryos were being flushed from oviduct or uterus, a disposable syringe filter with 0.22 mm pores, diameter 13 mm, was interposed between the flushing syringe and the flushing needle to sterilize the medium. When culture medium was being dispensed into culture dishes (either under an oil layer or without an oil layer), it was similarly sterilized with a syringe filter attached to the dispensing syringe. If it was being dispensed under oil, a 25 G sterile needle was attached to the filter unit. We always discarded the first 0.25–0.5 ml to rinse out contaminants.

9. Suitable pipettes may be drawn from glass Pasteur pipettes as described in [7]. Diameters of the drawn pipettes should fit the different sizes of various embryo stages being handled. This issue arises because (1) 1-cell rabbit embryos have a diameter about 160–170 μm, almost twice that of the mouse embryo, (2) deposition of the mucin coat in the oviduct can add 200 μm

or more to the total diameter of the embryo, and (3) intrauterine blastocyst growth can lead to diameters of 5–6 mm [77]. Such large blastocysts are probably best handled using sterile, individually wrapped plastic Pasteur pipettes. The tip of the pipette can be cut at a suitable point to fit a given size blastocyst and the bulb cut off at a suitable point to allow a 1 ml syringe to be inserted.

We have also used the following types of pipettes, all of which require a suitable type of adapter to connect them to a 1 ml syringe: (a) disposable glass capillary micropipettes (10 μl for day 1–3 embryos, 50 μl for day 4 embryos) and (b) pipettes prepared from sterile plastic insemination straws (day 5 embryos). An aspirator tube assembly for calibrated microcapillary pipettes can be used for attaching the capillary pipettes to a 1 ml syringe (Sigma-Aldrich, cat no. A5177). Alternatively one can use tubing supplied with the aspirator assembly for mouth pipetting control. The aspirator assembly will also work with the insemination straws provided one uses unplugged straws or cuts off the plugged end of the straw before use.

10. Embryos will readily develop from 1-cell to blastocyst in 20% oxygen but one will get superior growth with 5%. However, one can carry out informative experiments in 20% oxygen; also instead of an incubator with variable oxygen control, one can use a modular chamber filled with 5% oxygen within the incubator.

11. Mayer's albumin, also called albumin glycerol fixative is a mixture of 50% egg white and 50% glycerol and is readily available commercially. Aceto-orcein–HCl stain is prepared by adding 2 g orcein to 45 ml acetic acid at 100 °C, followed by 55 ml H_2O. Before use the stain is filtered and mixed 9:1 with concentrated HCl [69].

Acknowledgments

Thanks are due to Dr. G.E. Seidel, Jr., Colorado State University, Dr. L.R. Quinlan, National University of Ireland, Galway for very helpful criticism of the manuscript and to Ms. B. Coen, National University of Ireland, Galway for advice on equipment.

References

1. Brachet A (1912) Développement *in vitro* de blastodermes et jeunes embryons de mammifères. C R Acad Sci Paris 155:1191–1193
2. Lewis WH, Gregory PW (1929) Cinematographs of living developing rabbit-eggs. Science 69:226–229
3. Whitten WK (1957) Culture of tubal ova. Nature 179:1081–1082
4. McLaren A, Biggers JD (1958) Successful development and birth of mice cultivated *in vitro* as early embryos. Nature 182:877–878

5. Kane MT, Foote RH (1970) Culture of two-and four-cell rabbit embryos to the expanding blastocyst stage in synthetic media. Proc Soc Exp Biol Med 133:921–925
6. Carney EW, Foote RH (1991) Improved development of rabbit one-cell embryos to the hatching blastocyst stage by culture in a defined, protein-free medium. J Reprod Fertil 91:113–123
7. Hogan B, Costantini F, Lacy E (1986) Manipulating the mouse embryo: a laboratory manual. Cold Spring Harbor Laboratory, Cold Spring Harbor
8. Brinster RL (1967) Protein content of the mouse embryo during the first five days of development. J Reprod Fertil 13:413–420
9. Turner K, Goldstein DJ, Rogers AW (1992) Variation in the dry mass of mouse embryos throughout the preimplantation period. Hum Reprod 7:112–116
10. Denker H-W (1977) Implantation: the role of proteinases, and blockage of implantation by proteinase inhibitors. Adv Anat Embryol Cell Biol 53:1–123
11. Fujimoto S, Pahlavan N, Woody HD, Dukelow WR (1975) Cell numbers in rabbit pre-implantation blastocysts. Cytologia (Tokyo) 40:307–311
12. Daniel JC Jr (1964) Early growth of rabbit trophoblast. Am Naturalist 98:85–97
13. Morgan PM, Kane MT (1993) Protein content of rabbit embryos: one cell to peri-implantation blastocyst. J Reprod Fertil 97:101–106
14. Kane MT (1969) *In vitro* culture of two- to four-cell rabbit embryos to expanding blastocysts in serum extracts and synthetic media. PhD Thesis, Cornell University
15. Brinster RL (1963) A method for *in vitro* cultivation of mouse ova from two-cell to blastocyst. Exp Cell Res 32:205–208
16. Ham RG (1963) An improved nutrient solution for diploid Chinese hamster and human cell lines. Exp Cell Res 29:515–526
17. Hegele-Hartung C, Fischer B, Beier HM (1988) Development of preimplantation rabbit embryos after in-vitro culture and embryo transfer: an electron microscopic study. Anat Rec 220:31–42
18. Lindenau A, Fischer B (1996) Embryotoxicity of polychlorinated biphenyls (PCBS) for preimplantation embryos. Reprod Toxicol 10:227–230
19. Kane MT (1988) The effects of water soluble vitamins on the expansion of rabbit blastocysts *in vitro*. J Exp Zool 245:220–222
20. Kane MT (1989) Effects of the putative phospholipid precursors, inositol, choline, serine and ethanolamine, on formation and expansion of rabbit blastocysts *in vitro*. J Reprod Fertil 87:275–279
21. Fahy MM, Kane MT (1992) Inositol stimulates DNA and protein synthesis, and expansion by rabbit blastocysts *in vitro*. Hum Reprod 7:550–552
22. Warner SM, Conlon FV, Kane MT (2003) Inositol transport in preimplantation rabbit embryos: effects of embryo stage, sodium, osmolality and metabolic inhibitors. Reproduction 125:479–493
23. Fahy MM, Kane MT (1993) Incorporation of [^{3}H]inositol into phosphoinositides and inositol phosphates by rabbit blastocysts. Mol Reprod Dev 34:391–395
24. Kane MT, Foote RH (1971) Factors affecting blastocyst expansion of rabbit zygotes and young embryos in defined media. Biol Reprod 4:41–47
25. Kane MT (1987) Minimal nutrient requirements for culture of one-cell rabbit embryos. Biol Reprod 37:775–778
26. Kane MT (1972) Energy substrates and culture of single-cell rabbit ova to blastocysts. Nature 238:468–469
27. Naglee DL, Maurer RR, Foote RH (1969) Effect of osmolarity on in vitro development of rabbit embryos in a chemically defined medium. Exp Cell Res 58:331–333
28. Kane MT (1974) The effects of pH on culture of one-cell rabbit ova to blastocysts in bicarbonate buffered medium. J Reprod Fertil 38:477–480
29. Cross MH (1974) Rabbit blastocoele bicarbonate: accumulation rate. Biol Reprod 11:654–666
30. Kane MT (1975) Bicarbonate requirements for culture of one-cell rabbit ova to blastocysts. Biol Reprod 12:552–555
31. Hallden K, Li JM, Carney EW, Foote RH (1992) Increasing carbon dioxide from five percent to ten percent improves rabbit blastocyst development from cultured zygotes. Mol Reprod Dev 33:276–280
32. Farrell PB, Foote RH (1995) Beneficial effects of culturing rabbit zygotes to blastocysts in 5% oxygen and 10% carbon dioxide. J Reprod Fertil 103:127–130
33. Lindenau A, Fischer B (1994) Effect of oxygen concentration in the incubator's gas phase on the development of cultured preimplantation rabbit embryos. Theriogenology 41:889–898
34. Li J, Foote RH (1993) Culture of rabbit zygotes into blastocysts in protein-free

medium with one to twenty per cent oxygen. J Reprod Fertil 98:163–167

35. Seidel GE Jr, Bowen RA, Kane MT (1976) *In vitro* fertilization, culture and transfer of rabbit ova. Fert Steril 27:861–869
36. Denker HW (2000) Structural dynamics and function of early embryonic coats. Cells Tissues Organs 166:180–207
37. Fasano M, Curry S, Terreno E et al (2005) The extraordinary ligand binding properties of human serum albumin. IUBMB Life 57:787–796. https://doi.org/10.1080/15216540500404093
38. Westphal U (1970) Corticosteroid-binding globulin and other steroid hormone carriers in the blood stream. J Reprod Fertil Suppl 10:15–38
39. Blundell TL (1977) The binding of heavy metals to proteins. Chem Soc Rev 6:131–171
40. Kane MT (1983) Variability in different lots of commercial bovine serum albumin affects both cell multiplication and hatching of rabbit blastocysts in culture. J Reprod Fertil 69:555–558
41. Kane MT (1979) Fatty acids as energy sources for culture of one-cell rabbit ova to viable morulae. Biol Reprod 20:323–332
42. Kane MT, Headon DR (1980) The role of commercial bovine serum albumin preparations in the culture of one-cell rabbit embryos to blastocysts. J Reprod Fertil 60:469–475
43. Kane MT (1985) A low molecular weight extract of bovine serum albumin stimulates rabbit blastocyst cell division and expansion. J Reprod Fertil 73:147–150
44. Gray CW, Morgan PM, Kane MT (1992) Purification of an embryotrophic factor from commercial bovine serum albumin and its identification as citrate. J Reprod Fertil 94:471–480
45. Rorie RW, Miller GF, Nasti KB, McNew RW (1994) *In vitro* development of bovine embryos as affected by different lots of bovine serum albumin and citrate. Theriogenology 42:397–403
46. Wassarman PM, Litscher ES (2013) Biogenesis of the mouse egg's extracellular coat, the zona pellucida. Curr Top Dev Biol 102:243–266. https://doi.org/10.1016/B978-0-12-416024-8.00009-X
47. Denker H-W, Gerdes H-J (1979) The dynamic structure of rabbit blastocyst coverings. I. Transformation during regular preimplantation development. Anat Embryol 157:15–34
48. Lin SP, Lee RK, Tsai YJ (2001) *In vivo* hatching phenomenon of mouse blastocysts during implantation. J Assist Reprod Genet 18:341–345
49. Kane MT (1975) Inhibition of zona shedding of rabbit blastocysts in culture by the presence of a mucin coat. J Reprod Fertil 44:539–542
50. Murakami H, Imai H (1996) Successful implantation *of in vitro* cultured rabbit embryos after uterine transfer: a role for mucin. Mol Reprod Dev 43:167–170
51. Koerber S, Santos AN, Tetens F, Küchenhoff A, Fischer B (1998) Increased expression of NADH-ubiquinone oxidoreductase chain 2 (ND2) in preimplantation rabbit embryos cultured with 20% oxygen concentration. Mol Reprod Dev 49:394–399
52. Navarrete Santos A, Tonack S, Kirstein M, Pantaleon M, Kaye P, Fischer B (2004) Insulin acts via mitogen-activated protein kinase phosphorylation in rabbit blastocysts. Reproduction 128:517–526
53. Ramin N, Thieme R, Fischer S, Schindler M, Schmidt T, Fischer B, Navarrete Santos A (2010) Maternal diabetes impairs gastrulation and insulin and IGF-I receptor expression in rabbit blastocysts. Endocrinology 151:4158–4167. https://doi.org/10.1210/en.2010-0187
54. Liu Z, Foote RH, Simkin ME (1996) Effect of amino acids and α-amanitin on the development of rabbit embryos in modified protein-free KSOM with HEPES. Mol Reprod Dev 45:157–162
55. Bai Y, Wu C, Zhao J, Liu YH, Ding W, Ling WL (2011) Role of iron and sodium citrate in animal protein-free CHO cell culture medium on cell growth and monoclonal antibody production. Biotechnol Prog 27:209–219. https://doi.org/10.1002/btpr.513
56. Baker E, Baker SM, Morgan EH (1998) Characterisation of non-transferrin-bound iron (ferric citrate) uptake by rat hepatocytes in culture. Biochim Biophys Acta 1380:21–30
57. Keen MJ, Rapson NT (1995) Development of a serum-free culture medium for the large scale production of recombinant protein from a Chinese hamster ovary cell line. Cytotechnology 17:153–163. https://doi.org/10.1007/BF00749653
58. Graham RM, Morgan EH, Baker E (1998) Characterisation of citrate and iron citrate uptake by cultured rat hepatocytes. J Hepatol 29:603–613
59. Hanson RW, Ballard FJ (1968) Citrate, pyruvate, and lactate contaminants of commercial serum albumin. J Lipid Res 9:667–668
60. Krisher RL, Prather RS (2012) A role for the Warburg effect in preimplantation embryo development: metabolic modification to support rapid cell proliferation. Mol Reprod Dev

79:311–320. https://doi.org/10.1002/mrd.22037

61. Kane MT (1987) *In vitro* growth of preimplantation rabbit embryos. In: Bavister BD (ed) The mammalian preimplantation embryo: regulation of growth and differentiation *in vitro*. Plenum, New York, pp 193–217
62. Close B, Banister K, Baumans V et al (1996) Recommendations for euthanasia of experimental animals: Part 2. Lab Anim 31:1–32
63. Green CJ (1975) Neuroleptanalgesic drug combinations in the anaesthetic management of small laboratory animals. Lab Anim 9:161–178
64. Blandau RJ (1961) Biology of eggs and implantation. In: Young WC (ed) Sex and internal secretions, vol II, 3rd edn. The Williams and Wilkins Co., Baltimore, pp 797–882
65. Kladakis S (2014) Choosing sutures in small animal surgery. J Dairy Vet Anim Res 1:68–71. https://doi.org/10.15406/jdvar.2014.01.00015
66. Chen PH, White CE (2006) Comparison of rectal, microchip transponder, and infrared thermometry techniques for obtaining body temperature in the laboratory rabbit (*Oryctolagus cuniculus*). J Am Assoc Lab Anim Sci 45:57–63
67. Wenger RH, Kurtcuoglu V, Scholz CC, Marti HH, Hoogewijs D (2015) Frequently asked questions in hypoxia research. Hypoxia (Auckl) 3:35–43. https://doi.org/10.2147/HP.S92198
68. Pursel VG, Wall RJ, Rexroad CE, Hammer RE, Brinster RL (1985) A rapid whole-mount staining procedure for nuclei of mammalian embryos. Theriogenology 24:687–691
69. Grayson K (1978) An improved method for staining mammalian oocytes. Stain Technol 53:115–116
70. Giles JR, Foote RH (1995) Rabbit blastocyst: allocation of cells to the inner cell mass and trophectoderm. Mol Reprod Dev 41:204–211
71. Kennelly JJ, Foote RH (1965) Superovulatory response of pre- and post-pubertal rabbits to commercially available gonadotrophins. J Reprod Fertil 9:177–188
72. Varian NB, Maurer RR, Foote RH (1967) Ovarian response and cleavage rate of ova in control and FSH-primed rabbits receiving varying levels of luteinizing hormone. J Reprod Fertil 13:67–73
73. Daniel N, Renard JP (2010) Artificial insemination in rabbits. Cold Spring Harb Protoc 2010:pdb prot5358. https://doi.org/10.1101/pdb.prot5358
74. Bredderman PJ, Foote RH, Yassen AM (1964) An improved artificial vagina for collecting rabbit semen. J Reprod Fertil 7:401–403
75. Amann RP, Foote RH (2004) Artificial vagina for rabbits. J Androl 25:184
76. Flecknell PA, Mitchell M (1984) Midazolam and fentanyl-fluanisone: assessment of anaesthetic effects in laboratory rodents and rabbits. Lab Anim 18:143–146
77. Alliston CW, Pardee NR (1973) Variability of embryonic development in the rabbit at 19 to 168 hours after mating. Lab Anim Sci 23:665–670

Chapter 6

In Vitro Maturation, Fertilization, and Culture of Pig Oocytes and Embryos

Bethany K. Redel, Lee D. Spate, and Randall S. Prather

Abstract

Assisted reproductive technologies in the pig are critical for the production of genetically modified pigs as models of human disease and to improve production agriculture. Methods of oocyte maturation, fertilization, and culture all play an extremely important role in how the embryo, fetus, and offspring will develop. In this chapter, we discuss the historical methods and recent advances that have been essential in promoting efficient and competent embryo development. Here we describe the procedures that can be used to mature, fertilize, and culture pig embryos to the blastocyst stage.

Key words Porcine, Oocyte, Embryo, Maturation, Culture

1 Introduction

Pigs are an important species for use in biomedicine as they make excellent models for many human conditions [1, 2], as well as having an important role in agriculture [3]. Genome editing and somatic cell nuclear transfer techniques utilized to genetically engineer these pigs require the in vitro use of oocytes and embryos. Historically, the in vitro production of pig oocytes and embryos is a very inefficient process with low rates of subsequent embryonic development. Therefore, it is our goal to increase the efficacy of these techniques to produce more competent embryos that can be transferred to a recipient. This chapter will focus on the historical perspectives of these techniques and recent advances that have assisted in improving porcine oocyte and embryo competence in vitro.

1.1 Historical Perspective on Oocyte Maturation Systems

Pig oocyte maturation has been investigated for more than 50 years [4]. Compared to other species, pig oocyte maturation has been very inefficient in producing competent oocytes and embryos. The length of maturation is dependent on the species. In pigs, in vitro oocyte maturation lasts about 40–44 h, in cows maturation lasts

Jason R. Herrick (ed.), *Comparative Embryo Culture: Methods and Protocols*, Methods in Molecular Biology, vol. 2006, https://doi.org/10.1007/978-1-4939-9566-0_6, © Springer Science+Business Media, LLC, part of Springer Nature 2019

about 16–24 h [5], humans maturation lasts 24–36 h [6] and mouse maturation lasts about 15–16 h [7]. Due to the low efficiency of nuclear and cytoplasmic maturation in pigs, much research has been completed to determine how this in vitro environment is suboptimal. The most commonly used maturation medium is Tissue Culture Medium (TCM)-199, while a variety of other media have been used such as Waymouth MB 752/1, North Carolina State University (NCSU23), and modified Tyrode's solution (mTLP-PVA). A few obstacles that porcine maturation has had to overcome were (1) low rates of male pronuclear formation, (2) high rates of polyspermy, and (3) low development to the blastocyst stage. In the past, fetal calf serum was used in the medium and caused low male pronuclear formation [8, 9]. Improvements to oocyte maturation were made by removing fetal calf serum from the medium, supplementing with porcine follicular fluid [9], and by the addition of cysteine to the medium [10, 11]. With many maturation systems in vitro matured porcine oocytes have consistently high rates of polyspermic fertilization, which can often exceed 50% of the oocytes fertilized [12, 13]. Many methods have been implemented to decrease the levels of polyspermy in pig oocytes, whether it is finding the optimal sperm concentration [14], identifying the correct volume of fertilization medium and number of oocytes [15], or by decreasing the time of coincubation of sperm with oocytes [16]. One way of improving cytoplasmic maturation and subsequent embryo development was by the addition of epidermal growth factor (EGF) [17]. The presence of EGF in maturation did not affect nuclear maturation; however, it did increase the percentage of blastocysts that developed after fertilization. The addition of EGF also allowed for BSA, fetal bovine serum, and follicular fluid to be removed from the maturation system so that it was a chemically defined system. Many maturation systems are now serum-free to eliminate variability and increase repeatability among different laboratories [18].

1.2 Recent Advances to Porcine Oocyte Maturation

Recently, three cytokines have been identified to improve nuclear oocyte maturation and developmental competence [19]. Fibroblast growth factor 2 (FGF2), leukemia inhibitory factor (LIF), and insulin-like growth factor 1 (IGF1), together referred to as FLI, improve the nuclear maturation of immature pig oocytes and increase the efficiency of blastocyst production by twofold after in vitro fertilization. This is thought to occur as FLI matured cumulus oocyte complexes (COCs) display differences in MAPK activation in the cumulus cells, which causes an increase in cumulus cell expansion (Fig. 1). Another consistent finding in the FLI matured oocytes is that the perivitelline space is much larger than in oocytes matured without FLI (Fig. 2). One hypothesis is that FLI promotes the severance of transzonal projections from the cumulus cells into the oocyte and this may allow the oocyte to

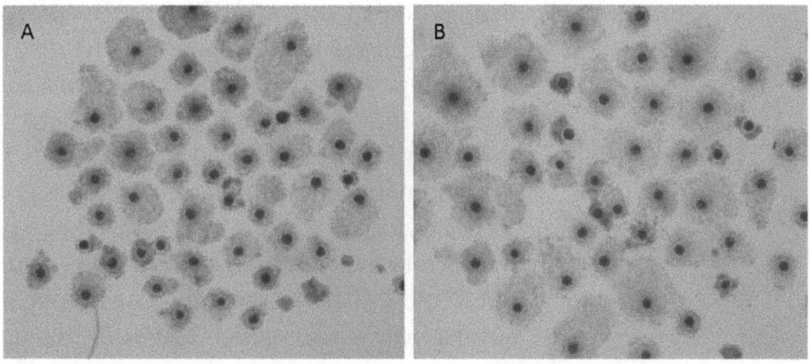

Fig. 1 Effects of FLI on cumulus cell expansion after 24 h of maturation. (**a**) COCs matured with control medium, (**b**) COCs matured with FLI medium

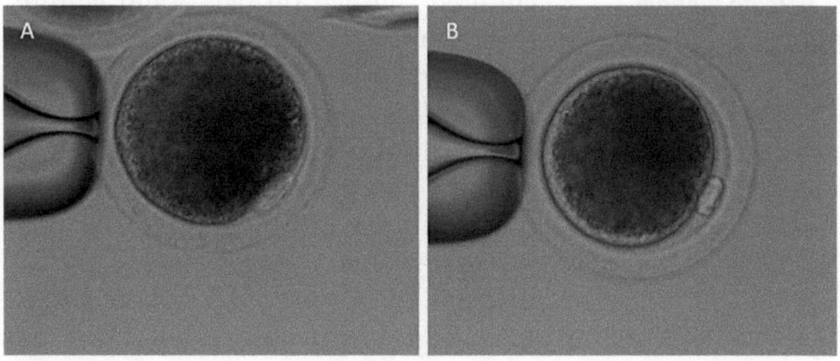

Fig. 2 Effects of FLI on the perivitelline space in metaphase II (MII) oocytes after 42 h of maturation. (**a**) MII oocyte matured in control medium, (**b**) MII oocyte matured in FLI medium

detach more completely from the zona pellucida, thus better simulating what is seen in in vivo-derived oocytes [19]. Following transfer, blastocysts produced from oocytes matured in the presence of FLI have about twice the developmental competence as embryos produced from oocytes not matured in the presence of FLI, thus doubling the litter size to about nine piglets per litter [19].

1.3 Historical Prospective on Embryo Culture

Prior to the application of modern culture media it was not possible to culture the pig embryo through the four-cell stage [20]. Zygotes and two-cell stage embryos would stop development or "block" at the four-cell stage. Embryos recovered at or beyond the eight-cell stage and placed in culture would continue development to the blastocyst stage [20]. The stage at which the "block" occurred was species-specific and corresponded with a major onset of transcription. The first report of successful culture through the "block" stage in the pig was with Whitten's medium [21]. Later a serendipitous observation led to using a Tyrode's Lactate HEPES buffered medium [22]. The observation was stumbled upon as extra zygotes were collected from a gilt on a Friday. The embryos were flushed in the Tyrode's Lactate HEPES buffered medium and then placed in a humidified warm air incubator. The next week when the incubator was being cleaned out, it was noticed that the zygotes had developed to the blastocyst stage (RSP, personal observation). Prior to the identification of culture conditions that were consistent with development through the "block," zygotes were transferred back to the oviduct of a pig, or even a sheep [23]. In this heterologous oviductal environment, the embryos of many species readily developed to the blastocyst stage. The composition of the media and the atmosphere under which the embryos are currently cultured continue to evolve as we understand more about the unique properties of the early embryo.

1.4 Recent Advances to Porcine Embryo Culture

By using transcriptional profiling of in vivo-derived and in vitro-cultured (flushed on day 2 and cultured until day 6) blastocyst-stage embryos, supplements have been identified to add to the culture medium. The first supplement identified was arginine, because the message for an arginine transporter, solute carrier family 7 member 1 (SLC7A1), was upregulated by 63-fold in embryos cultured in vitro compared to in vivo-derived blastocyst stage embryos. It was hypothesized that this transporter message was upregulated to compensate for a lack of arginine in the culture medium. When arginine was added at a final concentration of 1.69 mM, there was an increased number of blastocyst stage embryos with increased cell number and lower level of *SLC7A1* message [24]. The addition of arginine to culture medium became standard, and the medium was named MU1 medium. Recently, another compound, 5-(4-chloro-phenyl)-3-phenyl-pent-2-enoic acid (PS48), was added to the culture system, which improved the percentage of embryos that developed to the blastocyst stage and the number of nuclei in those blastocyst stage embryos [25]. In addition to arginine, we now routinely add PS48 to culture medium and it is referred to as MU2.

In this chapter, we present the optimal methods and techniques, which we have found to produce competent oocytes and embryos in our laboratory. We and others are continuing to investigate what is needed in our culture systems to promote higher-quality oocyte and embryo development and will likely continue to make improvements to our systems.

2 Materials

Prepare all media solutions by using sterilized glassware and sterile technique. Reagents are weighed by using either plastic weighing dishes or weighing paper on a balance. Glassware is sterilized in an oven heated to 225 °F for 2–4 h.

1. *TL-HEPES*: Combine 6.663 g NaCl (114 mM), 0.239 g KCl (3.2 mM), 0.168 g $NaHCO_3$ (2 mM), 0.041 g NaH_2PO_4 (0.4 mM), 1.868 g Na-lactate (10 mM), 0.102 g $MgCl_2{\cdot}6H_2O$ (0.5 mM), 2.383 g HEPES (10 mM), 0.294 g $CaCl_2{\cdot}2H_2O$ (2 mM), 0.100 g PVA (0.01%), 2.186 g sorbitol (12 mM), 0.022 g Na-pyruvate (0.25 mM), 1 mL 100× (10 μg/mL) gentamicin, and 0.010 g phenol red, q.s. with 1000 mL with culture-grade H_2O. Adjust the pH to 7.2–7.4 and the osmolarity to 295–310 mOsm. Filter by using a 0.22 μM filter and store at 4 °C for 2–3 weeks (*see* **Note 1**).
2. *Oocyte Maturation Medium Stock A*: To a 500 mL bottle of M199 which already contains 2.2 g/L sodium bicarbonate and does not contain HEPES, combine 0.500 g (0.1%) PVA, 0.150 g (3.6 mM) sodium bicarbonate, and 500 μL (10 μg/mL) gentamicin. This medium can be stored at 4 °C for 4 weeks.
3. *Oocyte Maturation Medium Stock B*: Combine 100 mL of stock A, 0.055 g (3.05 mM) glucose, and 0.001 g (908.8 μM) Na-pyruvate. Store at 4 °C for 1 week.
4. *Oocyte Maturation Medium*: Combine 40 mL of stock B, 10 ng/mL epidermal growth factor (EGF), 0.5 μg/mL luteinizing hormone (LH), 0.5 μg/mL follicle stimulating hormone (FSH), and 0.57 mM L-cysteine. Place 2 mL of oocyte maturation medium into two 35 mm dishes. Add 40 ng/mL FGF2, 20 ng/mL LIF, and 20 ng/mL IGF1 to one of the dishes. Place both dishes in the incubator at 38.5 °C (*see* **Note 2**).
5. *Cumulus Denuding Medium*: Combine 5 mL of TL-HEPES, 95 mL of culture-grade H_2O, 5.460 g mannitol (0.3 M), 0.001 g BSA (0.01%), and 0.030 g hyaluronidase (0.03%). Adjust the pH to 7.0–7.4. Filter with a 0.22 μm filter and place 750 μL aliquots in microfuge tubes, and store in −20 °C. A day or two before the hyaluronidase solution is needed place the microfuge tubes in a dry, non-CO_2 incubator at 38.5 °C.
6. *Sperm Thawing Medium*: Combine 0.03 g of BSA with 30 mL of PBS and add 10 μg/mL gentamicin. This medium can be stored in a dry, non-CO_2 incubator at 38.5 °C for up to 1 month.
7. *Sperm Washing Medium*: Combine 14.380 g (0.3 M) glucose, 0.625 g alpha-D-lactose (6.9 mM), 1.125 g Na-citrate·$2H_2O$ (15.3 mM), 0.875 g ethylenediaminetetraacetic acid (EDTA) (12 mM), 0.300 g $NaHCO_3$ (14.3 mM), 0.100 g KCl (5.4 mM), and 25 μL (10 μg/mL) gentamicin to 250 mL of culture-grade H_2O. Filter and store at 4 °C for 1 week. On the day of use remove from refrigerator and warm to room temperature. This sperm wash needs to be made fresh every week.

8. *Sperm Freezing Medium*: Androhep CryoGuard is purchased from MOFA Global (Verona, WI).
9. *Oocyte Manipulation Medium*: Combine 500 mL of M199, 0.025 g (0.6 mM) $NaHCO_3$, 0.375 g (2.9 mM) HEPES, 0.876 g (30 mM) NaCl, and 500 μL (10 μg/mL) gentamicin. Adjust the osmolarity to 295–310 mOsm and pH to 7.2–7.4 and filter by using a 0.22 μm filter. Every 2 weeks take 200 mL of the stock manipulation medium and add 0.600 g BSA and place in a dry non-CO_2 incubator at 38.5 °C.
10. *Modified Tris Buffered Medium (mTBM) for IVF*: For 100 mL mTBM stock solution, combine 0.661 g (113.1 mM) NaCl, 0.022 g (3.0 mM) KCl, 0.110 g (7.5 mM) $CaCl_2 \cdot 2H_2O$, 0.242 g (20.0 mM) Tris Base, 0.198 g (11.0 mM) glucose, 0.055 g (5.0 mM) Na-pyruvate, and 100 μL (10 μg/mL) gentamicin to 100 mL culture-grade H_2O and mix until dissolved and store 50 mL aliquots at 4 °C. Forty-eight hours before IVF, then take 50 mL and add 0.332 g (1.0 mM) caffeine and 0.100 g (0.2%) bovine serum albumin (BSA) to 50 mL of mTBM and store in a 5% CO_2 incubator at 38.5 °C to equilibrate. This is the fertilization medium. Any remaining mTBM can be stored at 4 °C for 1 week. On the day of use, make 50 μL microdroplets in a 35 mm petri dish, overlay with mineral oil treated with Whitten's salts (*see* **Note 3**), and place in 5% CO_2 incubator to equilibrate.
11. *Missouri University Medium 2 (MU2)*: Combine 6.311 g (108 mM) NaCl, 0.746 g (10 mM) KCl, 0.048 g (0.35 mM) KH_2PO_4, 0.099 g (0.40 mM) $MgSO_4$, and 10 μg/mL gentamicin to 1 L of culture-grade H_2O and mix. This PZM stock is then filtered by using a 45 mm neck top 0.22 μm filter into a 1 L filter unit and can be stored at 4 °C for 1 month.

A PS48 stock solution (10 mM) diluted in dimethyl sulfoxide (DMSO) can be prepared and stored at −80 °C until use, and can stay at 4 °C for 2 weeks.

The week of culture MU2 medium [25] is made by combining 98 mL of the PZM stock solution, 0.002 g (0.2 mM) Na-pyruvate, 0.062 g Ca-lactate (2.0 mM), 0.015 g (1.0 mM) L-glutamine, 0.055 g (5.0 mM) hypotaurine, 1 mL (100×) Basal Medium Eagle (BME) amino acids, 1 mL (100×) Minimum Essential Medium (MEM) amino acids, 0.300 g (3 mg/mL) fatty acid free BSA, 0.211 g (25.07 mM) $NaHCO_3$, 100 μL (10 μg/mL) gentamicin, 0.027 g (to a final concentration of 1.69 mM) arginine [24], and 5 μM PS48 [25]. On the day of culture, make one wash dish with 2 mL of MU2 medium and place 500 μL of MU2 in each well of a 4-well Nunc dish and overlay with mineral oil treated with Whitten's salts. Place in the incubator (5% CO_2 and 38.5 °C in air) to equilibrate at least 4 h before use (*see* **Notes 4** and **5**).

3 Methods

3.1 Sperm Freezing

1. Collect the boar's sperm-rich fraction in a semen collection cup (Minitube) and filter once through a sperm filter.
2. Subjectively examine the sperm motility and concentration.
3. Let the semen stand at room temperature for 1.5–2 h.
4. Transfer 25 mL of semen to a 50 mL conical tube and wash 1:1 (v:v) with 25 mL of sperm wash medium. Mix gently and centrifuge at 1000 × *g* for 7 min.
5. Remove supernatant and add 5 mL of Androhep CryoGuard Cooling Extender and mix semen well.
6. Calculate the sperm concentration by using a hemocytometer. For freshly collected semen, dilute 1:10,000 to count.
7. Adjust the sperm concentration to 5–10 × 10^8/mL by adding Androhep CryoGuard Cooling Extender.
8. Place the sperm suspension inside a cooling machine and allow the sperm to cool to 4 °C over a 1.5 h period. Or the sperm suspension can be placed inside a glass beaker containing room temperature water. Place beaker inside a 4 °C refrigerator and allow the suspension to cool down to 4 °C over a 1.5 h period.
9. Dilute the cooled sperm suspension 1:1 with an equal volume of Androhep CryoGuard Freezing Extender containing 4% glycerol (4 °C).
10. Gently mix and place the 50 mL conical tube on ice.
11. Freeze sperm in 100 μL drops on dry ice depressions. To make the impressions in dry ice:
    (a) Heat 0.5–1 cm ball bearings in boiling water.
    (b) Place ball bearings on dry ice to make a circular depression on the dry ice by using forceps.
    (c) Make about 25–50 depressions on a single block of dry ice.
    (d) Continue to mix sperm in between pipetting of the sperm suspension.
12. Allow the sperm pellets to stay on the dry ice for 1–2 min and then immediately place them into liquid nitrogen. To store the sperm pellets in liquid nitrogen:
    (a) Drill a small hole in a 50 mL conical tube and tie a string through the hole.
    (b) Pour the sperm pellets into a funnel placed in the 50 mL tube.
    (c) Tie the tube to the canister inside the liquid nitrogen tank.
13. The sperm pellets need to be completely submerged in liquid nitrogen.

3.2 In Vitro Maturation

1. To collect oocytes for in vitro maturation, immature pig ovaries are obtained from a slaughterhouse and transported back to the laboratory at room temperature in a cooler.
2. The ovaries are washed with 37 °C saline to remove excess blood. Follicular fluid is aspirated from 3–5 mm follicles with an 18 gauge needle and a 10 mL syringe. The fluid is then placed into a 50 mL conical tube.
3. The follicular fluid is allowed to settle for 10 min to allow the COCs to fall to the bottom of the tube. The supernatant of the follicular fluid is discarded and the pellet is then resuspended in 25–50 mL of PVA-TL-HEPES. This is repeated three times to remove blood and extra follicular debris.
4. The remainder of the clean follicular fluid is then placed into 100 mm petri dishes and searched by using a dissecting microscope.
5. The COCs with homogenous colored cytoplasm and at least two layers of cumulus cells are selected and washed in a dish containing maturation medium and then placed into a new 35 mm petri dish or one well of a four-well plate at a rate of no less than 8.3 μL maturation medium/COC.

3.3 In Vitro Fertilization

1. Forty to 44 h after the start of maturation the 35 mm dish or 4-well plate of maturation medium containing COCs is removed from the incubator.
2. COCs are transferred into a microfuge tube containing 0.03% hyaluronidase in TL-HEPES by using a 200 μL pipette in order to minimize medium crossover. To remove cumulus cells, vortex the tube for 2–3 min on a low setting.
3. Transfer the contents of the microfuge tube to an empty 35 mm petri dish and rinse the empty tube with Oocyte Manipulation Medium to collect any oocytes left behind (*see* **Note 6**). Gather all denuded oocytes and transfer them to a new 35 mm petri dish containing Oocyte Manipulation Medium. This dilutes any hyaluronidase that could potentially damage the oocyte.
4. Next, select oocytes based on the presence of an extruded polar body, marking a metaphase II (MII) oocyte. Further select oocytes that have a homogenous color of cytoplasm and that are not lysed.
5. Take the selected MII oocytes and wash in a 35 mm dish of 2 mL equilibrated mTBM. Wash and move groups of 30–50 oocytes to equilibrated 50 μL drops of mTBM covered in 3 mL light mineral oil in a 35 mm petri dish.
6. Thaw a sperm pellet in warmed 38.5 °C Sperm Thawing Medium.

7. Transfer the thawed semen directly on top of a 45% Percoll solution making a single layer and centrifuge at 550 × *g* for 30 min at room temperature (*see* **Note** 7).
8. After centrifugation, remove the sperm pellet and transfer it to a tube containing 6 mL equilibrated mTBM.
9. Mix gently and centrifuge again for 30 min at room temperature.
10. Carefully remove all of the supernatant leaving just the undisturbed pellet.
11. Resuspend the pellet with 100 μL mTBM.
12. Calculate the concentration of spermatozoa and dilute to 0.5×10^6 sperm/mL and add 50 μL of the diluted sperm to the pellet containing the oocytes for a final concentration of 0.25×10^6 sperm/mL.
13. Coincubate the sperm and oocytes for 4–6 h in 5% CO_2 with 100% humidity and at 38.5 °C.

3.4 In Vitro Culture

1. After coincubation with sperm, the presumptive zygotes are washed in a 35 mm dish containing 2 mL MU2 medium.
2. Culture groups of 50 embryos in 500 μL of MU2 with an overlay of 200 μL mineral oil in a four-well NUNC dish.
3. These embryos are cultured at 38.5 °C in a humidified atmosphere of 5% CO_2 overnight and then the next day (day 1) they are moved to a humidified atmosphere of low oxygen 5% O_2, 5% CO_2, and 90% N_2 until day 6 (144 h postinsemination).

4 Notes

1. Before mixing medium, add PVA to 100 mL of culture grade H_2O, stir while using moderate heat to get PVA to dissolve and go into solution.
2. 1000× LH and FSH, 1000× EGF, and 2400× FLI stock can be made and stored long term at −80 °C.
3. To treat mineral oil with Whitten's salts, combine an equal volume of Whitten's salts with mineral oil and shake well. Let stand at RT at least 1 week and up to 3 months.
4. To make a chemically defined culture system, PVA can be added instead of BSA. Add 0.10 g PVA to 100 mL of PZM working solution instead of fatty acid-free BSA.
5. Arginine is already supplied in BME at a 0.12 mM concentration, so add the difference to make a final concentration of 1.69 mM.

6. To visualize the polar body more effectively, the Oocyte Manipulation Medium contains a higher osmolarity (295–310 mOsm) than normal oocyte maturation medium (275–290 mOsm). This causes the oocyte's cytoplasm to shrink, allowing the polar body to be more distinct.
7. Some sperm may be sensitive to Percoll so a Percoll gradient where a 45% Percoll solution is placed over a 90% Percoll solution may be used.

Acknowledgments

Funding from Food for the 21st Century and the National Institutes of Health are acknowledged (U42 OD011140, R01 HD 080636).

References

1. Prather RS, Lorson M, Ross JW, Whyte JJ, Walters E (2013) Genetically engineered pig models for human diseases. Annu Rev Anim Biosci 1:203–219. https://doi.org/10.1146/annurev-animal-031412-103715
2. Redel BK, Prather RS (2016) Meganucleases revolutionize the production of genetically engineered pigs for the study of human diseases. Toxicol Pathol 44(3):428–433. https://doi.org/10.1177/0192623315613160
3. Whitworth KM, Rowland RR, Ewen CL, Trible BR, Kerrigan MA, Cino-Ozuna AG, Samuel MS, Lightner JE, McLaren DG, Mileham AJ, Wells KD, Prather RS (2016) Gene-edited pigs are protected from porcine reproductive and respiratory syndrome virus. Nat Biotechnol 34(1):20–22. https://doi.org/10.1038/nbt.3434
4. Edwards RG (1965) Maturation in vitro of mouse, sheep, cow, pig, rhesus monkey and human ovarian oocytes. Nature 208 (5008):349–351
5. Leibfried-Rutledge ML, Critser ES, Parrish JJ, First NL (1989) In vitro maturation and fertilization of bovine oocytes. Theriogenology 31 (1):61–74. https://doi.org/10.1016/0093-691X(89)90564-5
6. Cobo AC, Requena A, Neuspiller F, Aragon SM, Mercader A, Navarro J, Simon C, Remohi J, Pellicer A (1999) Maturation in vitro of human oocytes from unstimulated cycles: selection of the optimal day for ovum retrieval based on follicular size. Hum Reprod 14(7):1864–1868
7. Eppig JJ, Schroeder AC (1989) Capacity of mouse oocytes from preantral follicles to undergo embryogenesis and development to live young after growth, maturation, and fertilization in vitro. Biol Reprod 41(2):268–276
8. Niwa K (1993) Effectiveness of in vitro maturation and in vitro fertilization techniques in pigs. J Reprod Fertil Suppl 48:49–59
9. Naito K, Fukuda Y, Toyoda Y (1988) Effects of porcine follicular fluid on male pronucleus formation in porcine oocytes matured in vitro. Gamete Res 21(3):289–295. https://doi.org/10.1002/mrd.1120210310
10. Sawai K, Funahashi H, Niwa K (1997) Stage-specific requirement of cysteine during in vitro maturation of porcine oocytes for glutathione synthesis associated with male pronuclear formation. Biol Reprod 57(1):1–6
11. Yoshida M, Ishigaki K, Pursel VG (1992) Effect of maturation media on male pronucleus formation in pig oocytes matured in vitro. Mol Reprod Dev 31(1):68–71. https://doi.org/10.1002/mrd.1080310112
12. Nagai T (1996) In vitro maturation and fertilization of pig oocytes. Anim Reprod Sci 42 (1):153–163. https://doi.org/10.1016/0378-4320(96)01487-X
13. Wang WH, Abeydeera LR, Prather RS, Day BN (1998) Morphologic comparison of ovulated and in vitro-matured porcine oocytes, with particular reference to polyspermy after in vitro fertilization. Mol Reprod Dev 49(3):308–316. https://doi.org/10.1002/(SICI)1098-2795(199803)49:3<308::AID-MRD11>3.0.CO;2-S

14. Abeydeera LR, Day BN (1997) Fertilization and subsequent development in vitro of pig oocytes inseminated in a modified tris-buffered medium with frozen-thawed ejaculated spermatozoa. Biol Reprod 57(4):729–734
15. Gil MA, Abeydeera LR, Day BN, Vazquez JM, Roca J, Martinez EA (2003) Effect of the volume of medium and number of oocytes during in vitro fertilization on embryo development in pigs. Theriogenology 60(4):767–776
16. Coy P, Martinez E, Ruiz S, Vazquez JM, Roca J, Matas C, Pellicer MT (1993) In vitro fertilization of pig oocytes after different coincubation intervals. Theriogenology 39 (5):1201–1208
17. Abeydeera LR, Wang WH, Cantley TC, Rieke A, Prather RS, Day BN (1998) Presence of epidermal growth factor during in vitro maturation of pig oocytes and embryo culture can modulate blastocyst development after in vitro fertilization. Mol Reprod Dev 51(4):395–401. https://doi.org/10.1002/(SICI)1098-2795(199812)51:4<395::AID-MRD6>3.0.CO;2-Y
18. Abeydeera LR, Wang WH, Prather RS, Day BN (1998) Maturation in vitro of pig oocytes in protein-free culture media: fertilization and subsequent embryo development in vitro. Biol Reprod 58(5):1316–1320
19. Yuan Y, Spate LD, Redel BK, Tian Y, Zhou J, Prather RS, Roberts RM (2017) Quadrupling efficiency in production of genetically modified pigs through improved oocyte maturation. Proc Natl Acad Sci U S A 114(29): E5796–E5804. https://doi.org/10.1073/pnas.1703998114
20. Prather RS (2010) Whatever happened to the "cell-block" during mammalian embryogenesis? Mol Reprod Dev 77(5):NA. https://doi.org/10.1002/mrd.21182
21. Beckmann LS, Cantely TC, Rieke AR, Day BN (1990) Development aviability of one- and two-cell porcine embryos cultured throught the 'four-cell block'. Theriogenology 33:193. (abstract)
22. Hagen DR, Prather RS, Sims MM, First NL (1991) Development of one-cell porcine embryos to the blastocyst stage in simple media. J Anim Sci 69(3):1147–1150
23. Prather RS, Sims MM, First NL (1991) Culture of porcine embryos from one- and two-cell to blatocyst in sheep oviducts. Theriogenology 35:1147–1151
24. Redel BK, Tessanne KJ, Spate LD, Murphy CN, Prather RS (2015) Arginine increases development of in vitro-produced porcine embryos and affects the protein arginine methyltransferase-dimethylarginine dimethylaminohydrolase-nitric oxide axis. Reprod Fertil Dev. https://doi.org/10.1071/RD14293
25. Spate LD, Brown A, Redel BK, Whitworth KM, Prather RS (2015) PS48 can replace bovine serum albumin in pig embryo culture medium, and improve in vitro embryo development by phosphorylating AKT. Mol Reprod Dev 82(4):315–320. https://doi.org/10.1002/mrd.22474

Chapter 7

In Vitro Culture of Late Stage Pig Embryos in a Chemically Defined Medium, Porcine Blastocyst Medium (PBM)

Tomomi Mito and Hiroyoshi Hoshi

Abstract

In vitro production (IVP) of porcine preimplantation embryos is an important technique not only for basic and biomedical research purposes but also for animal biotechnology application such as transgenesis, cloning, and embryo transfer. In this chapter, we demonstrate a superior IVP procedure of porcine embryos derived from cumulus–oocyte complexes (COCs) of slaughtered pig ovaries which are cultured sequentially in different defined media. Porcine blastocyst medium (PBM) particularly designed for the late stage embryo culture could improve the potential of morulae or blastocysts to develop into hatching and hatched blastocysts with good quality.

Key words Pig, In vitro production, Defined medium, Blastocyst development, Embryo quality, Embryo transfer

1 Introduction

Optimization of various nutrients and factors in a culture medium is important for efficient production of high quality mammalian preimplantation embryos in vitro. Suzuki and Yoshioka [1] developed a chemically defined medium, Porcine Zygote Medium-5 (PZM-5) suitable for pig embryo culture in vitro. This medium composition is designed on the basis of inorganic compound and energy substrate concentrations present in porcine oviductal fluid [2]. Although PZM-5 better supports porcine embryo development after IVF than the conventional NCSU-23 medium [3], in vitro-derived embryos cultured in PZM-5 still have reduced developmental ability compared with in vivo-produced embryos [4]. PZM-5 medium contains pyruvate and lactate, but not glucose, as the major energy substrates [5]. It has been known that pyruvate and lactate are the preferred carbohydrates at the early developmental stages of porcine embryos in vitro [6]. Glucose consumption is low during the early stages of embryo development, but increases toward the morulae stage and finally reaches a

Jason R. Herrick (ed.), *Comparative Embryo Culture: Methods and Protocols*, Methods in Molecular Biology, vol. 2006, https://doi.org/10.1007/978-1-4939-9566-0_7, © Springer Science+Business Media, LLC, part of Springer Nature 2019

maximum level at the blastocyst stage [7]. In our previous study, the addition of glucose to PZM-5 for culture from day 5 to 7 after IVF greatly improved embryo viability, blastocyst development rate, and ATP content of blastocysts compared to those in glucose-free medium [8].

Oviduct and uterine fluids contain significant amounts of free amino acids and amino acid supplementation of the culture medium is beneficial for the embryo development of different mammalian species [9]. Glycine is one of the most abundant amino acids in porcine oviductal and uterine fluids [10]. Glycine has been reported to act as an intracellular stabilizer of osmotic pressure in early mouse embryos in vitro [11] and protect against apoptotic cell death in sinusoidal endothelial cells [12]. Glycine and alanine were found to be the major amino acids required at higher concentrations during in vitro culture of bovine embryos [13]. Glucose- and high glycine (10 mM)-containing PZM-5 medium (the so-called porcine blastocyst medium [PBM]) synergistically improved embryo development to the hatching and hatched blastocyst stages as well as embryo quality, as determined by an increase of blastocyst diameter and total cell numbers, and/or a reduction in apoptotic index [8]. Moreover, we have recently found that PBM is clearly beneficial for the production of high-quality porcine blastocysts in vitro that could develop full term after embryo transfer [14].

2 Materials

Formulation of various defined media for oocyte collection, maturation, fertilization, and embryo development of pigs is shown in Table 1. All media and stock solutions are stored at 2–6 °C for 4 months, unless otherwise noted.

1. Porcine oocyte/embryo collection medium: POE-CM (IFP1040P; Research Institute for the Functional Peptides, Yamagata, Japan).
2. Basic porcine oocyte maturation medium: POM (IFP1010P; Research Institute for the Functional Peptides).
3. Porcine fertilization medium: PFM (IFP1020P; Research Institute for the Functional Peptides).
4. Porcine zygote medium-5: PZM-5 (IFP0410P; Research Institute for the Functional Peptides).
5. Porcine blastocyst medium: PBM (IFP1030P; Research Institute for the Functional Peptides).
6. Concentrated dbcAMP solution-100× (100-fold, 100 mM): dbcAMP-100× (IFP1060P; Research Institute for the Functional Peptides).

Table 1
Formulation of different culture media for in vitro production of porcine embryos

Component (mM)	POE-CM	POM	PFM	PZM-5	PBM
NaCl	108.00	108.00	108.00	108.00	108.00
KCl	10.00	10.00	10.00	10.00	10.00
KH_2PO_4	0.35	0.35	0.35	0.35	0.35
$MgSO_4 \cdot 7H_2O$	0.40	0.40	0.40	0.40	0.40
$NaHCO_3$	5.00	25.00	25.00	25.00	25.00
Hepes	25.00	–	–	–	–
Glucose	–	4.00	1.00	–	5.00
Na-pyruvate	0.20	0.20	0.20	0.20	0.20
Ca-$(lactate)_2 \cdot 5H_2O$	2.00	2.00	4.00	2.00	2.00
L-Glutamine	–	2.00	–	2.00	2.00
Hypotaurine	–	5.00	–	5.00	5.00
Basal Medium Eagle amino acids (ml/l)[a]	–	20.00	–	20.00	20.00
Minimum Essential Medium nonessential amino acids (ml/l)[b]	–	10.00	–	10.00	10.00
Glycine	–	–	–	–	10.00
Theophylline	–	–	2.50	–	–
Adenosine (μM)	–	–	1.00	–	–
L-Cysteine (μM)	–	600	0.25	–	–
Gentamicin (μg/ml)	10.00	10.00	10.00	10.00	10.00
Polyvinyl alcohol (mg/ml)	3.00	3.00	3.00	3.00	3.00
Phenol red (mg/ml)	2.00	–	–	–	–

All media were adjusted to pH 7.3 at preparation
Abbreviations: *POE-CM* porcine oocyte/embryo collection medium, *POM* basic porcine oocyte maturation medium, *PFM* porcine fertilization medium, *PZM-5* porcine zygote medium-5, *PBM* porcine blastocyst medium
[a]50-fold stock solution
[b]100-fold stock solution

7. Concentrated sperm diluent for in vitro fertilization: SEM-5× (fivefold, IFP1050P; Research Institute for the Functional Peptides).
8. Saline solution: 0.9% NaCl in ultrapure water. Autoclave and store at a room temperature for 1 year.
9. Mineral oil (Cat. No. 107162; Merck, Darmstadt, Germany).
10. Equine chorionic gonadotropin (eCG) and human chorionic gonadotropin (hCG) stock solutions (each 1000 IU/ml):

Prepare 2000 IU/ml solutions of eCG and hCG in POM medium. Mix equal volumes of eCG and hCG stock solutions to prepare at the final concentration of 1000 IU/ml stock solution. Aliquots (70-μl) of the eCG-hCG stock solution are placed into vials and stored at −30 °C for 1 year.

11. Complete Oocyte Maturation Medium: Supplement POM medium with 1 mM dbcAMP, 10 IU/ml eCG and 10 IU/ml hCG just before use.
12. Percoll solution: Prepare 80% Percoll solution (add 0.4 ml SEM-5× to 1.6 ml Percoll) and 50% Percoll solution (add 1 ml Percoll to 0.4 ml SEM-5× stock solution containing 0.6 ml ultrapure water) just before use.

3 Methods

The protocol for in vitro embryo culture in defined media is briefly described in Fig. 1. All cultures are maintained at 38.5 °C in a humidified atmosphere with 5% CO_2, 5% O_2, and 90% N_2. Droplets of media are maintained in Reproplates (IFP9670; Research Institute for the Functional Peptides) under 150 μl of mineral oil. Handling of ovaries, COCs, and embryos outside of the incubator is carried out at room temperature (23–25 °C), unless otherwise mentioned. Media in petri dishes or Reproplates are left on warm plates and kept at 38.5 °C.

3.1 Oocyte Collection and In Vitro Maturation of Oocytes

1. Collect porcine ovaries at a local slaughterhouse and transport them in saline solution (approx. 25 °C) to the laboratory within 2–3 h.
2. Wash ovaries several times with saline solution.
3. Aspirate follicular fluid containing cumulus–oocyte complexes (COCs) from antral follicles (3–6 mm diameter) using an 18-gauge needle attached to a 10-ml disposable syringe.
4. Transfer the follicular fluid containing COCs to 50 ml centrifuge tubes and let stand for 20 min.
5. Discard the supernatant and dilute the precipitated material with POE-CM in a 100 × 15 mm square petri dish.
6. Select healthy COCs with a uniform ooplasm and more than three layers of compact cumulus cells under a stereomicroscope (Fig. 2).
7. Wash the selected COCs twice with 3 ml POE-CM in 35 × 10 mm petri dishes.
8. Wash COCs twice in 200-μl droplets of the complete oocyte maturation medium (POM with dbcAMP, eCG, and hCG) (*see* **Note 1**, Fig. 3).

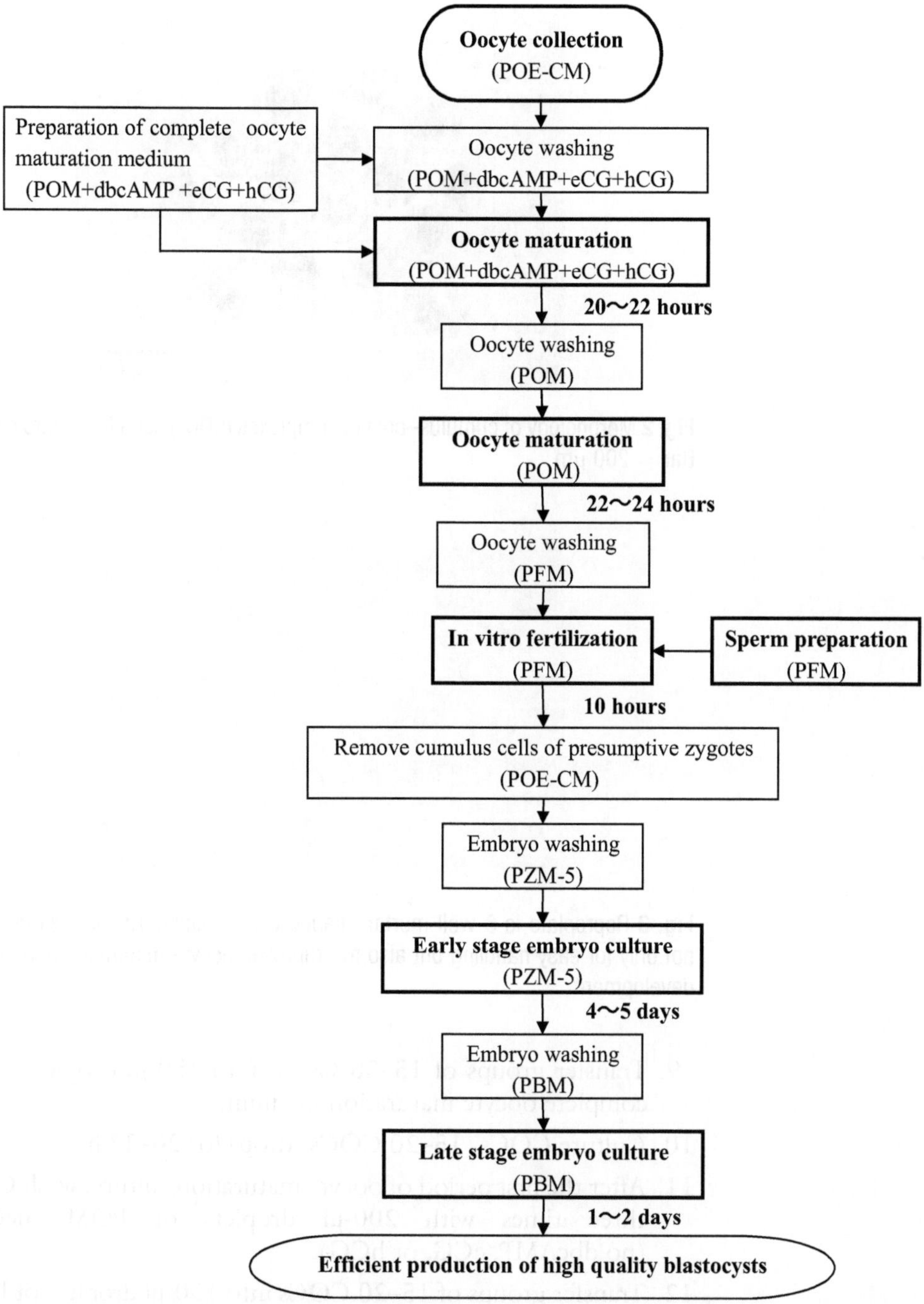

Fig. 1 Protocol of porcine embryo culture in defined medium. *Abbreviations*: *POE-CM* porcine oocyte/embryo collection medium, *POM* basic porcine oocyte maturation medium, *PFM* porcine fertilization medium, *PZM-5* porcine zygote medium-5, *PBM* porcine blastocyst medium

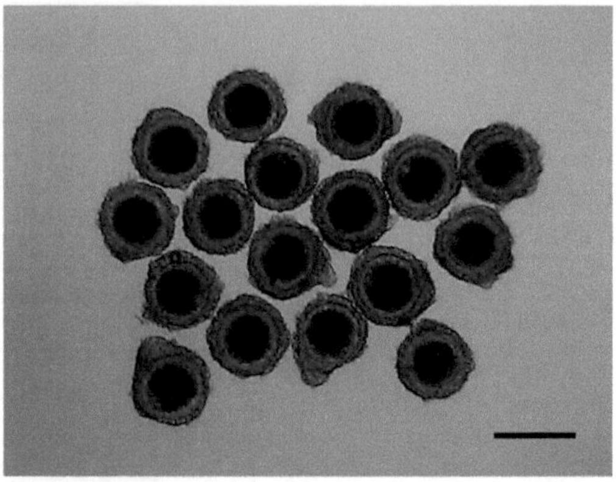

Fig. 2 Morphology of cumulus–oocyte complexes (COCs) used for in vitro culture. Bar = 200 μm

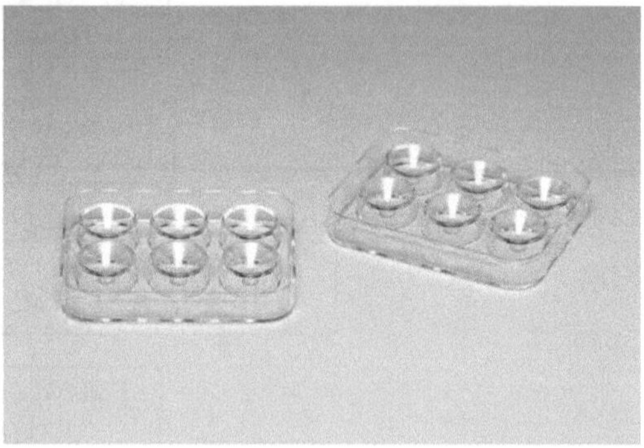

Fig. 3 Reproplate is 6-well mortar-shaped culture plate. This is well designed not only for easy handling but also for improved oocyte maturation and embryo development

9. Transfer groups of 15–20 COCs into 150-μl droplets of the complete oocyte maturation medium.
10. Culture COCs (15–20 COCs/drop) for 20–22 h.
11. After the first period of oocyte maturation culture, wash COCs three times with 200-μl droplets of POM medium (no dbcAMP, eCG, or hCG).
12. Transfer groups of 15–20 COCs into 150-μl droplets of POM medium.
13. Maturation of COCs is continued for an additional 22–24 h.

3.2 In Vitro Fertilization

1. Keep 20 ml PFM (pH 7.3) and 2 ml each of 80% and 50% Percoll solutions warm in a water bath at 38 °C.
2. Thaw frozen semen immediately in a water bath at 38 °C for 30–40 s.

3. Dilute thawed semen with 2 ml of 50% Percoll solution.
4. Gently overlay the sperm suspension on top of 2 ml of 80% Percoll solution.
5. Centrifuge at 700 × *g* for 20 min at 35 °C or room temperature.
6. Discard the supernatant and resuspend the sperm pellet with about 7 ml PFM and gently mix. Centrifuge sperm suspension at 500 × *g* for 5 min.
7. Repeat this procedure once again.
8. Discard the supernatant and resuspend the remaining pellet in 1 ml of PFM and gently mix.
9. Count the numbers of sperm with a hemacytometer (*see* **Note 2**).
10. Dilute sperm suspension with PFM to 2×10^6 spermatozoa/ml (*see* **Note 3**).
11. Add 50-μl of the sperm suspension to each droplet (50-μl) of gas-phase equilibrated PFM (pH 7.3) immediately before addition of COCs (final sperm concentrations: 1×10^6 spermatozoa/ml).
12. Wash COCs twice with 200-μl droplets of PFM.
13. Transfer matured COCs into the PFM droplets (15–20 COCs/drop) containing sperm suspension and coincubate for 10 h.

3.3 In Vitro Embryo Culture

1. After in vitro fertilization, wash the presumptive zygotes once with 300-μl POE-CM at 38 °C in Reproplate without mineral oil (*see* **Note 4**).
2. Transfer presumptive zygotes into 1 ml POE-CM in a 15 ml centrifuge tube and remove cumulus cells by vortexing for 4 min.
3. Wash the inside wall of centrifuge tube with 1 ml POE-CM and transfer the entire contents into a 35 × 10 mm petri dish.
4. Repeat this procedure.
5. Collect the denuded presumptive zygotes under a stereomicroscope.
6. Discard presumptive zygotes with strongly attached cumulus cells and/or degenerated (e.g., damaged cell membrane, shrunk ooplasm) cytoplasm.
7. Wash selected presumptive zygotes twice with 300-μl POE-CM in Reproplate without mineral oil.
8. Wash presumptive zygotes twice in 200-μl droplets of PZM-5.
9. Transfer groups of 20–30 presumptive zygotes into 50-μl droplets of PZM-5.

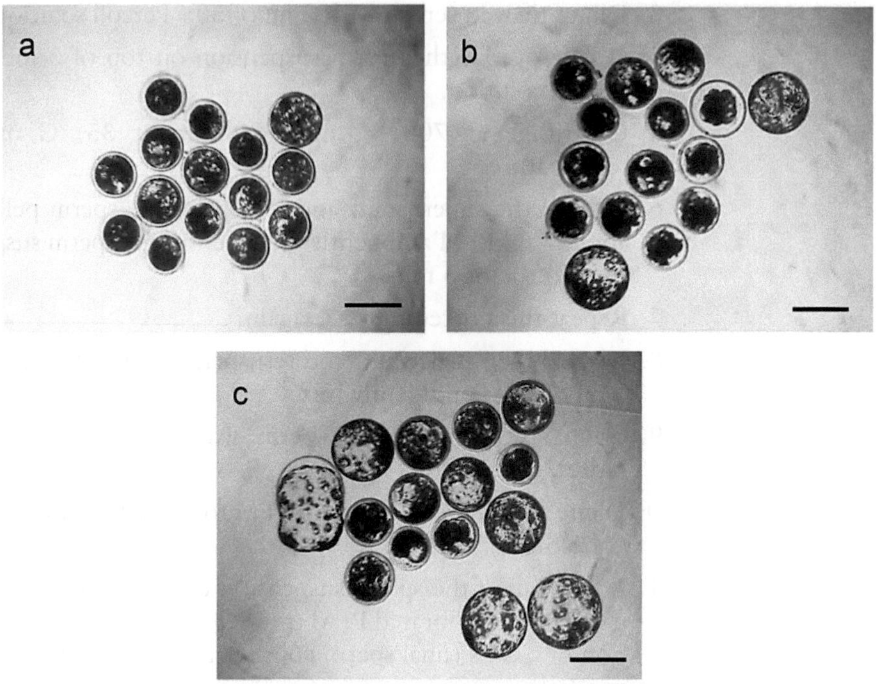

Fig. 4 Microscopic observation of porcine in vitro-derived embryos cultured in defined medium. (**a**) Day-5 embryos cultured in PZM-5 from Day-0. (**b**) Day-6 embryos cultured additionally for 24 h in fresh PZM-5 from Day-5. Expanded and degenerated blastocysts are seen. (**c**) Day-6 embryos cultured for 24 h in PBM from Day-5. Expanded and hatching blastocysts are seen. Bar = 200 μm

10. 4–5 days after the in vitro fertilization, embryos should develop to the morula and blastocyst stages. Wash morulae and blastocysts twice in 200-μl droplets of PBM.
11. Transfer the embryos into 50-μl droplets (10–15 embryos/drop) of PBM and culture them for an additional 1–2 days for development to expanded, hatching and hatched blastocysts.
12. The morphology of porcine in vitro-derived embryos cultured in various defined media was investigated under a stereomicroscope (Fig. 4). Greater numbers of brighter expanded and hatching blastocysts are observed following culture in PBM for an additional 24 h (Fig. 4c) compared to the dark blastocysts and expanded blastocysts produced without additional culture (Control, Fig. 4a) and/or culture in fresh PZM-5 for 24 h (Fig. 4b).

4 Notes

1. Prepare the given number of medium droplets in Reproplate (Fig. 3) and equilibrate in an incubator (38.5 °C, 5% CO_2, 5% O_2, 90% N_2) just before ($\geq$1 h) the culture.
2. To count sperm numbers, dilute the sperm suspension with 3% NaCl solution.
3. The optimal sperm concentration (final sperm concentrations: 1–10 $\times$ 10^6 spermatozoa/ml) and the incubation time (10–20 h) of insemination must be adjusted for every lot of semen used for each experiment.
4. Keep POE-CM warm in a 38 °C water bath. Do not warm in CO_2 incubator.

References

1. Suzuki C, Yoshioka K (2006) Effects of amino acid supplements and replacement of polyvinyl alcohol with bovine serum albumin in porcine zygote medium. Reprod Fertil Dev 18:789–795
2. Nicole R, Hunter RHF, Gardner DK, Leese HJ, Cooke GM (1992) Concentrations of energy substrates in oviductal fluid and blood plasma of pigs during the peri-ovulatory period. J Reprod Fertil 96:699–707
3. Petters RM, Wells KD (1993) Culture of pig embryos. J Reprod Fertil 48(Suppl):61–73
4. Yoshioka K, Suzuki C, Tanaka A, Anas IM, Iwamura S (2002) Birth of piglets derived from porcine zygotes cultured in a chemically defined medium. Biol Reprod 66:112–119
5. Yoshioka K, Suzuki C, Onishi A (2008) Defined system for in vitro production of porcine embryos using a single basic medium. J Reprod Dev 54:208–213
6. Sturmey RG, Leese HJ (2003) Energy metabolism in pig oocytes and early embryos. Reproduction 126:197–204
7. Kikuchi K, Onishi A, Kashiwazaki N, Iwamoto M, Noguchi J, Kaneko H, Akita T, Nagai T (2002) Successful piglet production after transfer of blastocysts produced by a modified in vitro system. Biol Reprod 66:1033–1041
8. Mito T, Yoshioka K, Yamashita S, Suzuki C, Noguchi M, Hoshi H (2012) Glucose and glycine synergistically enhance the in vitro development of porcine blastocysts in a chemically defined medium. Reprod Fertil Dev 24:443–450
9. Gardner DK (1988) Changes in requirements and utilization of nutrients durig mammalian preimplantation embryo development and their significance in embryo culture. Theriogenology 49:83–102
10. Li R, Whitworth K, Lai L, Wax D, Spate L, Murphy CN, Rieke A, Isom C, Hao Y, Zhong Z, Katayama M, Schatten H, Prather RS (2007) Concentration and composition of free amino acids and osmolarities of porcine oviductal and uterine fluid and their effects on development of porcine IVF embryos. Mol Reprod Dev 74:1228–1235
11. Dawson KM, Collins JL, Baltz JM (1998) Osmolarity-dependent glycine accumulation indicates a role for glycine as an organic osmolyte in early preimplantation mouse embryos. Biol Reprod 59:225–232
12. Zhang Y, Ikejima K, Honda H, Kitamura T, Takei Y, Sato N (2000) Glycine prevents apoptosis of rat sinusoidal endothelial cells caused by deprivation of vascular endothelial growth factor. Hepatology 32:542–546
13. Moore K, Bondioli KR (1993) Glycine and alanine supplementation of culture medium enhances development of in vitro matured and fertilized cattle embryos. Biol Reprod 48:833–840
14. Mito T, Yoshioka K, Noguchi M, Yamashita S, Misumi K, Hoshi T, Hoshi H (2015) Birth of piglets from in vitro-produced porcine blastocysts vitrified and warmed in a chemically defined medium. Theriogenology 84:1314–1320

Chapter 8

Production and Culture of the Bovine Embryo

Paula Tríbulo, Rocío Melissa Rivera, Martha Sofia Ortega Obando, Elizabeth Ann Jannaman, and Peter J. Hansen

Abstract

A protocol for production of bovine embryos from oocytes collected from ovaries obtained from an abattoir is described. The protocol includes methods for in vitro maturation of oocytes, capacitation of sperm, fertilization, and development of the resultant embryos to the blastocyst stage. The protocol can be easily modified to use oocytes collected by ultrasound-guided follicular aspiration.

Key words Embryo, In vitro fertilization, In vitro maturation, Culture, Bovine

1 Introduction

The first live calf born as a result of transfer of an in vitro-produced embryo was achieved in 1981 [1]. Since that time, in vitro production (IVP) of bovine embryos has become commonplace. In part, the procedure is extensively used because of the importance of the IVP embryo for cattle production. Moreover, the biological material needed, that is, oocytes from ovaries collected at abattoirs and frozen sperm packaged and distributed as part of the artificial insemination industry, is widely available. In addition, requirements for in vitro maturation (IVM) of oocytes, in vitro capacitation of sperm, in vitro fertilization (IVF) of oocytes, and in vitro culture (IVC) of the resultant embryos are relatively simple. Indeed, the bovine embryo can develop to the blastocyst stage and achieve pregnancy after transfer into recipients when cultured in a protein-free medium containing salts, amino acids, the dipeptide alanyl-glutamine, saccharides, and polyvinylpyrrolidone to control surface tension [2].

The relative ease of producing bovine embryos in vitro makes them an important research tool for advancing our understanding of key biological processes that take place during oocyte maturation and the period of embryonic development between fertilization and blastocyst formation. The IVP bovine embryo is also becoming

Jason R. Herrick (ed.), *Comparative Embryo Culture: Methods and Protocols*, Methods in Molecular Biology, vol. 2006, https://doi.org/10.1007/978-1-4939-9566-0_8,

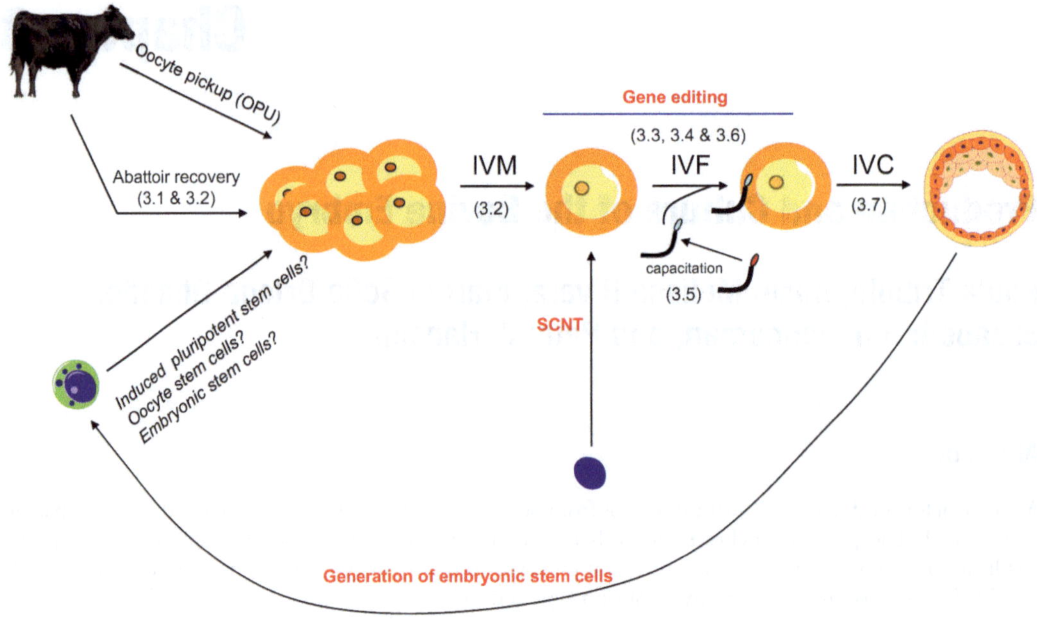

Fig. 1 Flowchart for production of bovine embryos in vitro to illustrate (1) methods described in the chapter and (2) importance of specific procedures for in vitro production for other reproductive technologies. Details of procedures for collection of oocytes from abattoir derived ovaries, in vitro oocyte maturation (IVM), sperm capacitation, in vitro fertilization (IVF) and in vitro embryo culture (IVC) are located in sections delineated underneath each procedure (i.e., sperm are capacitated in procedures described in Subheading 3.5). Also shown are some other reproductive technologies that depend on the in vitro produced embryo including generation of embryos using oocytes derived from stem cells, somatic cell nuclear transfer (SCNT) and gene editing

increasingly important economically for the cattle industry. Production of embryos in vitro can be used to increase the number of offspring from genetically superior females and reduce the generation interval [3]. Transfer of embryos produced in vitro or in vivo can also increase fertility in specific populations of subfertile animals [4, 5]. Moreover, as shown in Fig. 1, the mastery of procedures related to IVP is required for several established and emerging biotechnologies including somatic cell nuclear transfer, gene editing, and production of stem cells.

It is important to bear in mind that embryos produced in vitro are different in terms of morphology, gene expression, and competence to establish pregnancy after transfer than embryos produced in vivo [6]. Moreover, typically only about 20–40% of oocytes subjected to IVM and IVF become blastocysts, so a large proportion of embryos collected at any one stage are fated to cease development before the blastocyst stage.

The key steps in any procedure for IVP of the bovine embryo, illustrated in Fig. 1, are in vitro maturation of the oocyte, capacitation of sperm, fertilization of mature oocytes with capacitated sperm, and culture of the resultant embryos to the blastocyst stage. While there is near uniformity in the overall flow in IVP, there is a great deal of variation between laboratories in media, incubation times, etc. The procedures for IVP in this protocol are based on an amalgam of procedures developed largely in laboratories at the University of Wisconsin (Neal First), University of Missouri (R. Michael Roberts), and University of Guelph (W. Allan King), as well as modifications developed in our own laboratory or borrowed from the literature.

The protocol itself is geared toward producing embryos for research purposes. Oocytes are collected from ovaries obtained from an abattoir (an inexpensive and often abundant source). The medium used for IVC does not contain serum to allow examination of cell signaling molecules on embryonic development without interference from growth factors in serum. Although the procedure is designed for the developmental biologist, it can be used as well for commercial IVP. In this case, the source of female gametes will be oocytes harvested from living cows using a procedure called transvaginal, ultrasound-guided aspiration of follicles (oocyte pickup or OPU for short) (Fig. 1) [7]. The culture medium for IVC in many commercial laboratories also includes serum, which increases the yield of blastocysts [8, 9] but may predispose the resultant calves to experience large offspring syndrome (*see* discussion by [10, 11]).

2 Materials

2.1 Specialized Items of Equipment

1. Stereomicroscope with magnification capacity up to 40×.
2. An incubator set at 38.5 °C (body temperature of the cow) capable of maintaining a gaseous environment of 5% (v/v) CO_2 for IVM and IVF.
3. An incubator set at 38.5 °C capable of maintaining a gaseous environment of 5% (v/v) CO_2 and 5% (v/v) O_2 in humidified air (*see* **Note 1**).
4. Warming dish set at 38.5 °C.
5. Laminar flow hood.
6. Water bath or warming oven set at 38.5 °C. Fill the water bath with thermal beads instead of water to reduce risk of mold growth.
7. Water bath (37 °C) for thawing semen samples.
8. Hemocytometer.

2.2 Glassware

Glassware for media preparation and IVP is exclusively used for this purpose. The glassware used in the IVP laboratory must be thoroughly washed with a phosphate-free detergent and completely

Table 1
Composition of media used for oocyte manipulation and maturation[a,b]

Medium	Ingredient	Concentration
Oocyte collection medium (OCM)	Tissue culture medium 199 (with Hank's salts, L-glutamine and without bicarbonate)	
	$NaHCO_3$	4.16 mM
	HEPES	10.0 mM
	Alanyl-glutamine	1 mM
	Bovine steer serum	2% (v/v)
	Heparin	2 U/ml
	Penicillin	100 U/ml
	Streptomycin	0.1 mg/ml
Oocyte maturation medium (OMM)	Tissue culture medium 199 (with Earle's salts, L-glutamine and sodium bicarbonate)	
	Bovine steer serum	10% v/v
	Estradiol-17β	2 μg/ml
	Porcine purified follicle-stimulating hormone	25 μg/ml
	Sodium pyruvate	22 μg/ml
	Gentamicin	50 μg/ml
	Alanyl-glutamine	1 mM

[a]Adapted from Ortega et al. [12]
[b]Also *see* **Notes 2** and **3**

rinsed. It is essential that there is no soap residue left in any glassware because it will harm the embryos. Rinse glassware five times with tap water followed by five times with distilled water and sterilize by autoclaving.

2.3 Media and Other Reagents

1. Transport saline: 0.9% (v/v) NaCl, 10,000 units penicillin, and 10,000 μg streptomycin per ml in tissue culture grade H_2O.
2. Mineral oil produced specifically for embryo culture.
3. Media for oocyte manipulation and maturation are described in Table 1. These can be stored at 4 °C up to 1 month (*see* **Notes 2** and **3**).
4. 80% (v/v) and 40% (v/v) solutions of PureSperm (Nidacon, Mölndal, Sweden) or Isolate® (Irvine Scientific, Santa Ana, CA, USA) (*see* **Note 4**).
5. Media for sperm capacitation and fertilization. Media used in our laboratory are described in Table 2. Heparin is the key ingredient for induction of capacitation. Optimal concentrations vary from bull to bull and can be determined experimentally if one bull is to be used frequently.
6. Penicillamine stock: 2 mM penicillamine in 0.9% (w/v) NaCl in water.

Table 2
Composition of media composition used for sperm preparation, capacitation, and fertilization[a,b,c]

Medium	Ingredient	Concentration
HEPES-TALP	HEPES	10.0 mM
	$CaCl_2{\cdot}2H_2O$	2.0 mM
	$MgCl_2{\cdot}6H_2O$	0.5 mM
	$NaH_2PO_4{\cdot}H_2O$	0.4 mM
	KCl	3.2 mM
	NaCl	114.0 mM
	$NaHCO_3$	2.0 mM
	Na-lactate	10.0 mM
	Fraction V BSA	3.0 mg/ml
	Sodium pyruvate	0.20 mM
	Gentamicin	7.5 μg/ml
IVF-TALP	$CaCl_2{\cdot}2H_2O$	2.0 mM
	$MgCl_2{\cdot}6H_2O$	0.5 mM
	$NaH_2PO_4{\cdot}H_2O$	0.4 mM
	KCl	3.2 mM
	NaCl	114.0 mM
	$NaHCO_3$	25.0 mM
	Na-lactate	10.0 mM
	Essentially fatty acid-free BSA	6.0 mg/ml
	Sodium pyruvate	0.2 mM
	Gentamicin	5 μg/ml
	Heparin	20 μg/ml

[a]Adapted from Ortega et al. [12]
[b]Abbreviations: *BSA* bovine serum albumin, *HEPES-TALP* Hepes-buffered Tyrode's albumin–lactate–pyruvate, *IVF-TALP* Tyrode's albumin–lactate–pyruvate for fertilization
[c]Also *see* **Notes 2** and **3**

7. Hypotaurine stock: 1 mM hypotaurine in 0.9% (w/v) NaCl in water.
8. Epinephrine stock: 250 μM epinephrine in a lactate–metabisulfite solution [77 μl of a 98% (w/w) sodium lactate syrup and 50 mg sodium metabisulfite in 50 ml water].
9. PHE (penicillamine–hypotaurine–epinephrine). Combine 10 ml penicillamine stock, 10 ml hypotaurine stock, 4 ml of epinephrine stock, and 16 ml of 0.9% (w/v) NaCl in water. Store at −20 ° C for 12 months.
10. Media for embryo manipulation and culture are described in Table 3. Store at 4 °C up to 1 month.
11. Water. The choice of water depends upon local availability of highly purified water. We make up stock solutions using commercially available water for tissue culture or embryo culture. For all other media, deionized water is further purified through a water purification system.

Table 3
Composition of synthetic oviduct fluid medium—bovine embryo 2 (SOF-BE2) medium used for embryo culture[a,b,c,d]

Ingredient	Concentration
NaCl	100 mM
KCl	6.64 mM
KH_2PO_4	1.10 mM
$CaCl_2{\cdot}2H_2O$	1.09 mM
Sodium lactate	4.92 mM
$NaHCO_3$	23.27 mM
$MgCl_2{\cdot}6H_2O$	0.45 mM
Alanyl-glutamine	0.93 mM
Sodium pyruvate	0.37 mM
Sodium citrate tribasic dihydrate	0.46 mM
Myoinositol	2.56 mM
Gentamicin sulfate	23.3 μg/ml
Nonessential amino acids (100×)	9.3 μl/ml
Essential amino acids (50×)	18.6 μl/ml
Essentially fatty acid-free BSA	3.7 mg/ml

[a]Note that concentrations of amino acids are slightly more diluted than would be expected for a 100x (nonessential) or 50x (essential) solution. This is a deliberate deviation from using amino acids at a final concentration of 1× because, in our hands, development is better when the dilution is performed as described in the table

[b]Adapted from Ortega et al. [12]. Note that there are slight differences in the concentrations of the first seven ingredients from that published earlier by Ortega et al. [12] because of alterations in how stock solutions are added to make the final medium (e.g., NaCl is 100 mM instead of 92 mM)

[c]Abbreviations: *BSA* Bovine serum albumin, *SOF-BE2* Synthetic oviduct fluid—bovine embryo 2

[d]Also *see* **Notes 2** and **3**

12. Media designed for use in room air (transport saline, OCM, HEPES-TALP, PureSperm, and PHE) should be warmed to 38.5 °C in a tightly sealed container at least 2–3 h prior to use.
13. Media designed for use in a 5% (v/v) CO_2 environment (OMM, IVF-TALP, and SOF-BE2) should be equilibrated in an incubator for at least 2 h before use. To facilitate equilibration of the medium with the CO_2, it is important to loosen the lid of containers of medium placed in the incubator. Leaving an opened container of CO_2-equilibrated medium outside the incubator will result in an increase in pH that could affect embryonic viability and development. Care should also be

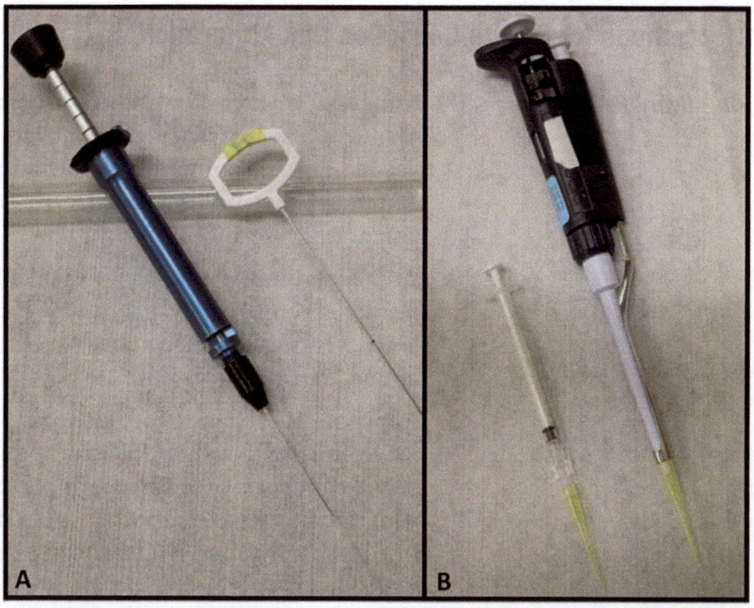

Fig. 2 Examples of instruments used for manipulation of cumulus–oocyte complexes (COC) and embryos. Panel A: instruments of small diameter used for manipulation of immature COC and embryos. Panel B: Instruments of larger diameter used to manipulate matured (i.e., expanded) COC

taken not to let the medium sit too long in the incubator as evaporation can change the molarity of components and pH.

14. Semen is purchased as frozen straws that are kept in liquid nitrogen (−196 °C) until use. Using the procedures described here, one straw should provide enough spermatozoa for 800–1000 oocytes. Unless the choice of sire is critical, we typically pool semen from three bulls (1 straw per bull) to enhance the probability that spermatozoa from at least one sire will perform well.

2.4 Instruments and Disposable Supplies

1. Device for handling and manipulating cumulus–oocyte complexes and embryos (Fig. 2, *see* **Notes 5** and **6**).
2. 60 × 15 mm and 35 × 10 mm petri dishes, untreated polystyrene.
3. Sterile hemostat, scalpel blades, and forceps.
4. 6–12 ml syringes and 18G needles.
5. Petri dish marked on the bottom with a grid.
6. X-dish (Fisherbrand™ compartmentalized petri dishes).
7. 100 μm cell strainer.
8. Sterile 1.5 ml microcentrifuge tubes.
9. Sterile 15 ml tubes.

3 Methods for Oocyte Collection and Maturation

3.1 Ovary Retrieval

1. Immediately after extraction of internal organs from the carcass, ovaries are removed from the reproductive tract and placed in plastic containers with transport saline. We use plastic containers that hold about 45 ovaries and fill each container with about 500 ml transport saline.
2. After all ovaries have been collected, excess blood from the ovaries is removed by massaging the ovaries within the container. Ovaries are then transferred to a second container containing transport saline.
3. Transport the ovaries to the lab as soon as possible. Optimal oocyte quality is achieved by keeping ovaries at about 25 °C. Avoid exposing ovaries to temperature extremes of heat or cold, such as can occur when placed in the trunk of a car. If possible, transport buckets of ovaries in a temperature-controlled cooler set at 22 °C.
4. Upon return to the lab, classify the ovaries as good [large number of small (<5 mm in diameter) and medium (5–10 mm) sized follicles, no very large follicles, ovaries not discolored] or marginal (others). The number of follicles on the ovary is an indicator of oocyte competence [13]. Marginal ovaries should only be used if the desired number of oocytes is not obtained from the good ovaries.
5. Wash ovaries (by massaging within the plastic container) 2–3 times with warm (38.5 °C) transport saline until the majority of the blood has been washed away from the ovaries. Following washing, place ovaries in a container containing fresh transport saline and store at room temperature until time of oocyte collection.

3.2 Collection of Cumulus–Oocyte Complexes (COC) and Oocyte Maturation

1. At least 2 h before needed, prepare 60 × 15 mm dishes with enough microdrops (50-μl) of oocyte maturation medium (OMM) for the expected number of COC to be collected (10 oocytes/microdrop) (Fig. 3, *see* **Notes 6** and 7). Cover the microdrops with 9 ml mineral oil and maintain in the incubator (5% CO_2 with humidified air; 38.5 ° C) until needed (*see* **Note 8**).
2. Add 100 ml warm (38.5 °C) oocyte collection medium (OCM) to a sterile 400 ml beaker.
3. Attach a hemostat to the base of the ovary to hold the ovary firmly in place. Cut the excess tissue from the ovarian stalk if needed. Hold ovary above the beaker and make 2–3 mm incisions with a scalpel blade across all visible follicles <10 mm in diameter (Fig. 4, *see* **Note 9**). Submerge ovary into OCM and

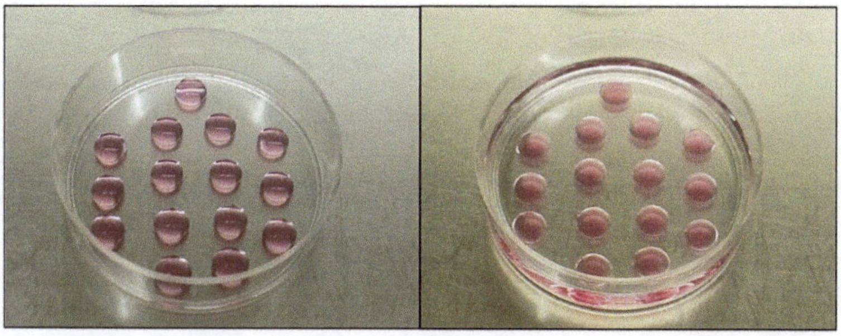

Fig. 3 Preparation of microdrops. Shown here are 15 50-μl drops on the bottom of a 60 mm petri dish before (left) and after (right) being covered with 9 ml mineral oil

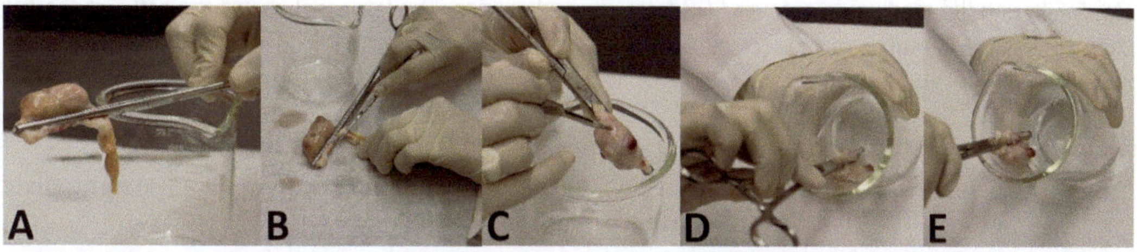

Fig. 4 Slashing of ovaries to recover oocytes. In panel A, a hemostat is attached to the base of the ovary to hold the ovary firmly in place. In panel B, the excess tissue from the ovarian stalk is removed. Panel C shows how the ovary is held above the beaker and 2–3 mm incisions are made in a downward direction with a rapid but firm movement across follicles. In panel D, the slashed ovary is being swirled in oocyte collection medium. In panel E, the ovary is being pressed against the side of the beaker to allow drainage

swirl vigorously. Repeat this process until the desired number of ovaries has been processed (Fig. 4).

4. Once all ovaries have been processed, set the beaker in the water bath (38.5 °C) while setting up for the next step.
5. Fill a 12 ml syringe with warm OCM (for later use in rinsing the oocytes from filter) and pour ~2 ml OCM into a petri dish marked on the bottom with a grid. Also, pour ~3 ml warm OCM into three wells of an X-dish.
6. Use forceps to hold a 100 μm cell strainer over a 400 ml beaker (Fig. 5). Filter the medium containing COC through the cell strainer (waste fluid should be caught in the beaker below the cell strainer). Continually swirl the beaker containing the fluid with the COC so that the oocytes do not stick to the bottom of the glass beaker during pouring.
7. When all the oocyte-containing fluid has passed through the filter, immediately turn the filter upside down over the grid dish and rinse the COC into the petri dish with a grid on the bottom using a 10 ml syringe fitted with an 18G needle and

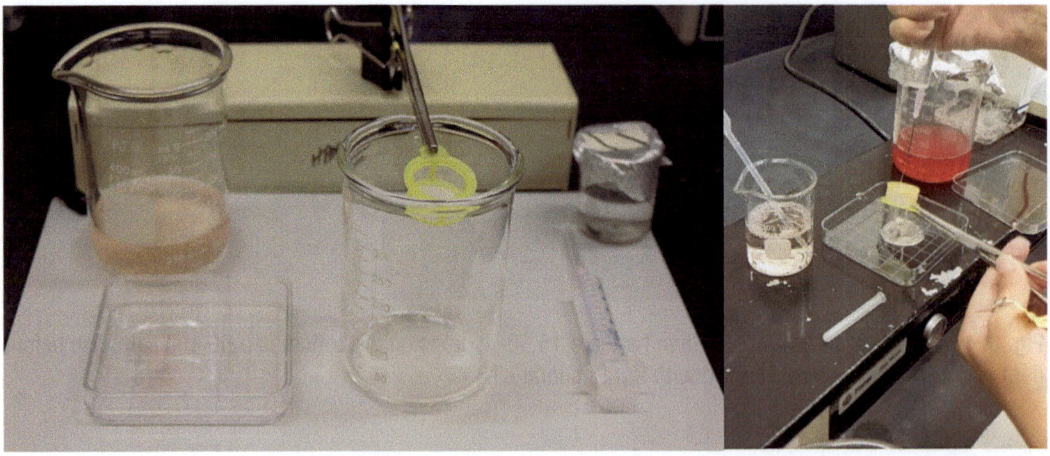

Fig. 5 Filtering the medium to recover COC from slashed ovaries. The left panel shows the setup of the cell strainer used to filter medium and recover COC. The right panel shows rinsing of oocytes from the cell strainer into a grid petri dish

filled with OCM (Fig. 5). Maintain the dish on a warming plate until ready for searching.

8. Under a stereomicroscope, use the instrument of your choice (Fig. 2, *see* **Note 5**) to harvest COC from the dish. Collect COC as fast as possible to prevent adverse effects of cold shock. Only COC that have at least three layers of compact cumulus cells and an evenly granulated cytoplasm with no clear spaces should be used for subsequent steps (*see* **Note 10**).
9. Place retrieved COC into the first well of the X-dish containing OCM.
10. When all COC have been collected, transfer COC through two additional wells containing warm OCM, leaving all debris behind.
11. Transfer groups of ten COC to 50 μl microdrops of OMM.
12. For oocyte maturation, incubate COC in OMM for 18–22 h at 38.5 °C and 5% CO_2 with humidified air.

3.3 Preparation for In Vitro Fertilization

The following procedures are performed a minimum of 2–3 h before fertilization, so that all supplies and media are warm and, where necessary, equilibrated with CO_2.

1. Add 1700 μl of IVF-TALP to 35 × 10 mm dishes (1 dish per 300 COC) and place in a 5% CO_2 incubator at 38.5 °C.
2. For each fertilization dish (up to 300 COC/dish), warm 20 ml HEPES-TALP (in tubes with caps tightened) in an oven or water bath at 38.5 °C. Medium in these tubes will be used for washing the COC before and after fertilization.

3. Add 10 ml HEPES-TALP to another 15 ml conical tube and label it "Wash." Tighten the cap and warm to 38.5 °C in an oven or water bath.
4. Add 3 ml IVF-TALP to a tube and label it "Sperm Diluent." Leave the cap loose and place in a 5% CO_2 incubator at 38.5 °C.
5. Place 1.5 ml of 80% PureSperm® and 1.5 ml 40% PureSperm® in the warming oven or water bath (*see* **Note 4**).
6. Prepare water bath (37 °C) for thawing semen.
7. Place aliquots of PHE (80 μl per fertilization dish) in the warming oven and cover with aluminum foil to reduce exposure to light.
8. Place 2–3 centrifuge carriers for 15 ml conical tubes in the oven to warm to 38.5 °C.

3.4 Preparation of Oocytes for In Vitro Fertilization

1. Place an X-dish on the slide warmer and add ~3 ml of prewarmed HEPES-TALP to one well per fertilization dish.
2. Transfer matured COC from microdrops of OMM to one corner of a well of the X-dish containing HEPES-TALP. Repeat as necessary until there are ~300 COC per well. Wash the COC by moving them from one corner to the next within the same well, completing a total of three washes.
3. Transfer a group of no more than 300 washed COC from the X-dish to a dish containing preequilibrated IVF-TALP (1700 μl/dish).
4. Return the dish with the COC to the incubator until fertilization.

3.5 Preparation of Sperm for In Vitro Fertilization

Spermatozoa are very sensitive to changes in temperature, especially cold temperatures. A space heater in front of the area where sperm preparation will be performed can aid in preventing cold shock of the sperm cells (use caution to not allow the area to get overly hot; about 35 °C). Also, make sure that all media used for sperm are warmed to 38.5 °C before use. Procedures described here fit best for conventional semen (i.e., not sex-sorted). *See* **Note 11** for modifications for sex-sorted semen.

1. Place 1.5 ml of 80% PureSperm (bottom layer) into a 15 ml conical tube. Slowly layer 1.5 ml 40% PureSperm (top layer) on top of the 80% PureSperm using a sterile plastic Pasteur pipet.
2. Thaw straws of semen for 45–60 s in the 37 °C. Note that 1 straw of semen should provide enough semen for 800–1000 oocytes.
3. Wipe the straw dry, cut the tip of the straw with scissors, and expel contents of the straw onto the top of the PureSperm

gradient. Care must be taken so that the gradient is not disturbed and the semen remains on top of the 40% layer.

4. Place the conical tube containing the semen and sperm purification gradient into a warm centrifuge carrier and centrifuge at $1000 \times g$ for 10 min.
5. After centrifugation, aspirate sperm pellet from the bottom of the conical tube with as little PureSperm as possible using a plastic Pasteur pipette.
6. Place the sperm pellet into the 15 ml conical tube containing 10 ml HEPES-TALP (labeled "wash") and centrifuge for 5 min at $200 \times g$ in another warm centrifuge carrier. Note that when adding the sperm pellet to the wash tube, care should be taken to not immerse the plastic pipette into the washing medium.
7. Remove the supernatant with a Pasteur pipette without disturbing the pellet. This step must be done quickly because motile sperm will swim out of the pellet. If the pellet is accidentally disturbed, stop the procedure and recentrifuge.
8. Determine the concentration of spermatozoa using a hemocytometer and dilute to 17×10^6/ml using prewarmed, equilibrated IVF-TALP. This will produce a final concentration of 1×10^6 spermatozoa/ml in the fertilization dish.
9. Examine one drop of diluted semen under the microscope to ensure that the spermatozoa are motile (at least 70% motile is desirable but sometimes motility for sex-sorted spermatozoa is much less and will still be used).

3.6 In Vitro Fertilization

1. Add 120 μl of the sperm suspension and 80 μl PHE into each fertilization dish.
2. Return dishes to the incubator. Fertilization should proceed for at least 8–10 h. Fertilization can proceed longer (up to 18–20 h) if convenient.

3.7 In Vitro Culture of Embryos

1. Prepare culture dishes by making 50 μl microdrops of SOF-BE2 for each group of 30 zygotes in petri dishes and cover with mineral oil. For smaller groups of embryos (<15 per drop), we typically prepare 25 μl microdrops. Place the dishes in the incubator (5% CO_2 and 5% O_2; 38.5 °C with humidity) to equilibrate for at least 2 h before removing zygotes from the fertilization dish.
2. Place an X-dish on the slide warmer and add ~3 ml of HEPES-TALP to three wells.
3. Remove one fertilization dish from the incubator.
4. Transfer putative zygotes from the fertilization dishes to a microcentrifuge tube containing 700 μl HEPES-TALP. Up to 300 embryos can be loaded in each microcentrifuge tube.

5. Allow zygotes to settle to the bottom of the tube by placing the tube in a warm oven for a few minutes.
6. Remove all but 100–200 μl of HEPES-TALP from each tube. Leaving a small volume prevents zygotes from going into the lid of the tube during vortexing. To minimize loss of zygotes during this process, expel the removed supernatant onto the lid of the dish and examine for zygotes prior to discarding it.
7. Remove cumulus cells from zygotes by vortexing the tube for 5 min.
8. Rinse the tube three times using ~1 ml warm HEPES-TALP, transferring the contents of each rinse to an X-dish.
9. Transfer presumptive zygotes to a second well of the X-dish containing warm HEPES-TALP, leaving cumulus cells and debris behind. Repeat this step by transferring the presumptive zygote to a third well of the X-dish.
10. Transfer groups of 25–30 presumptive zygotes into equilibrated 50 μl drops of SOF-BE2 and return to the incubator (5% CO_2 and 5% O_2; 38.5 °C with humidity).
11. At 72 h from the time sperm were added to the fertilization dishes, assess cleavage rate of embryos by determining the number of embryos that have cleaved and dividing by the number of putative zygotes initially placed into the microdrops (a putative zygote is an oocyte exposed to sperm). Return dishes to the incubator as quickly as possible.
12. Seven days after addition of sperm to the fertilization dishes, assess development of embryos to the blastocyst stage. Determine the number of embryos classified as a blastocyst and divide by the number of putative zygotes placed in the microdrops or by the number of cleaved embryos present on day 3. Return dishes to the incubator as quickly as possible or harvest for other purposes. Typically, embryos are not cultured for more than 8 days. One should expect that the percentage of oocytes that cleave is ≥70% and the percentage of oocytes that become blastocyst is ≥20%.

4 Notes

1. One option is to purchase a trigas incubator containing CO_2 and O_2 sensors that uses nitrogen gas to regulate oxygen content. An alternative is to use one of the benchtop incubators sold by several suppliers such as WTA (São Paulo, SP, Brazil) and Cook Medical (Bloomington, IN, USA) that deliver premixed gas to the chamber in which dishes are cultured. The inexpensive option is to culture embryos in cell-culture environmental chambers, which are airtight containers that

are perfused with gas mixture, sealed, and then placed in an incubator or oven.

2. For each medium, the pH is adjusted to 7.3–7.4 and the medium is sterile-filtered using a 0.2 μm surfactant-free cellulose acetate filter. For HEPES-TALP, IVF-TALP and SOF-BE2, osmolarity is determined after filtration and the medium only used if the osmolarity is within acceptable limits (HEPES-TALP, 260–270 mOsm; IVF-TALP, 280–300 mOsm; and SOF-BE2, 260–270 mOsm).
3. Note that commercial media are also available. For example, BoviPRO oocyte washing medium with BSA from MOFA (Verona, WI, USA) can substitute OCM, BO-IVM from IVF Bioscience (Falmouth, UK) can substitute OMM, and BO-IVC from IVF Bioscience can substitute SOF-BE2.
4. One can also make sperm purification media based on 90% (v/v) and 45% (v/v) Percoll® solutions suspended in medium used for sperm preparation. Note that sperm cells from some bulls are sensitive to these solutions. In such a case, alternative methods such as swim up or glass wool purification may be used.
5. Hand-held pipettes or microdispensers are used for handling and manipulations of cumulus oocyte complexes (COC) and embryos (Fig. 2). When transferring oocytes or embryos from one medium to another, transfer the oocytes or embryos in as little medium as possible. A video demonstrating oocyte and embryo handling procedures is available at https://www.youtube.com/watch?v=vbpLv1r0TnU.
6. Pipet tips may contain toxins or other substances that might inhibit development of embryos. As a precaution, always fill and empty pipet tip at least once before using, especially before adding new medium or serum to a microdrop containing embryos.
7. Microdrops for culture of COC and embryos are made by pipetting drops of medium into the bottom of a dish and then carefully flooding drops with mineral oil. When pipetting medium to make the drops, the pipettor should be held perpendicular to the dish. Oil is added at a point near the side of the dish. See video at https://www.youtube.com/watch?v=2D0CE1LnZzo for an example of how to make microdrops.
8. Due to repeated openings of the incubator door, temperature at the front of the incubator fluctuates. Thus, place dishes at the back of the incubator to reduce exposure to changes in temperature.
9. See video at https://www.youtube.com/watch?v=Rfi80TUjbG4 for a demonstration of COC recovery.

10. See video at https://www.youtube.com/watch?v=6UCwVkPKEQg for an example of COC selection.

11. Straws of sex-sorted semen will be sufficient for many fewer oocytes (~150) because the quantity and quality of the spermatozoa in these straws is reduced. Procedures for IVF using sex-sorted semen are similar, but working volumes are less. For **step 1**, each layer is made with 200 μl in a 1.5 ml tube. For **step 4**, a microcentrifuge is used (1600–2600 × *g* for 5 min). For **step 6**, transfer 100 μl from the bottom of the tube (the pellet may not be visible) to an 1.5 ml tube containing 500 μl of prewarmed IVF-TALP and centrifuge at 400–600 × *g* for 3 min. Use 100 μl from the bottom of the tube to fertilize. Note that we do not count the sperm but instead inseminate 60 μl drops containing 30 COC with 20 μl of sperm and 3.5 μl of PHE. We coincubate these gametes for 18–20 h.

References

1. Brackett BG, Bousquet D, Boice ML, Donawick WJ, Evans JF, Dressel MA (1982) Normal development following in vitro fertilization in the cow. Biol Reprod 27:147–158
2. Block J, Bonilla L, Hansen PJ (2009) Effect of addition of hyaluronan to embryo culture medium on survival of bovine embryos in vitro following vitrification and establishment of pregnancy after transfer to recipients. Theriogenology 71:1063–1071
3. Bouquet A, Juga J (2013) Integrating genomic selection into dairy cattle breeding programmes: a review. Animal 7:705–713
4. Baruselli PS, Ferreira RM, Sales JN, Gimenes LU, Sá Filho MF, Martins CM, Rodrigues CA, Bó GA (2011) Timed embryo transfer programs for management of donor and recipient cattle. Theriogenology 76:1583–1593
5. Hansen PJ (2017) Physiological approaches to improving fertility during heat stress. In: Beede DK (ed) Large dairy herd management, 3rd edn. American Dairy Science Assn, Champaign, IL, pp 7–42
6. Hansen PJ, Block J, Loureiro B, Bonilla L, Hendricks KE (2010) Effects of gamete source and culture conditions on the competence of in vitro-produced embryos for post-transfer survival in cattle. Reprod Fertil Dev 22:59–66
7. Galli C, Duchi R, Colleoni S, Lagutina I, Lazzari G (2014) Ovum pick up, intracytoplasmic sperm injection and somatic cell nuclear transfer in cattle, buffalo and horses: from the research laboratory to clinical practice. Theriogenology 81:138–151
8. Leivas FG, Brum DS, Fialho SS, Saliba WP, Alvim MT, Bernardi ML, Rubin MI, Silva CA (2011) Fetal calf serum enhances in vitro production of *Bos taurus indicus* embryos. Theriogenology 75:429–433
9. Pinyopummintr T, Bavister BD (1994) Development of bovine embryos in a cell-free culture medium: Effects of type of serum, timing of its inclusion and heat inactivation. Theriogenology 41:1241–1249
10. Farin PW, Piedrahita JA, Farin CE (2006) Errors in development of fetuses and placentas from in vitro-produced bovine embryos. Theriogenology 65:178–191
11. Thompson JG, Allen NW, McGowan LT, Bell AC, Lambert MG, Tervit HR (1998) Effect of delayed supplementation of fetal calf serum to culture medium on bovine embryo development in vitro and following transfer. Theriogenology 49:1239–1249
12. Ortega MS, Wohlgemuth S, Tribulo P, Siqueira LG, Null DJ, Cole JB, Da Silva MV, Hansen PJ (2017) A single nucleotide polymorphism in *COQ9* affects mitochondrial and ovarian function and fertility in Holstein cows. Biol Reprod 96:652–663
13. Brevini-Gandolfi TA, Favetta LA, Mauri L, Luciano AM, Cillo F, Gandolfi F (1999) Changes in poly(A) tail length of maternal transcripts during in vitro maturation of bovine oocytes and their relation with developmental competence. Mol Reprod Dev 52:427–433

Chapter 9

In Vitro Embryo Production in Sheep

S. Mondal, A. Mor, I. J. Reddy, S. Nandi, P. S. P. Gupta, and A. Mishra

Abstract

The recent advances in biotechnological research have led to development of many advanced reproductive techniques and biological tools which are set to revolutionize the productive efficiency of livestock species. The development of technology for sequencing of whole genomes and mass screening of gene regulation has widened our approach to genetic profiling and mapping, as well as furthering our understanding of underlying physiological mechanisms. The newer biotechnologies of gene transfer, in vitro fertilization and embryo production, cloning, and stem cell technology have been developed and are being refined with efficiencies suitable for use in animal farming. Efficient in vitro systems for maturing oocytes, fertilizing, and developing embryos have resulted in commercial in vitro production of embryos. Here we describe in vitro maturation, in vitro fertilization, embryo production, embryo culture, and quantitation of gene expression in sheep embryos.

Key words In vitro maturation, In vitro fertilization, In vitro embryo production, Embryo culture, Embryonic gene expression

1 Introduction

Sheep are said to have been one of the first species to be domesticated and hence have been closely associated with man from early times [1]. These livestock are important around the world, especially in less developed countries, which are known to have limited land and other natural resources. Besides meat production, sheep provide milk, organic manure to the crop sector, and hides, skins, bones, blood, and fiber that are used in industries. Thus, they are an important source of income and employment, helping to alleviate poverty and smooth the income distribution among small landholders and the landless, who represent the bulk of rural populations and the majority of livestock owners.

Multiple ovulation and embryo transfer (MOET) and in vitro embryo production (IVEP) are assisted reproductive technologies aimed at increasing the diffusion of female genetics. These technologies have been effectively utilized to multiply the characteristics of superior animals in animal breeding systems in many

Jason R. Herrick (ed.), *Comparative Embryo Culture: Methods and Protocols*, Methods in Molecular Biology, vol. 2006,
https://doi.org/10.1007/978-1-4939-9566-0_9, © Springer Science+Business Media, LLC, part of Springer Nature 2019

developed countries. In vitro embryo production is important in sheep as it has shown the potential of producing sheep embryos even during the nonbreeding season through in vitro maturation and in vitro fertilization techniques [2]. In vitro embryo production is a multistep methodology and consists of the following steps: (1) in vitro maturation of oocytes recovered directly from the follicles, (2) in vitro fertilization (IVF) or coincubation of capacitated spermatozoa with the in vitro matured oocytes, and (3) in vitro culture (IVC) of zygotes up to the blastocyst stage. The natural in vivo process of fertilization and early embryonic development involves innumerable factors that interact to affect each event of the developmental process, including the first cleavage division, the timing of which is known to be an indicator of subsequent developmental potential of the embryo, activation of embryonic genome [3], compaction of morulae, and formation of blastocysts. These morphological and physiological transitions in preimplantation embryo development are known to be accompanied and regulated by differential expression of developmentally important genes [4, 5]. A better knowledge about gene expression patterns during early embryo development would give us detailed information about the molecular pathways controlling early development [6, 7]. We have optimized in vitro fertilization and embryo production protocol in sheep using FGF2 for the first time [8–11]. Maturation and cleavage rates were significantly higher ($P < 0.05$) in the oocytes cultured in TCM-199 + 10% FBS + FSH (10 μg/ml) supplemented with both 20 ng/ml ITS and 20 ng/ml FGF2 as compared to the control indicating that the supplementation of ITS and FGF2 in maturation medium was beneficial for improving maturation and cleavage rates of sheep oocytes. This chapter focuses on oocyte collection and evaluation, in vitro maturation, in vitro capacitation of sperm, in vitro fertilization, in vitro embryo culture, and gene expression quantitation.

2 Materials

2.1 Culture Media for In Vitro Oocyte Maturation, Fertilization, and Embryo Culture

1. Saline: 9.0 mg/ml NaCl in culture-grade H_2O.
2. Mineral oil.
3. 1 mg/ml hyaluronidase in TCM199.
4. Aspiration Medium (10 ml): 5 ml TCM-199 (25 mM HEPES, with Earle's salts and L-glutamine), 5 ml Dulbecco's phosphate-buffered saline, 3.0 mg/ml bovine serum albumin (BSA), 10 μg/ml of heparin, and 10 μg/ml gentamicin.
5. In Vitro Maturation Medium (1 ml): 885 μl TCM-199, 100 μl fetal bovine serum (FBS, 10%), 1 μg/ml FSH, and 50 μg/ml gentamicin.

Table 1
Composition of Brackett and Oliphant (BO)-A medium

Chemicals	Amount/100 ml	mM
Sodium chloride	808 mg	138.35
Potassium chloride	39.50 mg	5.3
Calcium chloride	43.50 mg	3.9
Sodium dihydrogen phosphate	17 mg	1.4
Magnesium chloride hexahydrate	14 mg	0.7
Sodium bicarbonate	310.46 mg	37
HEPES	60 mg	2.5
H_2O	100 ml	–
0.5% Phenol red	20 µl	–

6. Brackett and Oliphant (BO)-A Medium (Table 1, *see* **Note 1**).
7. BO-B Capacitation Medium (50 ml): 50 ml BO-A Medium, Na-pyruvate (1.25 mM), 10 µg/ml heparin, and 50 µg/ml gentamicin.
8. BO-C Fertilization Medium (4 ml): 4 ml BO-B Medium, 20 mg (5.0 mg/ml) fatty acid-free BSA, and caffeine (9.3 mM).
9. Embryo Culture Medium (1 ml): 895 µl TCM-199, 100 µl FBS (10%), and 50 µg/ml gentamicin.

2.2 Supplies for Molecular Biology

1. TRIzol LS (AMBION, Life Technologies, USA).
2. RNeasy Mini Kit (Qiagen, USA).
3. Qubit® 2.0 Fluorometer (Invitrogen, USA).
4. 0.2% chloroform.
5. SuperScript™ III First-Strand Synthesis kit, Invitrogen, USA.
6. 50 µM oligo(dT)20.
7. 50 ng/µl random hexamers.
8. 10 mM dNTPs mix.
9. DEPC-treated, RNase-free H_2O.
10. DNA Synthesis Mix: 2 µl 10× RT buffer (4 µl 25 mM $MgCl_2$, 2 µl 0.1 M DTT, 1 µl RNase OUT (40 U/µl)); and 1 µl Superscript III RT (200 U/µl).
11. RNaseH.
12. qRT-PCR Reaction Mixture (10 µl): 5 µl SYBR master mix, primers (7.5 pM), cDNA (100–200 ng), and bring to a final volume of 10 µl with DEPC-treated H_2O.

13. STEP ONE PLUS real-time PCR system (Applied Bio systems, USA).
14. Real-Time RT-PCR (ABI Prism 7300).

3 Methods

3.1 Oocyte Collection and Evaluation

1. Collect ovaries from nonpregnant slaughtered sheep from a local slaughterhouse and transport to the laboratory in saline (37 °C).
2. Trim off extra tissues from the ovaries and wash three times in saline (37 °C).
3. Aspirate visible follicles with a diameter of 1–6 mm using a 20G needle attached to a 5 ml syringe containing 0.5–1.0 ml of aspiration medium.
4. Grade cumulus–oocyte complexes by the morphological appearance of the cumulus cells investments and homogeneity of ooplasm under a stereo zoom microscope (magnification 100×). Grade 1: compact cumulus–oocyte complexes with an unexpanded cumulus mass having $\geq$5 layers of cumulus cells and with homogenous cytoplasm; grade 2: cumulus–oocyte complexes with $\leq$4 layers of cumulus cells and with homogenous cytoplasm; and grade 3: oocytes without cumulus cells and with irregular (shrunken) cytoplasm. Only oocytes having more than five layers of unexpanded cumulus cells and granular homogenous ooplasm (grade 1) will be recovered and selected for in vitro maturation.

3.2 In vitro Maturation of Oocytes

1. Wash all the selected oocytes for in vitro maturation (IVM) once with aspiration medium and three times with IVM medium, then culture the oocytes (10–15 oocytes/50 μl drop) in the maturation medium covered with mineral oil.
2. Culture oocytes at 38.5 °C, 5% CO_2 and 95% humidity for 24 h.
3. Determine the degree of cumulus cell expansion after 24 h of IVM. Only oocytes whose cumulus cells have expanded to at least twice the diameter of the oocyte will be considered expanded.
4. Cumulus cells are then removed by 0.1% hyaluronidase.

3.3 Sperm Preparation

1. Collect fresh semen from a ram using electronic ejaculator.
2. For capacitation of spermatozoa, 100 μl of semen is diluted with 14 ml of BO-B medium and centrifuged for 6 min at 2500 rpm (500 × *g*).

3. Remove the supernatant, resuspend the pellet in 1 ml BO-B, and centrifuge again for 6 min at 2500 rpm (500 × *g*).
4. After the second centrifugation, remove the supernatant and resuspend the spermatozoa in 1 ml BO-C medium.

3.4 In Vitro Fertilization

1. Before transfer to fertilization drops, wash the oocytes four times in 50 μl drops of BO-C medium in order to remove the cumulus cells adhered to the ZP and residual cumulus cells in the medium.
2. Oocytes with an expanded cumulus cell mass to at least 2 diameters away from the zona pellucida and with an extruded first polar body in the perivitelline space were considered as matured and used for IVF.
3. After the matured oocytes have been denuded, keep them in 50 μl of BO-C medium at 38.5 °C in 5% CO_2.
4. Carry out in vitro fertilization by coculturing denuded oocytes and capacitated spermatozoa.
5. Add 5.0 μl aliquot of sperm suspension, containing $1–2 \times 10^6$ sperm/ml into the fertilization drop (10–15 oocytes per 50 μl BO-C), and incubate at 38.5 °C under 5% CO_2 in humidified air for 2–4 h.

3.5 In Vitro Embryo Culture

1. After 4 h of coincubation, transfer the oocytes to 50 μl embryo culture medium and incubate for 18–20 h.
2. After 18 h, wash oocytes in embryo culture medium.
3. Culture the embryos in the embryo culture medium, which is similar to the maturation medium, except FSH in the maturation medium, for further development at 38.5 °C, 5% CO_2 and 95% humidity and check the cleavage after 42–48 h post insemination.
4. Culture the embryos further 6–7 days in order to produce a complete developmental series from the 2-cell to blastula stage of embryonic development.

3.6 Expression Profiling of Developmentally Important Genes in Embryos

1. Examine presumptive matured oocytes under inverted microscope (magnification) for the presence of the first polar body. Examine all embryos at all stages of development carefully for uniform size and quality.
2. Denude immature and mature oocytes ($n = 100$) with 100 μl 0.1% hyaluronidase in aspiration medium and keep at 38.5 °C under 5% CO_2 for 2 min in a 100 μl droplet, then vortex for 2–3 min.
3. After denudation of the oocytes, wash them extensively and verify the absence of cumulus cells at 100× magnification.

4. Collect pools of immature and mature oocytes ($n = 100$), 2-cell ($n = 50$), 4-cell ($n = 25$), 8-cell ($n = 13$), 16-cell ($n = 6$), morulae ($n = 5$), and blastulae ($n = 3$) at the appropriate stages of development in 100 μl TRIzol LS and store at −80 °C.

3.7 Isolation of Total RNA

Extract total RNA from the pooled oocytes and embryos using TRIzol LS (AMBION, life technologies, USA) and RNeasy Mini Kit. In the TRIzol LS (AMBION, Life Technologies, USA) method after separation of the aqueous phase, transfer total RNA to the RNeasy column according to the manufacturer's instructions. Check the quality and integrity of the purified RNA by agarose gel electrophoresis and quantify RNA using Qubit® 2.0 Fluorometer as per manufacturer's protocol. The purified RNA can be preserved at −80 °C until use. Different steps involved in RNA isolation are as follows:

1. Collect embryos and oocytes in 200 μl TRIzol LS and bring the total volume to 1 ml with additional TRIzol LS. Homogenize the sample with the use of syringe and keep the sample at RT for 5 min.
2. After incubation, add 0.2% chloroform and centrifuge for 10 s at 8000 × *g* for proper mixing. Again, keep it at RT for 2 min and centrifuge at 12,000 rpm (8000 × *g*) for 15 min.
3. After centrifugation, separate the aqueous phase in a new tube and add an equal volume of 70% ethanol to the supernatant and mix well by pipetting. Do not centrifuge.
4. Transfer up to 700 μl of the sample, including any precipitate, to an RNeasy Mini spin column and place in a 2 ml collection tube and centrifuge for 15 s at ≥8000 × *g*. Discard the flow-through.
5. Add 700 μl Buffer RW1 to the RNeasy spin column and centrifuge for 15 s at ≥8000 × *g*. Discard the flow-through.
6. Add 500 μl Buffer RPE to the RNeasy spin column and centrifuge for 15 s at ≥8000 × *g*. Discard the flow-through.
7. Add 500 μl Buffer RPE to the RNeasy spin column and centrifuge for 2 min at ≥8000 × *g*. Optional: Place the RNeasy spin column in a new 2 ml collection tube (supplied). Centrifuge at full speed for 1 min to dry the membrane.
8. Place the RNeasy spin column in a new 1.5 ml collection tube (supplied). Add 30–50 μl RNase-free H_2O directly to the spin column membrane and centrifuge for 1 min at ≥8000 × *g* to elute the RNA.

3.8 Synthesis of First-Strand cDNA

cDNA synthesis is performed in the first step using either total RNA or poly(A) + -selected RNA primed with oligo(dT), random primers, or a gene-specific primer. In the second step, PCR is performed in a separate tube using primers specific for the gene of interest. The following procedure is designed to convert 1 pg to 5 μg of total RNA or 1 pg to 500 ng of poly(A) + RNA into first-strand cDNA.

1. Mix and briefly centrifuge (30 s at 8000 × *g*) each component before use.
2. Combine up to 5 μg of total RNA, 1 μl of required primer (50 μM oligo(dT)20, 2 μM of gene specific primers, or 50 mg/ml random hexamers), and 1 μl of 10 mM dNTPs mix, and bring the final volume to 10 μl with DEPC-treated H_2O in a 0.2 or 0.5 ml tube.
3. Incubate the tube at 65 °C for 5 min, then place on ice.
4. Add 10 μl of cDNA synthesis mix to each RNA–primer mixture, mix gently, and collect by brief centrifugation.
5. Incubate as follows:

   Oligo (dT)20 or GSP primed: 50 min at 50 °C.

   Random hexamer primed: 10 min at 25 °C, followed by 50 min at 50 °C 1 min.
6. Terminate the reactions by 85 °C for 5 min.
7. Chill on ice.
8. Collect the reaction by brief centrifugation (30 s at 8000 × *g*).
9. Add 1 μl of RNaseH to each tube and incubate for 20 min at 37 °C.
10. cDNA synthesis reaction can be stored at −20 °C or used for PCR immediately.

3.9 Real-Time Quantitative Polymerase Chain Reaction

1. Primers are designed using the Primer3 program [12] with an annealing temperature of 60 °C and amplification size of less than 250 bp using sequences found in NCBI (*see* **Note 2**).
2. Perform PCR analyses with three replicates per sample for each gene using the STEP ONE PLUS real time PCR system.
3. Confirm primer efficiencies (98–103%) by tenfold serial dilution of cDNA [13].
4. Carry out thermal cycling as per manufacturer's protocol (Table 2).
5. Determine the specificity of each PCR product by a melt curve analysis and amplicon size determination by agarose gel (1%). Negative controls which consisted of all components of qRT-PCR mix except cDNA will be used for all primers.
6. The relative quantification of gene expression changes will be recorded after normalizing for β-actin gene expression computed by using the $2^{-\Delta\Delta CT}$ method [14] in which *CT* value from controls served as calibrator.

Table 2
Real-time PCR temperature and time profile

	Stage-I	Stage-II	Stage-III	Stage-IV
Cycle	1	40	45	1
Temperature	95 °C	95 °C and 60 °C	95 °C and 60 °C	95 °C and 60 °C
Time	20 s	95 °C–3 s 60 °C–30 s	95 °C–15 s 60 °C–1 min	95 °C–15 s 60 °C–1 min 95 °C–15 s

7. Real-Time RT-PCR was performed with SYBR green chemistry with cDNA as template.
8. The specificity of amplification during real time RT-PCR will be monitored by evaluating the melting curve, running some of the products on an agarose gel. The ΔCt method [14] will be used to determine the relative amounts of RNA transcripts of the genes of interest that are present in the targeted and mock-transfected cells.

4 Notes

1. First, dissolve all the chemicals in 50 ml culture-grade H_2O except $CaCl_2$ and $MgCl_2$. Dissolve $MgCl_2$ in 2–3 ml of culture-grade H_2O and add drop by drop to the prepared solution. Repeat the same process with $CaCl_2$ and subsequently make the volume to 100 ml with culture-grade H_2O.
2. Tips for Primer Design.
   (a) Primer length: The optimal length of PCR primers is 18–22 bp. This length is long enough for adequate specificity and short enough for primers to bind easily to the template at the annealing temperature.
   (b) Primer melting temperature: Primer melting temperature (T_m) is the temperature at which one half of the DNA duplex will dissociate to become single stranded and indicates the duplex stability. Primers with melting temperatures in the range of 52–58 °C generally produce the best results.
   (c) Primer annealing temperature: The primer melting temperature is the estimate of the DNA–DNA hybrid stability and critical in determining the annealing temperature. Too high Ta will produce insufficient primer–template hybridization, resulting in low PCR product yield. Too low Ta may possibly lead to nonspecific products caused by a high number of base pair mismatches. Mismatch

tolerance is found to have the strongest influence on PCR specificity.

(d) GC content: The GC content (the number of G's and C's in the primer as a percentage of the total bases) of primer should be 40–60%.

(e) GC clamp: The presence of G or C bases within the last five bases from the 3′ end of primers (GC clamp) helps promote specific binding at the 3′ end due to the stronger bonding of G and C bases. More than three G's or C's should be avoided in the last five bases at the 3′ end of the primer.

(f) Primer secondary structures: The presence of the primer secondary structures produced by intermolecular or intramolecular interactions can lead to poor or no yield of the product. They greatly reduce the availability of primers to the reaction.

- Hairpins: It is formed by intramolecular interaction within the primer and should be avoided.
- Self dimer: A primer self-dimer is formed by intermolecular interactions between the two (same sense) primers, where the primer is homologous to itself.
- Cross dimer: Primer cross dimers are formed by intermolecular interaction between sense and antisense primers, where they are homologous.

(g) Repeats: A repeat is a dinucleotide occurring many times consecutively and should be avoided because they can misprime. For example: ATATATAT. The maximum number of dinucleotide repeats acceptable in an oligo is four dinucleotides.

(h) Runs: Primers with long runs of a single base should generally be avoided as they can misprime. For example, AGCGGGGGATGGGG has runs of base "G" of value 5 and 4. The maximum number of runs accepted is 4 bp.

(i) 3′ End Stability: It is the maximum ΔG value of the five bases from the 3′ end. An unstable 3′ end (less negative ΔG) will result in less false priming.

(j) Avoid Template Secondary Structure: Single-stranded nucleic acid sequences are highly unstable and fold into conformations (secondary structures). The stability of these template secondary structures depends largely on their free energy and melting temperatures (T_m). Consideration of template secondary structures is important in designing primers, especially in qPCR.

(k) Avoid Cross Homology: To improve specificity of the primers it is necessary to avoid regions of homology.

References

1. Shelton M (1995) Harnessing the biological potential of sheep in providing protein for growing world population. J Anim Sci 73:243. (Abstract)
2. Pugh A, Fukui Y, Tervit HR, Thompson JG (1991) Developmental ability on in vitro matured sheep oocytes collected during non-breeding season and fertilized in vitro with frozen ram semen. Theriogenology 36:771–778
3. Memili E, Dominko T, First NL (1998) Onset of transcription in bovine oocytes and preimplantation embryos. Mol Reprod Dev 51:36–41
4. Zimmermann JM, Schultz RM (1994) Analysis of gene expression in the preimplantation mouse embryo: use of mRNA differential display. Dev Biol 91:5456–5460
5. Schultz RM, Davis W, Stein P, Svoboda P (1999) Reprogramming of gene expression during preimplantation development. J Exp Zool 285:276–282
6. Mor A, Mondal S, Reddy IJ, Soumya NP (2015d) Expression of cell growth regulatory factor IGF-2 mRNA in sheep embryos. J Vet Sci Tech 6(6):61
7. Mor A, Mondal S, Reddy IJ, Nandi S, Gupta PSP (2018) Molecular cloning and expression of FGF2 gene in preimplantation developmental stages of in vitro produced sheep embryos. Reprod Domest Anim 53(4):895–903
8. Mondal S, Mor A, Reddy IJ, Nandi S, Gupta PSP (2015a) Effect of fibroblast growth factor 2 (FGF2) and insulin transferrin selenium (ITS) on in vitro maturation, fertilization and embryo development in sheep. Brazilian Arch Biol Tech 58(4): 521–525
9. Mondal S, Mor A, Reddy IJ, Gupta PSP, Nandi S (2015b) Influence of Season on Quality and In Vitro Maturation of Sheep Oocytes. Int J Environ Sci 6(2):162–166
10. Mondal S, Mor A, Reddy IJ, Nandi S, Gupta PSP (2015c) Selection of developmentally competent ovine oocytes by BCB staining improves blastocyst development rate. J Biotech Biomater 5(6):58
11. Mor A, Mondal S, Reddy IJ (2017) Production of sheep embryos in vitro. Aust Endocrinol Diabet Case Rep 2(1):1011
12. Rozen S, Skaletsky H (2000) Primer3 on the WWW for general users and for biologist programmers. In: Kravetz S, Minsener S (eds) Methods in molecular biology. Humana Press, Totown, NJ, pp 365–386
13. Svec D, Tichopad A, Novosadova V, Pfaffl MW, Kubista M (2015) How good is a PCR efficiency estimate: recommendations for precise and robust qPCR efficiency assessments. Biomol Detect Quantif 3:9–16
14. Livak KJ, Schmittgen TD (2001) Analysis of relative gene expression data using real-time quantitative PCR and the $2^{-\Delta\Delta CT}$ method. Methods 25:402–408

Chapter 10

In Vitro Culture of Embryos from LOPU-Derived Goat Oocytes

Joanna Maria G. Souza-Fabjan, Ribrio Ivan T. P. Batista, Vicente José F. Freitas, and Pascal Mermillod

Abstract

A high oocyte quality is the prerequisite for in vitro embryo production. Goat cumulus–oocyte complexes (COC) are mainly collected from slaughterhouse ovaries or by laparoscopic ovum pickup (LOPU) from live animals. Several features can influence the availability of good quality oocytes recovered by the LOPU technique. Interestingly, slaughterhouse and LOPU oocytes have different in vitro maturation kinetics and requirements, and thus, the IVP system must be adapted regarding the oocyte origin. Overall, the use of undefined media in the different steps makes interpretation of results more difficult, hampers their reproducibility, and introduces a sanitary risk. Thus, there is an effort worldwide to use simpler conditions for goat IVP. Although the success of IVP rates is relatively high, in vitro embryos differ from in vivo-derived ones in many aspects, resulting in lower viability. Therefore, strategies to improve in vitro embryo quality are crucial, such as the use of oviductal epithelium cells for coculture. Here we describe the main steps and culture media which can be utilized to produce embryos in vitro from LOPU or slaughterhouse oocytes in goats.

Key words Caprine, IVF, IVP, Goat COC, Laboratory protocol, OPU

1 Introduction

Animal biotechnology is one of the fields of science that has contributed to the most remarkable achievements in the last two decades: from the sheep Dolly to transgenic pigs that can donate organs for humans, and the generation of animal bioreactors producing human therapeutic proteins in milk [1]. The potential application of in vitro embryo production (IVP), cloning, or transgenesis depends on the acquirement of reliable, repeatable, and efficient techniques for recovery of good quality *cumulus*–oocyte complexes (COC), in vitro maturation (IVM), fertilization (IVF), and embryo development (IVD) [2, 3]. Currently, goat COC are mostly collected from slaughterhouse ovaries or, less often, by laparoscopic ovum pickup (LOPU) from live animals. Slaughterhouse ovaries may provide a cheap and relatively abundant source of

Jason R. Herrick (ed.), *Comparative Embryo Culture: Methods and Protocols*, Methods in Molecular Biology, vol. 2006,
https://doi.org/10.1007/978-1-4939-9566-0_10,

COC, which are helpful for research. However, the use of IVP for genetic improvement or diffusion requires collecting COC from given females of high economic or genetic merit. Soon, the National Center of Production of Goat Semen in France (Cap-Gene) will switch to genomic selection based on SNP array and decrease progeny testing in dairy goats. As was the case for bovine, this will probably increase the use of MOET or IVP to accelerate the selection process on the female line.

1.1 LOPU

LOPU procedure is less stressful and less invasive, and takes less (each session takes between 10 and 20 min per goat) time than laparotomy and can be repeated at short intervals without distressing animals or altering COC developmental competence [4]. Our group has demonstrated that goats raised in a tropical climate maintained a consistent ovarian response (quantitatively and qualitatively) to repeated hormonal treatments followed by LOPU [5]. It is noteworthy that after repeated LOPU in sheep (similar technique as in goats), no complication such as adhesions and fibroses occurred, and normal histology of ovaries was observed [6], demonstrating that laparoscopy is a minimally invasive procedure.

Although the development of LOPU started in the 1990s, there is still little information available on immature oocytes collected from live goats and their subsequent IVP. In addition, despite heavy research efforts, the variability of the number and quality of COC collected still limits the large-scale use of this technology. Under the current conditions, oocyte recovery rates (the number of oocytes recovered by puncturing follicles $\times$ 100) range from 40 to 80%, and the number of harvested COC per female ranges between 4 and 14 in different laboratories (Reviewed by [7]).

Different schemes of hormonal stimulation have been evaluated in order to improve the quantity and quality (developmental competence) of COC from does submitted to LOPU [4, 8]. It is well described that FSH stimulation prior to oocyte recovery increases oocyte production efficiency, with significantly more follicles punctured and COC retrieved from each female [9]. A few studies have indicated that FSH has an adverse effect on cleavage rate, although it positively affects blastocyst development rate. Possibly, some COC recruited by FSH treatment may be of low quality and unable to cleave. The positive or negative effect of FSH may be related to its dose and/or administration scheme (personal observation). Besides the use of hormonal treatments, several other features can influence the availability of COC presenting great quality in goats, such as individual characteristics, age, season, breed, nutrition, and the interval of hormonal treatment to LOPU procedure. Moreover, it is important to highlight that there are few important aspects for achieving a high recovery rate, such as follicle size to be punctured, the type and length of the needle, the diameter of the tubing, aspiration pressure, and flow rate. The detailed procedure is well described elsewhere (Reviewed by [7]).

1.2 IVM

Although many studies aim at searching for molecular biomarkers of COC quality, their morphologic evaluation (CC, cumulus cell number and appearance, homogeneity of oocyte cytoplasm) remains the only noninvasive way to select homogenous COC population before IVM. A grade classification based on the number of CC layers is currently used by many laboratories and, commonly, only oocytes with at least one complete layer of CC are selected for IVM. Those found denuded at collection (DOC, grade 3) are considered not suitable and routinely discarded. Nonetheless, the assessment of COC morphology before IVM has proved to be a poor predictor of oocyte developmental competence, resulting in a great variability in the production of embryos (Reviewed by [2]).

Our team has designed some experiments in order to evaluate the potential of the use of goat DOC to generate blastocysts, applying different approaches. Similar maturation, fertilization, and embryo development rates were obtained when using complex, semidefined, or simplified IVM media in slaughterhouse DOC (19–23% blastocyst/DOC). Parthenogenetically activated DOC presented higher development capacity compared with DOC submitted to IVF (59% versus 28% blastocyst/DOC, respectively) [10]. A total of 45% of oocytes reached the blastocyst stage when COC and DOC underwent IVM and IVF together in a similar proportion (1:1) [11]. It is reasonable to assume that the coculture of DOC with COC during IVM has a beneficial effect on the development rate of DOC. The mechanisms by which CC improve oocyte maturation are yet poorly understood. We believe that the presence of CC is not essential for meiotic maturation, if at the time of removal of the oocyte from follicular environment, they already acquired meiotic competence. Nevertheless, CC may significantly contribute to the acquisition of developmental competence acquisition during IVM. Although DOCs are less competent for development than COC, they can result in an additional number of embryos from genetically valuable females collected by LOPU or after culling [10].

Oocyte maturation includes meiotic resumption and progression to the fertilizable stage of metaphase II, after emission of the first polar body, and related events in oocyte cytoplasm. Overall, the result of IVM depends on the intrinsic quality of immature COC, but the maturation conditions can also widely modulate the final competence of IVM oocytes. In goats, in vitro maturation is commonly performed using tissue culture medium (TCM) 199 medium supplemented with FSH (porcine or ovine origin), LH, estradiol, and 10% fetal calf serum (FCS). However, although serum exerts beneficial effects on oocyte maturation, it makes more difficult the interpretation of results, hampers their reproducibility, and introduces a sanitary risk [2]. For these reasons, there is a trend to use defined or at least semidefined maturation media.

To make IVM simpler and more repeatable, we proposed a simplified maturation medium containing only tissue culture medium (TCM 199), supplemented with epidermal growth factor (EGF) and cysteamine. Good results were obtained in embryo development using slaughterhouse oocytes, with no difference when using semidefined (EGF, hCG, eCG, IGF, FGF, ITS, L-cysteine, β-mercaptoethanol, vitamin C, glycine, glutamine, and pyruvate) and undefined (10% fetal calf serum, cysteamine, and oFSH) IVM media (85–92% of MII rate, 66–77% of cleavage and 45–46% of blastocyst/initial COC). However, using the same simplified and semidefined media for LOPU-oocytes, significantly lower rates of MII (66–72%), cleavage after IVF (38–39%) and blastocyst/initial COC (25–28%) were found [12], whereas high development rate was obtained for LOPU oocytes matured in complex medium (51% of blastocyst development rate/initial COC) [13]. LOPU and slaughterhouse oocytes have similar intrinsic quality, reaching similar developmental competence in our simplified system when parthenogenetic activation was used instead of IVF, whereas LOPU oocytes were less competent for fertilization after IVM in simplified medium [12]. Moreover, slaughterhouse and LOPU oocytes may have different IVM kinetics, with the latter requiring more time, depending on the IVM medium [12]. These data clearly indicate that oocytes' requirements during IVM may differ according to their origin. It remains unclear if their difference in terms of response to IVM treatments may be related to FSH stimulation prior to LOPU session or to postmortem changes in oocyte responsiveness in slaughterhouse group.

Regarding IVM physical conditions, literature does not vary considerably, and it is usually performed incubating COC in large groups (30–50 COC per 500 μL of medium per well in four-well plates) under 5% CO_2 in air at 38–39 °C with maximum humidity for 22–24 h [2].

1.3 IVF

Either fresh or frozen-thawed spermatozoa (sptz) can be used to fertilize matured oocytes. In any case, it is essential to improve live sperm cells proportion, usually by the use of a Percoll gradient or swim-up separation. We have previously demonstrated that Percoll density gradient could be successfully used for frozen-thawed goat sperm selection and was much more efficient than swim-up in recovering spermatozoa [14]. In general, it is suggested that Percoll is preferred when working with frozen goat sperm, whereas swim-up is used for fresh sperm [15]. Some sheep IVF data showed that in contrast to other species (cattle and pig) sex-sorted spermatozoa achieve equal or greater cleavage and blastocyst development rates than nonsorted ones [16]. A large-scale commercial program for IVP in cattle using sexed sperm has been performed [17]. However, despite its interesting applicability, up to now there are no reports related to the use of sexed spermatozoa in goat IVF.

Overall, the procedures used in bovine IVF can be applied in goats after few modifications, such as the reduction in the centrifugation speed and the addition of estrous sheep serum in IVF medium for spermatozoa capacitation [18]. Sperm should undergo capacitation and acrosome reaction to be able to fertilize, causing the release of proteolytic enzymes that may assist sperm penetration through the zona pellucida surrounding the oocyte. The importance of capacitating agents for the success of IVF has been already described in many species. Heparin has been shown to increase fertilization rate in cattle [19], and is thus widely used for sperm capacitation in this species. Similarly, there are some reports confirming the role of heparin in goats and sheep. However, some studies indicated an adverse effect on fertilization or reduced embryo development and quality [20]. We have demonstrated that higher numbers of blastocysts were obtained from COC fertilized in the presence of 5 μg/mL heparin, suggesting that its addition to the IVF medium improves sperm capacitation of frozen-thawed goat semen.

The appropriate time of cumulus cells (CC) removal from COC during IVP also remains controversial. In some studies, oocytes were denuded either immediately after IVM or just after IVF. Our laboratory has demonstrated the importance of CC during the IVF process, and currently goat COC are denuded just after IVF. Indeed, the association of oocytes with CC, either separate or in intimate contact, during IVF improved the blastocyst development rate [11].

The most widely used IVF medium is the modified synthetic oviduct fluid (SOF), supplemented with serum, antibiotics, and other substances, varying according to the laboratories. Regarding IVF conditions for goats, sperm concentrations vary from 1 to 3×10^6 cells/mL, and large groups of oocytes (30–50 COC per 500 μL of medium) are coincubated for 16–20 h at 38–39 °C in a humidified atmosphere of 5% CO_2 in air [2].

1.4 IVD

The last step of embryo IVP is the culture of the presumptive zygotes to reach the blastocyst stage at 6–8 days after IVF in ruminants. The success of IVD depends on the various factors such as osmolarity and ionic composition, temperature, pH, CO_2, oxygen, carbohydrates, amino acids, lipids, proteins, growth factors, and cytokines. Any deviation from the appropriate environment could lead to embryonic arrest at any stage of development [21]. Different culture media have been successfully used for small ruminant embryo development such as TCM 199 [22], B2 [23], and "Sydney IVF Blastocyst" medium [24]. However, the most widely used medium is the synthetic oviduct fluid (SOF), which was based on the composition of ovine oviduct fluid. Tervit et al. [25] were the first to report successful culture of ruminant zygotes to the blastocyst stage in vitro using SOF medium. Subsequently, changes to the original composition have been made. Some laboratories routinely supplement SOF medium with 5–10% FCS [26, 27], but others only add BSA [28]. However, goat and sheep presumptive zygotes are routinely cultured for 8 days after insemination in a humidified atmosphere of 5% O_2, 5% CO_2, and 90% N_2 using SOF medium plus serum [21].

The success of IVD is relatively high (40–50% of the initial oocytes reach the blastocyst stage) [7, 11]; nevertheless, the resulting embryos differ from in vivo derived ones in many aspects (gene expression, metabolism, morphology), resulting in a lower viability [3], especially a lower resistance to different cryopreservation methods. The composition of the mammalian embryo's environment in vivo changes from the early to late cleavage stages. Embryonic genomic activation, which occurs at the 8- to 16-cell stage in ruminants, coincides with an increase in metabolic activity, oxygen consumption and uptake of carbohydrates [29] to the blastocyst stage. Thus, early embryos cultured in vitro fail to develop past the 8- to 16-cell stages in traditional culture media. This block occurs around the time of activation of the embryonic genome. Serum and cells have been added to the culture to avoid this block. Thus, since a research group [30] showed that coculture with oviductal epithelial cells (OEC) could support sheep embryos past the 8- to 16-cell stage, a large number of teams have used this undefined system, which does not require a special incubator to reduce the O_2 concentration to 5%. Indeed, cocultured cells reduce O_2 tension to less toxic levels for the embryos. Moreover, somatic cells used in coculture may produce unknown embryo growth promoting factors or delete embryo toxic factor from culture medium (reviewed by [31]). Although a wide variety of different cell types have been used, the most widely employed are the OEC [21].

In goats, Rodríguez-Dorta et al. [32] have compared the effect of two culture conditions, SOF in a humidified atmosphere of 5% O_2, 5% CO_2, and 90% N_2 vs. goat OEC (GOEC) coculture in a 5% CO_2 in air, on the pregnancy rate, embryo survival rate and offspring development after direct transfer of vitrified-warmed caprine IVP embryos. The percentage of blastocyst in SOF (28%) was higher than in GOEC (20%). However, the percentage of pregnancy and kids born were significantly higher in embryos from GOEC (56% and 33% respectively) than SOF (14% and 9% respectively). These results indicate that embryos cultured with GOEC acquired a better quality in terms of cryotolerance and embryo survival rate than embryos developed in SOF under low oxygen.

2 Materials

Prepare all solutions using ultrapure water (commercial or prepared by reverse osmosis device) and analytical grade reagents (embryo tested when available). Prepare and store all reagents at room temperature (unless indicated otherwise).

2.1 Oocyte Collection, Screening, and Washing

1. Oocyte Screening Medium: 500 mL of TCM199 HEPES, 1 mL BSA Fraction V (stock, 0.2 g/mL; final, 0.4 mg/mL), and 1.25 mL gentamycin (stock, 10 mg/mL; final, 25 μg/mL).
2. Oocyte Collection Medium: Add 100 μL heparin (5000 IU/mL) to 50 mL of oocyte screening medium.
3. Oocyte Washing Medium: 10 mL of TCM199 with bicarbonate and 40 μL of gentamycin (stock, 10 mg/mL; final, 40 μg/mL).

2.2 IVM

2.2.1 For Slaughterhouse-Derived Oocytes

Supplement 10 mL of TCM199 with bicarbonate with 100 μL human EGF (stock, 1 μg/mL; final 10 ng/mL) and 100 μL cysteamine (stock, 10 mM; final, 100 μM).

2.2.2 For LOPU-Derived Oocytes

Supplement 9 mL of TCM199 with bicarbonate, 100 μL Na pyruvate (stock 20 mM; final, 200 μM), 40 μL gentamycin (stock,10 mg/mL; final, 40 μg/mL), 1 mL of fetal calf serum (10%), 100 μL human EGF (stock, 1 μg/mL; final, 10 ng/mL), 100 μL FSH/LH (stock, 2 mg/mL; final, 20 μg/mL), 20 μL estradiol (stock, 500 μg/mL; final, 1 ng/mL), and 100 μL cysteamine (stock, 10 mM, final, 100 μM).

2.3 IVF (for Oocytes of Both Origins)

1. SOF: Dissolve 126 mg $CaCl_2 \cdot 2H_2O$ (final, 1.71 mM) and 49 mg $MgCl_2 \cdot 6H_2O$ (final, 0.49 mM) into 100 mL of ultrapure water. Separately, dissolve 3147 mg NaCl (final, 107.7 mM), 267 mg KCl (final, 7.16 mM), 81 mg KH_2PO_4 (final, 1.19 mM), 1053 mg $NaHCO_3$ (final, 25.07 mM), 141 μL Na-lactate (final, 3.3 mM), 17 mg Na-pyruvate (final, 0.3 mM), and 140 μL Phenol Red (stock, 5 mg/mL; final, 1.4 μg/mL) into 200 mL of ultrapure water. Combine both solutions when all ingredients have completely dissolved. Add 2.5 mL glutamine (stock, 200 mM; final, 1 mM), 10 mL essential AA (BME stock, 50×; final 1×), and 5 mL nonessential AA (MEM stock, 100×; final 1×). Mix well and adjust the volume to 500 mL. Measure the osmotic pressure (270–-285 mOsmoles) and pH (7.2–7.4). Store at 4 °C. This solution can be stored for up to 2 weeks at 4 °C.
2. IVF-SOF: 9 mL of SOF, 1 mL estrous sheep serum (10% v/v), 40 μL gentamycin (stock, 10 mg/mL; final, 40 μg/mL), and

50 μL heparin (stock, 1 mg/mL; final, 5 μg/mL). Filter (0.22 μm) and store for up to 7 days at 4 °C.

3. Sptz-SOF: 10 mL of SOF, 24.9 mg HEPES (final, 10 mM), and 40 μL gentamycin (stock, 10 mg/mL; final, 40 μg/mL). Filter (0.22 μm) and store for up to 7 days at 4 °C.
4. Percoll gradient: Add 5 mL of 0.9% NaCl to 45 mL of 100% Percoll to obtain 90% Percoll (store for up to 30 days at 4 °C). Percoll 45% is prepared by adding 750 μL of Percoll 90% and 750 μL of Sptz-SOF medium. For use, add 1.5 mL of 90% Percoll and 1.5 mL of 45% Percoll into a 15 mL tube. Place it into a CO_2 incubator.

2.4 IVD

2.4.1 System with Low Tension of Oxygen and No Coculture

1. SOF-IVD: 10 mL of SOF and 30 mg BSA Fraction V (final, 3 mg/mL). Filter (0.22 μm) and store for up to 15 days at 4 °C. At the time of cleavage evaluation, add 10% of fetal calf serum directly to the droplets.
2. Medium for CC removal: 10 mL of SOF, 24.9 mg HEPES (final, 10 mM) and 20 μL of BSA (stock, 0.2 g/mL; final, 0.4 mg/mL). Filter (0.22 μm) and store for up to 7 days at 4 °C.

2.4.2 Coculture with Oviduct Cells

1. Washing Medium: 500 mL of TCM 199 HEPES, 1 mL BSA Fraction V (stock, 0.2 g/mL; final, 0.4 mg/mL), and 1.25 mL gentamicin (stock, 10 mg/mL; final, 25 μg/mL).
2. Culture Medium: 9 mL of TCM 199 HEPES, 1 mL fetal calf serum (10%), and 80 μL gentamycin (stock, 10 mg/mL; final 80 μg/mL).

3 Methods

Carry out all procedures at room temperature. Media are placed in incubator at 38.8 °C, for at least 2 h before their use. IVM, IVF, and IVD are performed using four-well dishes.

3.1 Oocyte Collection (Slaughterhouse and LOPU), Screening and IVM

1. Prepare four-well dishes containing 500 μL of IVM medium for washing each group and place them into the incubator in air (5% CO_2).
2. For slaughterhouse derived oocytes, the time from the slaughter to the arrival in the laboratory should not exceed 3 h. Follicles >3 mm and <6 mm are punctured with a 19 G short bevel needle attached to a 10 mL-syringe (previously rinsed with collection medium) or vacuum pump. After the last ovary, remove the needle from the syringe and transfer contents into a 50 mL tube. Rinse the syringe and the vacuum collection system with medium to recover all the COC. The

COC remain stored in a 50 mL tube with 5 mL of collection medium in a water bath at 35 °C for 10 min to allow COC to settle to the bottom. The aspiration time should not exceed 1 h (slaughterhouse) or 2 h (LOPU-oocytes).

3. Deposit 5–10 mL of oocyte screening medium in a 90 mm petri dish divided in ±1 cm squares by lines on the bottom. Aspirate the pellet from the collection tube with a Pasteur pipette, transfer into the dish, and shake gently (each tube must be aspirated twice, change the petri dish when the medium is too cloudy). Observe each square and aspirate COC of Grade 1 and 2 (*see* **Note 1**). Place the COC in a 35 mm petri containing 2–3 mL of oocyte screening medium. At the end of the selection, pass the COC through a second dish containing oocyte screening medium and distribute them in groups of 50 (maximum 60, eliminate Grade 3 and 4 COC, if remaining).
4. Wash each group of COC through the wells on the left of the four-well IVM dishes (each containing 500 μL of IVM medium) and transfer them to the wells on the right side of the dish (also containing 500 μL of IVM medium) for maturation. At this stage, the medium is buffered with bicarbonate, so it is necessary to be fast in order to avoid pH variations. Place the dishes into the incubator (5% CO_2 in air, 38.8 °C) for 22–24 h (IVM).

3.2 IVF

1. Prepare four-well dishes containing 500 μL of SOF-IVF medium for washing each group of 50 COC (through four successive wells). Place into the incubator (5% CO_2 in air).
2. After IVM, matured COC are washed and then are transferred to wells containing 450 μL of SOF-IVF medium (groups of ~50 COC). Place these dishes into the incubator (5% CO_2 in air), until sperm processing.
3. Frozen sperm straws must be thawed by immersing them in a beaker of water at 37 °C for 30 s. Dry it with a paper towel, cut the end, and empty the contents of the straw by allowing it to gently flow along the wall of the tube into a 15 mL conical tube containing a 45% and 90% Percoll gradient. The tube is centrifuged for 15 min at 700 × *g*. Discard the supernatant and add 600 μL of Sptz-SOF medium, centrifuge again for 5 min at 100 × *g*. After centrifugation, remove the supernatant, leaving approximately 100 μL of pellet and resuspend the pellet. Aliquots can be taken at this time to evaluate total motility and vigor under the microscope.
4. Dilute 5 μL of pellet into 195 μL of water into a 1.5 mL microcentrifuge tube. Fill a Thoma cell for counting the sperm. Four out of 16 squares are counted randomly. The

formula for sperm concentration is [*C*] = Total number of sperm counted in four squares × 4 × 40 (dilution factor) × 15,625 (sptz/mL). Calculate the volume of the sperm suspension and the volume of SOF-IVF medium needed to obtain a concentration of 20×10^6 sperm/mL. Then, 50 μL of this solution is added to each IVF well (450 + 50 = 500 μL) (*see* **Note 2**). Sperm and oocytes are coincubated for 16–18 h in the incubator (5% CO_2 in air).

3.3 IVD

3.3.1 System with 5% Oxygen and No Coculture

1. Prepare four-well dishes containing 500 μL of IVD medium for washing each group of 25 presumptive zygotes (transfer the zygotes of each group through the four wells successively). Place 12.5 μL of SOF-IVD medium in the middle of a well, cover the drop with 700 μL mineral oil, and then complete the droplet with an additional 12.5 μL of SOF-IVD medium. Place all dishes into the incubator containing 5% O_2, 5% CO_2, and 90% N_2.
2. 16–20 h after fertilization (*see* **Note 3**), the cumulus cells around the presumptive zygotes and the remaining spermatozoa should be eliminated. Transfer all structures to a 15 mL-tube containing 2 mL of washing medium and vortex for 2 min at a "medium (strength)" position. Recover the embryos by washing several times with the washing medium (*see* **Note 4**).
3. Groups of 25 embryos are washed four times in 500 μL SOF-IVD and then transferred to the 25 μL SOF-IVD droplets under mineral oil (*see* **Note 5**). The dishes will remain for seven to 8 days in the incubator (38.8 °C, 5% CO_2, 5% O_2) (*see* **Note 6**).
4. One day after the start of IVD (Day 2 postinsemination), cleavage is observed and 2.5 μL (10%) of fetal calf serum is added into each droplet. Embryo development (morula and blastocysts) can be observed at Day 7 and 8 of culture.

3.3.2 Coculture with Oviduct Cells

1. In the slaughterhouse, both oviducts coming from the same genital tract and corresponding ovary are transported to the laboratory in a dry sterile tube at room temperature. In the laboratory, the oviducts are classified based on ovarian morphology as early follicular, late follicular, early luteal and late luteal phase. Only oviducts classified as late follicular (preovulatory follicle) or early luteal (recent ovulation) in the same genital tract are used.
2. In a petri dish, isolate the oviduct by removing the connective tissue. Then, disinfect the oviduct by immersing it briefly in 70% alcohol, and rinse three times in the washing medium.
3. With a flamed glass slide, compress the oviduct in a square petri dish from the uterotubal junction to push the epithelium

toward the fimbria. Finally, expel the mucosa into a clean, sterile petri dish.

4. Recover the epithelial cells with a Pasteur pipette and place them in a 15 mL tube containing 5 mL of wash medium.
5. Vortex the tube for 15 s and place the tube in the incubator at 38 °C, 5% CO_2 for 5 min to allow the cells to settle.
6. Remove the supernatant and add 5 mL of washing medium again. Repeat this procedure three times.
7. The washed cells are diluted 1/100 by adding 50 μL of the pellet obtained after the last settling to 5 mL of culture medium and mixing well.
8. 500 μL of culture medium plus cells (obtained in **step** 7) is deposited in four-well dishes. Mix throughly by pipette before aspirating cells.
9. Incubate at 38.8 °C, 5% CO_2, 20% O_2.
10. After 48 h of culture, the medium is completely renewed (500 μL). The concentration of gentamycin (final 80 μg/mL) in the culture medium is then reduced to 4 μL/mL (final 40 μg/mL).
11. Every 2 days, half of the culture medium (250 μL) is renewed (Be careful not to touch the carpet of cells with the pipette tip while removing and adding medium).
12. Allow the cells to grow to confluence (typically 6–7 days).
13. Renew the entire volume of culture medium 24 h prior to the start of coculture with the embryo development medium (wash the monolayer three times with the same medium before).

4 Notes

1. Grade 1: more than three complete and compact layers of cells; Grade 2: one to three full layers of cells; Grade 3: incomplete layer, with homogeneous ooplasm and Grade 4: Expanded cumulus or denuded oocyte with heterogeneous ooplasm [10]. Depending on the aim of the experiment, it is possible to use Grade 3 oocytes for IVM [10, 11].
2. Immediately after performing the insemination (IVF), it is recommended to check under the microscope if all the wells have an approximate number of sperm and mostly alive.
3. At the end of IVF, before starting IVD, it is essential to observe under the microscope if all the wells (perhaps different treatments) still have live sperm. It is reasonable to assume that at least some sperm must be alive at the end of IVF. When this does not happen, usually cleavage and blastocyst development rates are low.

4. In our routine, we observed that although it is acceptable to loose around 5% of presumptive zygotes in the vortex procedure for denuding, this strategy is preferred due to its repeatability. When the presumptive zygotes are repeatedly pipetted, the technician may apply different strength or hands position and thus an adverse effect can occur. At this time, we expect to have around 10% of structures already cleaved.
5. It is important to calculate after IVF how many presumptive zygotes per treatment we have before preparing IVD droplets. Consider that we may have 1 μL/embryo. Thus, if there are around 40 per treatment, two droplets of 20 μL can be prepared. Care must be observed when putting embryos into the droplets. Each embryo should be washed well in the IVD medium to avoid carryover of components from a different medium.
6. The use of incubator with low tension of oxygen (5% O_2) for IVD appears to be essential in order to achieve a high blastocyst development rate in goats. In case it is not possible to work with two different incubators in the laboratory, it is possible to use small "bags" containing each four-well dish under 5% O_2, 5% CO_2, and 90% N_2, sealed and placed into the incubator to control the temperature.

Acknowledgments

The authors wish to thank CAPES-COFECUB bilateral framework for financial support of collaboration between the State University of Ceará and INRA. J.M.G. Souza-Fabjan and R.I.T.P. Batista conducted their postdoctoral studies under the auspices of CAPES (Brasília, Brazil), J.M.G. Souza-Fabjan is a fellow of FAPERJ, and V.J.F. Freitas is a fellow of CNPq.

References

1. Duszewska AM, Trzeciak P, Compa A et al (2010) Selected issues concerning biotechnology of farm animals breeding—a review. Anim Sci Pap Rep 28:295–306
2. Souza-Fabjan JMG, Panneau B, Duffard N et al (2014a) In vitro production of small ruminant embryos: late improvements and further research. Theriogenology 81:1149–1162
3. Paramio MT, Izquierdo D (2016) Recent advances in in vitro embryo production in small ruminants. Theriogenology 86:152–159
4. Baldassarre H, Karatzas CN (2004) Advanced assisted reproduction technologies (ART) in goats. Anim Reprod Sci 82–83:255–266
5. Sanchez DJD, Melo CHS, Souza-Fabjan JMG et al (2014) Repeated hormonal treatment and laparoscopic ovum pick-up followed by in vitro embryo production in goats raised in the tropics. Livest Sci 165:217–222
6. Teixeira PPM, Padilha LC, Oiveira MEF et al (2011) Laparoscopic ovum collection in sheep: gross and microscopic evaluation of the ovary and influence on oocyte. Anim Reprod Sci 127:169–175
7. Souza-Fabjan JMG, Locatelli Y, Freitas VJF et al (2014b) Laparoscopic ovum pick up (LOPU) in goats: from hormonal treatment to oocyte possible destinations. R Bras Ci Vet 21:3–11

8. Gibbons A, Pereyra Bonnet F, Cueto MI et al (2008) Recovery of sheep and goat oocytes by laparoscopy. Acta Sci Vet 36:223–230
9. Katska-Ksiazkiewicz L, Rynska B, Gajda B et al (2004) Effect of donor stimulation, frozen semen and heparin treatment on the efficiency of in vitro embryo production in goats. Theriogenology 62:576–586
10. Souza-Fabjan JMG, Locatelli Y, Duffard N et al (2016) Intrinsic quality of goat oocytes already found denuded at collection for in vitro embryo production. Theriogenology 86:1989–1998
11. Souza JMG, Duffard N, Bertoldo MJ et al (2013) Influence of heparin or the presence of cumulus cells during fertilization on the in vitro production of goat embryos. Anim Reprod Sci 138:82–89
12. Souza-Fabjan JMG, Locatelli Y, Duffard N et al (2014c) In vitro embryo production in goats: Slaughterhouse and laparoscopic ovum pick up–derived oocytes have different kinetics and requirements regarding maturation media. Theriogenology 81:1021–1031
13. Souza-Fabjan JMG, Pereira AF, Melo CHS et al (2013) Assessment of the reproductive parameters, laparoscopic oocyte recovery and the first embryos produced in vitro from endangered Canindé goats (Capra hircus). Reprod Biol 13:325–332
14. Olivares CCS, Fonseca JF, Camargo LSA et al (2015) Comparison of different methods of goat sperm selection and capacitation for optimization of assisted reproductive technologies. Small Rumin Res 127:44–49
15. Palomo MJ, Izquierdo D, Mogas T et al (1999) Effect of semen preparation on IVF of prepubertal goat oocytes. Theriogenology 51:927–940
16. De Graaf SP, Beilby KH, Underwood SL (2009) Sperm sexing in sheep and cattle: the exception and the rule. Theriogenology 71:89–97
17. Pontes JHF, Silva KCF, Basso AC (2010) Large-scale in vitro embryo production and pregnancy rates from Bos taurus, *Bos indicus*, and *indicus-taurus* dairy cows using sexed sperm. Theriogenology 74:1349–1355
18. Cox JF, Alfaro V (2007) In vitro fertilization and development of OPU derived goat and sheep oocytes. Reprod Domest Anim 42:83–87
19. Parrish JJ, Susko-Parrish J, Winer MA et al (1988) Capacitation of bovine sperm by heparin. Biol Reprod 38:1171–1180
20. Malik RK, Lohan IS, Dhanda OP et al (1997) Test for the acrosomal reaction of goat spermatozoa treated with heparin. Small Rumin Res 26:163–166
21. Paramio MT, Izquierdo D (2014) Current status of in vitro embryo production in sheep and goats. Reprod Domest Anim 49(Suppl. 4):37–48
22. Wani AR, Khan MZ, Sofi KA et al (2012) Effect of cysteamine and epidermal growth factor (EGF) supplementation in maturation medium on in vitro maturation, fertilization and culturing of embryos in sheep. Small Rumin Res 106:160–164
23. Katska-Ksiazkiewicz L, Opiela J, Rynska B (2007) Effects of oocyte quality, sêmen donor and embryo co-culture system on the efficiency of blastocyst production in goats. Theriogenology 68:736–744
24. Beilby KH, de Graaf SP, Evans G et al (2011) Quantitative mRNA expression in ovine blastocysts produced from X- and Y chromosome bearing sperm, both in vitro and in vivo. Theriogenology 76:471–481
25. Tervit HR, Whittingham DG, Rowson LE (1972) Successful culture in vitro of sheep and cattle ova. J Reprod Fertil 30:493–497
26. Catalá MG, Izquierdo D, Rodríguez-Prado M (2012) Effect of oocyte quality on blastocyst development after in vitro fertilization (IVF) and intracytoplasmic sperm injection (ICSI) in a sheep model. Fertil Steril 97:1004–1008
27. Hammami S, Morató R, Romaguera R et al (2013) Developmental competence and embryo quality of small oocytes from pre-pubertal goats cultured in IVM medium supplemented with low level of hormones, insulin–transferrin–selenium and ascorbic acid. Reprod Domest Anim 48:339–344
28. Wang Z, Lin P, Yu S (2013) Effects of ghrelin on developmental competence and gene expression of in vitro fertilized ovine embryos. Theriogenology 79:695–701
29. Thompson JG, Partridge RJ, Houghton FD (1996) Oxygen uptake and carbohydrate metabolism by in vitro derived bovine embryos. J Reprod Fertil 106:299–306
30. Gandolfi F, Moor RM (1987) Stimulation of early embryonic development in the sheep by co-culture with oviduct epithelial cells. J Reprod Fertil 81:23–28
31. Bavister BD (1995) Culture of preimplantation embryos: facts and artifacts. Hum Reprod Update 1:91–148
32. Rodríguez-Dorta N, Cognié Y, González F et al (2007) Effect of coculture with oviduct epithelial cells on viability after transfer of vitrified in vitro produced goat embryos. Theriogenology 68(6):908–913

Chapter 11

Small Ruminants: Prepubertal Oocyte Donors

Dolors Izquierdo, María-Gracia Catalá, and María-Teresa Paramio

Abstract

In vitro embryo production using oocytes of prepubertal females is named JIVET (juvenile in vitro embryo technologies). The aim of the JIVET is to increase genetic gain by shortening the generation interval in breeding programs. In this chapter we describe the methodology currently used in our laboratory for producing in vitro embryos from prepubertal sheep (90–120 days old) and goats (30–45 days old).

Briefly, we obtain cumulus–oocyte complexes (COCs) from slaughtered female ovaries. These COCs are in vitro matured in TCM199 with 10% (v/v) fetal bovine serum (FBS) at 38.5 °C in a humidified air atmosphere with 5% CO_2 for 24 h. The next day we collect ejaculates from males of proven fertility and select the most motile spermatozoa by density gradient. Following coincubation with mature COCs, presumptive zygotes and embryos are cultured in SOF for 8 days at 38.5 °C, 90% N_2, 5% CO_2, 5% O_2, and 100% humidity to obtain blastocysts.

Key words Juvenile in vitro embryo technology, Prepubertal, Goat, Sheep, Oocyte, Embryo, In vitro maturation, In vitro fertilization, In vitro culture

1 Introduction

In vitro embryo production (IVEP) is a multistep methodology comprising the following procedures: (1) in vitro maturation (IVM) of oocytes recovered directly from the follicles, (2) in vitro fertilization (IVF) of oocytes, and (3) in vitro culture (IVC) of zygotes up to the blastocyst stage. The success of IVEP in small ruminants are lower than in other livestock species because of the reduced number of research teams working in these species compared to cattle and pigs. However, small ruminants are economical important in developing countries and they are interesting in transgenic studies as animals to express recombinant proteins in milk.

In vitro embryo production using oocytes of prepubertal females is named JIVET (juvenile in vitro embryo technologies). The aim of the JIVET is to increase genetic gain by shortening the generation interval in the breeding programs (reviewed in [1]). Oocytes from prepubertal small ruminants present the following characteristics compared to those form adult females: (a) The

Jason R. Herrick (ed.), *Comparative Embryo Culture: Methods and Protocols*, Methods in Molecular Biology, vol. 2006, https://doi.org/10.1007/978-1-4939-9566-0_11,

number of oocytes recovered per ovary is significantly higher than in adult ovaries [2]; (b) the ovaries of prepubertal goat and sheep are small, soft and with small follicles with a diameter < 3 mm. Thus, oocytes must be recovered by slicing these ovaries. In adult females, oocytes are conventionally recovered by follicle aspirations. It is well known the positive and direct relationship between follicle size and oocyte competence (reviewed in [3]), but selection by follicle size is very difficult in ovaries of kids and lambs; (c) oocytes recovered by slicing are heterogeneous in diameter and atresia stage. Thus, after oocyte liberation, oocytes must be selected by morphological criteria such as oocyte diameter, homogeneity and color of cytoplasm, and number of cumulus cell layers surrounding the oocyte; (d) oocytes from prepubertal females are less competent to develop up to blastocyst stage after IVF than those from adult goat and sheep (reviewed in [4, 5]).

In our laboratory we have compared the ultrastructure of oocytes from prepubertal goats with those of the adults and we have observed that the distribution patterns of cortical granules [6] and mitochondria [7] are different between both types of oocytes. Moreover, we have detected alteration in total RNA content, cyclin B1 and p34 (cdc2) expression, and maturation promoting factor (MPF) activity [8, 9]; and disorganization of microtubule and microfilament [10] in the oocytes from prepubertal goat and higher number of embryos with abnormal chromosome set [11].

It should be noted that a good selection of oocytes is essential for JIVET success because oocyte quality is the key factor to produce blastocysts. In our conditions, we have observed that the best oocyte for JIVET must have a diameter bigger than 135 μm [8], homogeneous cytoplasm and be surrounded by more than three layers of cumulus cells.

In the last decades we have carried out several studies to evaluate the effect of different maturation, fertilization, and embryo culture media, as well as different supplements (hormones concentrations, kinds of sera, thiol compounds, activin A, insulin–transferrin–selenium, ...) on embryo development and the quality of obtained blastocysts from prepubertal goat and sheep oocytes (reviewed in [5]). In this chapter we describe the methodology currently used in our laboratory for producing in vitro embryos from prepubertal sheep (90–120 days old) and goat (30–45 days old).

2 Materials

1. Phosphate buffered saline (PBS).
2. Slicing Medium: TCM199 (with Earle's salts, phenol red, and HEPES) supplemented with 2.2 g/L (26 mM) $NaHCO_3$ and

50 mg/L gentamicin. This medium contains 25 mM HEPES that maintains the correct pH for a long time (20 min). Dilute each compound one by one. After all compounds are completely dissolved, check osmolarity. It should be between 292 and 297 mOsm. If not, discard the medium and do it again. Filter medium using a 0.22 μm filter. It can be a syringe adapted filter (change filter every 200 mL medium) or 1 L filter that comes with 1 L sterile container. Keep medium in the refrigerator (5 °C) until use. This medium is stable at 5 °C for 45 days.

3. Heparin powder.
4. In vitro Maturation Medium: TCM-199 (With Earle's salts, L-glutamine, and sodium bicarbonate) supplemented with 5 μg/mL LH, 5 μg/mL FSH, 1 μg/mL 17β-estradiol, 10 ng/mL epidermal growth factor (EGF), 0.2 mM sodium pyruvate, 1 mM glutamine, 100 μM cysteamine, 10% (v/v) fetal bovine serum (FBS), and 5 μL/mL antibiotic–antimycotic (10,000 units penicillin, 10 mg streptomycin, and 25 μg amphotericin B per mL). In a 15 mL, combine 12.7 mL of TCM199, 50 μL sheep FSH (1.5 mg/mL stock solution) 42 μL sheep LH (1.79 mg/mL stock solution), 150 μL 17-β-estradiol (0.1 mg/mL stock solution), 75 μL murine EGF (2 μg/mL stock solution), 1.5 mL fetal bovine serum (FBS), 150 μL antibiotic–antimycotic (with 10,000 units penicillin, 10 mg streptomycin, and 25 μg amphotericin B per mL), 150 μL cysteamine (10 mM stock solution), 150 μL glutamine (200 mM stock solution), 33 μL sodium pyruvate (0.2 mM stock solution). Prepare this medium immediately before making drops and placed it in the incubator.
5. Bovipure Nidacon EVB S.L.: density gradient and washing solution.
6. TALP Stock: 250 mL water, 1.665 g (115 mM) NaCl, 59.5 mg (3.2 mM) KCl, 73.5 mg (2 mM) $CaCl_2{\cdot}2H_2O$, 6.8 mg (0.2 mM) $NaH_2PO_4{\cdot}H_2O$, 12.8 mg (0.25 mM) $MgCl_2{\cdot}6H_2O$, 7.5 mg gentamicin, and 2.5 mg phenol red. Place the medium in the refrigerator (5 °C). It is stable for 45 days.
7. Goat-IVF Medium: 50 mL TALP Stock, 105 mg (25 mM) $NaHCO_3$, 45 mg (5 mM) glucose, 1 mg (0.2 mM) pyruvic acid, 300 mg BSA—fatty acid free, and 85 μL (of 60% solution, 3.5 mM final) sodium D,L-lactate solution. Measure and adjust osmolarity to 292–297 mOsm by adding culture-grade H_2O. Using a 0.22 μm syringe filter, filter the medium and keep it in the refrigerator (5 °C) in a 50 mL sterile tube. This medium is stable in the refrigerator for 15 days.

8. Heparin Stock Solution: 100 μg heparin in 1 mL Goat-IVF Medium.
9. Hypotaurine powder.
10. Glutathione powder.
11. SOF Medium: 50 mL water, 314.65 mg (108 mM) NaCl, 27 mg (7.3 mM) KCl, 12.39 mg (1.69 mM) $CaCl_2 \cdot 2H_2O$, 8.1 mg (1.19 mM) KH_2 PO_4, 5 mg (830.8 mM) $MgSO_4$, 105.48 g (25.1 mM) $NaHCO_3$, 0.45 mg phenol red, 18.15 g (3.3 mM) sodium pyruvate, 72 μL (of a 60% solution, 3.5 mM final) sodium D,L-lactate, 0.4 g BSA—fatty acid free, 100 μL BME amino acids solution 50×, 100 μL MEM nonessential Amino (100×), 5 mg (0.34 mM) sodium citrate tribasic $2H_2O$, 25 mg (2.78 mM) myoinositol, 1.25 mL fetal bovine serum, and 0.5 mL antibiotic–antimycotic (with 10,000 units penicillin, 10 mg streptomycin, and 25 μg amphotericin B per mL). Measure and adjust molarity to 269–279 mOsm by adding culture-grade H_2O. Using a 0.22 μm syringe filter, filter the medium and keep it in the refrigerator (5 °C) in 15 mL sterile tubes. This medium is stable in the refrigerator for 1 month.
12. Estrous sheep serum (ESS) (*see* **Note 1**).
13. Sheep-IVF Medium: SOF Medium supplemented with 10% ESS.
14. Appropriate attire for slaughterhouse and farm.
15. Thermos.
16. Scissors.
17. Laminar flow hood.
18. Slide warmer.
19. Fifteen milliliter and 50 mL centrifuge tubes.
20. Thirty-five millimeter and 60 mm sterile embryo tested petri dishes.
21. Syringe filter (0.2 μm) and 0.2 μm bottle-top filter.
22. Micropipettes (1000, 200, 100, and 10 μL) and 3 mL plastic sterile Pasteur pipettes.
23. Sterile urine containers and tubes.
24. Scalpel blade and tweezers.
25. Precision bores with plunger (5 μL Wiretrol®; Drummond Scientific Company).
26. Powder-free gloves.
27. Artificial vagina for small ruminants.

3 Methods

3.1 Obtaining Ovaries from Slaughterhouse

1. Prepare a container, such as a 500 mL coffee thermos, which can maintain temperature for long periods of time.
2. Fill the container with warm (between 35 and 38 °C) PBS. A microwave can be used to heat PBS, but do NOT exceed 38 °C (*see* **Note 2**).
3. In the slaughterhouse stay near the place where animals are eviscerated and quickly, using scissors, cut the ovaries from the uterus and placed them into the container.
4. Transport ovaries within the shortest possible time to the laboratory.
5. Measure the temperature inside the container and rinse the ovaries with PBS at the same temperature. Example, if the temperature of the PBS inside the container arriving to the laboratory is 30 °C, you need to make three washes using PBS at 30 °C.
6. Transfer clean ovaries into a small container with PBS (could be a urine bottle) and put it inside a heater at 38 °C. In this way, the temperature of the ovaries will gradually increase to 38 °C (*see* **Note 3**).

3.2 Oocyte Retrieval Techniques

All supplies and media should be ready when we arrive to the laboratory with ovaries so as procedure can be initialized immediately.

Two types of retrieval can be done depending on the quality of the ovaries or type of oocytes needed. If they have a small number of antral follicles, slicing is the best option to recover the maximum number of COCs. However, if you need to select the biggest follicles, aspiration is suggested, but a lower number of COCs will be recovered.

Both techniques are recommended to be performed in a laminar flow hood while maintaining the temperature of the ovaries and oocytes as close as possible to 38 °C. A heating plate can be used for this purpose.

3.2.1 Slicing

1. Perform this technique with Slicing Medium that was warmed to 38 °C for at least 1 h.
2. Before starting to slice, add 100 μg/mL of heparin to the slicing medium (*see* **Note 4**).
3. Put two or three ovaries into a 60 mm petri dish containing approximately 4–5 mL of slicing medium. Using a scalpel blade make fine cuts on the ovary surface to release the oocytes from follicles.
4. Search for cumulus–oocyte complexes (COCs) and transfer them to a 35 mm petri dish with slicing medium (*see* **Note 5**).

5. Make a new 35 mm petri dish with slicing medium, select the best quality COCs (*see* **Note 6**).
6. Make a final wash in a new 35 mm petri dish with three microdrops of 100 μL IVM Medium covered with 3 mL mineral oil. Transfer COCs from one drop to another, to rinse residual slicing medium.

3.2.2 Aspiration

1. Aspirate follicles using an 18 or 19 gauge needle (*see* **Note 7**). Load the 5 mL syringe with 1 mL medium (slicing Medium or PBS). You can aspirate several follicles at a time without removing the needle from the ovary.
2. Remove the needle from the ovary and transfer the contents of the syringe into a 15 mL tube.
3. Leave the tube in the heater at 38 °C for 10–15 min until a pellet is formed.
4. Discard the supernatant.
5. Transfer the pellet to a new dish with slicing medium and repeat **steps 4–6** described for the slicing method (Subheading 3.2.1).

3.3 In Vitro Maturation Goat and Sheep Oocytes

1. Morphologically selected COCs will be cultured for 24 h in IVM Medium at 38.5 °C in a humidified air atmosphere with 5% CO_2 in a 4-well chamber with 35–40 COC's in 500 μL of maturation medium covered with 200 μL mineral oil or 35 mm petri dish with 25–30 COCs in 100 μL drops (3 drops/dish) covered with mineral oil (approx. 3 mL).
2. It is important to prepare the culture dishes at least 3 h before culture so the medium equilibrates in the gas atmosphere and reaches the correct pH (*see* **Note 8**).

3.4 Fresh Semen Selection for Goat and Sheep IVF

1. Using one female and at least two males of proven fertility collect fresh semen with an artificial vagina (*see* **Note 9**).
2. Ejaculates will be recovered in 15 mL tubes and maintained at 30–35 °C until selection in the laboratory (*see* **Note 10**).
3. Before starting selection it is important to check motility. Only pool ejaculates with at least 50% of motility. Discard the others (*see* **Note 11**).
4. Selection of the most motile sperm will be done by density gradient (Bovipure density gradient Nidacon EVB S.L.). Place 200 μL of ejaculate into the top of the 15 mL density gradient tube (2 mL bottom solution +2 mL top solution). Centrifuge at 300 × *g* for 25 min. Subsequently discard the supernatant and add 2 mL of washing solution (Bovipure density gradient Nidacon EVB S.L.) and centrifuge again for 5 min at 300 × *g*. Discard supernatant.

3.5 Goat Sperm Capacitation

1. Capacitate selected goat spermatozoa for 30 min in a 1:1 dilution with Heparin Stock Solution in a 15 mL tube in the CO_2 incubator (*see* **Note 12**).
2. Meanwhile sperm concentration should be determined with a Neubauer chamber. Calculate the volume required for a final concentration of 400,000 sperm per 100 μL IVF drop (4×10^6 spermatozoa/mL).

3.6 Sheep Sperm Capacitation

1. After gradient selection, determine the sperm concentration in the pellet using a Neubauer chamber.
2. Calculate the volume of sperm suspension required for a final concentration of 200,000 sperm per 100 μL IVF drop (2×10^6 spermatozoa/mL).
3. Sheep sperm is directly capacitated in the IVF drop with oocytes by adding estrous sheep serum (ESS).

3.7 Goat In Vitro Fertilization

1. Prepare 100 μL drops of Goat-IVF Medium (3 drops/35 mm petri dish) consisting of IVF-TALP Medium supplemented with 1 mg/mL hypotaurine and 0.3 mg/mL L-glutathione under 3 mL mineral oil (equilibrate for at least 3 h in CO_2 incubator). Prepare one or two extra dishes to rinse oocytes.
2. Wash in vitro matured COCs three times in Goat-IVF Medium and transfer 25 oocytes into 100 μL Goat-IVF drops.
3. Place the necessary volume of sperm suspension calculated (typically 0.5–2 μL) into each drop.
4. The culture dish containing oocytes and spermatozoa will be kept for 20 h in the incubator containing 5% CO_2 air and 100% humidity at 38.5 °C.

3.8 Sheep In Vitro Fertilization

1. Prepare 100 μL drops of Sheep-IVF Medium (3 drops/35 mm petri dish) cover with 3 ml mineral oil (*see* **Note 13**). Also, prepare one or two extra dishes to wash COCs.
2. Equilibrate for at least 3 h in a 90% N_2, 5% CO_2, and 5% O_2 incubator.
3. Transfer in vitro matured COCs into sheep-IVF drops (25 oocytes/100 μL SOF Medium + ESS drop).
4. Place the necessary volume of sperm (typically 0.5–2 μL) into each drop.
5. The culture dish containing oocytes and spermatozoa will be kept for 20 h at 38.5 °C in the incubator containing 90% N_2, 5% CO_2, and 5% O_2 with 100% humidity.

3.9 Goat and Sheep In Vitro Embryo Culture

1. Prepare culture dishes with ten drops of 10 μL of SOF Medium and one or two extra dishes with 100 μL SOF Medium for washing (3 mL mineral oil for 35 mm dishes). Keep these

dishes in the incubator for at least 3 h in 38.5 °C, 90% N_2 5% CO_2, 5% O_2, and 100% humidity.

2. After 20 h of IVF, denude presumptive zygotes by gently pipetting, and wash them twice before placing them into SOF drops.
3. Culture the presumptive zygotes in a ratio of 1 zygote/1 μL SOF medium for 8 days at 38.5 °C, 90% N_2, 5% CO_2, 5% O_2, and 100% humidity.
4. Assess embryo cleavage after 48 h of IVC.
5. At day 5 of IVC, transfer the embryos to fresh medium. For this, prepare and equilibrate dishes as at the initiation of embryo culture section and change oocytes from one dish to the other.
6. Assess embryo development at day 7 and 8 of IVC.

4 Notes

1. Heparin is added to avoid agglutination due to fibrin present in follicular fluid.
2. To prepare ESS, inject a sheep with 0.5 mL of estrogen or choose a female which is naturally in estrous. After 48 h of estrogen injection, extract 40 mL of blood. To avoid bubbles, blood must drain slowly down the tube wall. Allow blood to clot for 12 h at room temperature without moving the tube. Centrifuge (1 min at 250 × *g*) the tube to separate the clot. Serum will rest in the top of the tube. With a syringe, aspirate the clot, taking care not to break the clot, and centrifuge the remaining serum for 5 min at 250 × *g* (2 times). Keep the final supernatant. Maintain the recovered serum recovered for 30 min at 56 °C to deactivate the complement and finally filter the serum. Make 500 μL aliquots and keep them at −20 °C until use. It is recommended to collect serum from as many females as possible and then test each one to identify the ones that produce the greatest amount of embryos.
3. Ovaries can remain in the container between 25 °C and 38 °C for up to 6 h without causing degeneration of the oocytes. Higher temperatures will degenerate almost all oocytes and NO fertilization will occur.
4. It is important to change temperature gradually. Sudden changes in temperature will cause oocytes to deteriorate.
5. This dish can stay outside of the incubator as long as the color of the medium is orange-pink. When it turns fuchsia you should immediately return it to the incubator with CO_2 to reequilibrate to the correct pH.

6. At each stage of washing, bad oocytes (without cumulus cells or expanded cumulus) are discarded, so that the final IVM dish contains a pool of the best quality COCs (homogeneous cytoplasm, three or more complete cumulus layers).
7. Place the needle with the bevel pointing toward the ovary.
8. Do NOT change the pH of the medium with HCl or $NaHCO_3$. The CO_2 in the incubator will do this during equilibration.
9. One male of proven fertility provides enough spermatozoa for insemination of all oocytes. However, it is suggested to pool semen from two or three bucks to enhance penetration.
10. To maintain the temperature of the tubes during transport from the farm to the laboratory, you can keep them inside a pocket close to the body.
11. In the case of sheep, keep sperm suspension at room temperature for 1.5 h prior to insemination to help capacitation.
12. Remember to leave the cap of the tube open so as to allow exchange of CO_2 during capacitation.
13. This percentage may vary according to semen and ESS characteristic. If 10% does not work you can try other percentages between 2 and 20% of ESS.

References

1. Morton KM (2008) Developmental capabilities of embryos produced in vitro from prepubertal lamb oocytes. Reprod Domest Anim 43:137–143
2. Koeman J, Keefer CL, Baldassarre H, Downey BR (2003) Developmental competence of prepubertal and adult goat oocytes cultured in semi-defined media following laparoscopic recovery. Theriogenology 60:879–889
3. Paramio MT, Izquierdo D (2016) Recent advances in *in vitro* embryo production in small ruminants. Theriogenology 86:152–159
4. Armstrong DT (2001) Effects of maternal age on oocyte developmental competence. Theriogenology 55:1303–1322
5. Paramio MT, Izquierdo D (2014) Current status of *in vitro* embryo production in sheep and goats. Reprod Domest Anim 249:37–48
6. Velilla E, Izquierdo D, Rodriguez-Gonzalez E et al (2004) Distribution of prepubertal and adult goat oocyte cortical granules during meiotic maturation and fertilisation: ultrastructural and cytochemical study. Mol Reprod Dev 68:507–514
7. Velilla E, Rodriguez-Gonzalez E, Vidal F et al (2006) Mitochondrial organization in prepubertal goat oocytes during *in vitro* maturation and fertilization. Mol Reprod Dev 73:617–626
8. Anguita B, Jiménez-Macedo AR, Izquierdo D et al (2007) Effect of oocyte diameter on meiotic competence, embryo development, p34 (cdc2) expression and MPF activity in prepubertal goat oocytes. Theriogenology 67:526–536
9. Anguita B, Paramio M, Jiménez-Macedo A et al (2008) Total RNA and protein content, Cyclin B1 expression and developmental competence of prepubertal goat oocytes. Anim Reprod Sci 103:290–303
10. Velilla E, Rodríguez-González E, Vidal F, Paramio MT (2005) Microtubule and microfilament organization in immature, *in vitro* matured and *in vitro* fertilized prepubertal goat oocytes. Zygote 13:55–165
11. Villamediana P, Vidal F, Paramio MT (2001) Cytogenetic analysis of caprine 2- to 4-cell embryos produced *in vitro*. Zygote 9:193–199

Chapter 12

In Vitro Production of Bison Embryos

Jennifer P. Barfield

Abstract

Bison are an iconic species of cultural, conservation, and commercial interest. Various assisted reproductive technologies have been tested in bison over the last few decades (e.g., superovulation and embryo transfer), but their efficiencies are low. Since 2007, several methods for producing bison embryos in vitro have been published. All of these methods are based on cattle embryo production models and have varying degrees of success with regard to embryo production rates. In this chapter, a brief summary of these reports is presented followed by a detailed protocol that has been successfully used to produce bison embryos in vitro and live offspring following embryo transfer.

Key words Bison, IVF, Culture media

1 Introduction

The iconic North American bison (*Bison bison*) is widely recognized for its cultural significance to Native American communities and its representation of the American West. Having been recently named the National Mammal, this species was the first to be protected by the US government ushering in modern day legislative approaches to animal conservation. Despite the bison's tragic history and resurgence from the brink of extinction, it remains near threatened according to the IUCN red list [1]. Very little research has been conducted on assisted reproductive technologies (ART) in this species. However, with the recent identification of valuable genetic lineages within the bison population and disease risks associated with moving animals between herds, there has been a surge of interest in using in vitro-produced embryos in conservation strategies for bison, particularly as it pertains to mitigating diseases such as brucellosis. Additionally, there is a strong commercial bison industry in North American for which some ART may provide economic or practical benefit. The following chapter begins with

Jason R. Herrick (ed.), *Comparative Embryo Culture: Methods and Protocols*, Methods in Molecular Biology, vol. 2006, https://doi.org/10.1007/978-1-4939-9566-0_12, © Springer Science+Business Media, LLC, part of Springer Nature 2019

a history of in vitro embryo production in both plains (*Bison bison bison*) and wood bison (*Bison bison athabascae*) followed by a description of methods that can be used to produce bison embryos in the lab.

1.1 History of In Vitro Embryo Production in Bison

The first report of in vitro production of bison embryos was in wood bison in 2007 [2]. These methods were based on cattle embryo production protocols and utilized supplemented base media of TCM-199 for maturation, Sperm-TALP for fertilization, and SOF for embryo culture. Results varied depending on whether frozen-thawed semen or chilled epididymal spermatozoa were used for fertilization; chilled epididymal sperm yield better results (cleavage 90–95% vs. 29–75%, morula production 20–30% vs. 7.5–12.5%). Blastocyst rates were low for all treatments (10% or less). Similar rates were achieved in a follow-up study by the same group where cleavage rates of 71–73%, morulae rates of 39–46%, and blastocysts production rates of 6–8% were reported [3]. The generation of IVF embryos was done as an assay to evaluate semen from different sources and treatments.

Bison are a seasonal species, typically breeding in late summer or early autumn in North America. Seasonality has been demonstrated to affect the outcomes of IVF for bison in both the fertilizing ability of the sperm (82% breeding season, 50% nonbreeding season—based on a heterologous IVF assay with cattle oocytes) and the developmental potential of oocytes collected during the breeding vs. nonbreeding season (cleavage rates 84% vs 74%, respectively). However, blastocyst production rates were not different between seasons (18% breeding vs. 16% nonbreeding; [4]).

More recently, IVF has been attempted with oocytes collected via transvaginal aspiration [5]. In a study designed to determine the best time for collection of in vivo-matured oocytes for IVF, blastocyst rates from oocytes that were matured for 30 h in vivo and fertilized immediately after collection (13%) were compared to oocytes that were cultured for 4 h prior to fertilization (24%; $n = 38$ per treatment). By extending the in vivo maturation time to 34 h followed by immediate fertilization, a blastocyst rate of 34% was achieved ($n = 74$). No statistically significant seasonal differences were observed in embryo production rates possibly due to low sample numbers, but breeding season rates tended to be higher. Another study by this same group found no seasonal differences in the maturation capacity of wood bison oocytes [5]. The IVF system used in this study was very similar to previous studies, but with the use of Bracket-Oliphant fertilization medium and CR1aa (Charles Rosenkran's amino acid medium) for embryo culture [6]. Otherwise the timing and basic steps are similar to what is described in this chapter.

The quality of the oocyte that goes into an IVF system is of the utmost importance. In order to evaluate the value of morphological assessment of oocytes collected by transvaginal aspiration on wood bison IVF outcomes, Cervantes et al. [7] grouped oocytes according to the appearance of their ooplasm and the number of cumulus cell layers of immature oocytes. Using the same media system as described above, an overall day 7 blastocyst production rate of 27% was achieved ($n = 159$), though when evaluated for quality, oocytes with more than three layers of compact cumulus and homogeneous cytoplasm yielded 46% blastocyst rate ($n = 52$).

The development of bison-specific in vitro embryo production protocols has improved bison embryo production rates when compared to traditional bovine in vitro embryo production systems [8]. Day 7 blastocyst rates per oocyte and per eight-cell stage, respectively, were 16 and 27% for cattle embryos and 7 and 9% for bison embryos using an unmodified cattle embryo production protocol. However, the addition of 5% fetal calf serum (FCS) to the culture medium after embryos had reached the eight-cell stage resulted in improved blastocyst rates per oocyte and per eight-cell stage of in vitro-produced bison embryos to 13 and 30%, more closely resembling in vitro embryo production rates seen in cattle. In vitro embryo production rates in bison can vary due to a number of confounding factors including but not limited to seasonality, donor parameters such as age or pregnancy status, and differences in in vitro culture systems. Ongoing research continues to investigate if these effects can be mitigated in vitro through further modification of production systems.

2 Materials

2.1 Media

Media used for culture of bison embryos are based on media used in our lab for cattle embryo production [9]. Table 1 lists the stock solutions and the concentrations from which the media preparations are made (Table 2). Listed pH values are target pH values at media preparation not postequilibration. All media mentioned henceforth can be referenced in Table 2.

On the day of maturation, IVM medium is prepared for equilibration by addition of hormones and supplements according to Table 2. Hormones and supplements are stored at −80 °C in amounts aliquoted for a single round of in vitro embryo production to avoid repeated freeze–thaw cycles. Cysteamine is stored at −20 °C. All hormones and supplements are thawed immediately before addition to IVM for equilibration. Preparation of medium and plates for culture is always done in a hood. Once the hormones and supplements have been added, the medium is mixed by several gentle inversions of the tube. It is important not to vortex the medium as the bovine serum albumin (BSA) will trap air and cause the medium to become foamy. A 4-well culture dish is prepared with 1 mL of the IVM medium + supplements in each

Table 1
Stock solutions

Stock	mM or % (w/v)	Filter (Y/N)	Storage time	Storage temp
Alanyl-glutamine (Ala-Glu)	100	Y	2 weeks	4 °C
$NaHCO_3$ (B)[a]	250	Y	1 month	4 °C
Bovine serum albumin (BSA)	10%	Y	2 weeks	4 °C
$CaCl_2 \cdot 2H_2O$ (C)	200	Y	3 months	4 °C
Caffeine (Cafn)	50	Y	1 month	4 °C
Cysteamine (Cys)[b]	10	N	3 months	−20 °C
CDM:		Y	3 months	4 °C
NaCl	710			
KCl	60			
KH_2PO_4	10			
Na-citrate·$2H_2O$	5			
BME AA sol 50× (EAA)	N/A		1 year	4 °C
$C_{10}H_{16}N_2O_8$ (EDTA)[c]	1	Y	3 months	4 °C
D-Fructose (F)	200	Y	1 month	4 °C
D-Glucose (G)	350	Y	1 month	4 °C
Fetal calf serum (FCS)	N/A	N/A		−20 °C
Gentamicin	N/A	N		4 °C
Glycine (Glycn)	490	Y	1 month	4 °C
H-1:1:[a]				
HEPES free ac	100			
HEPES Na salt	100			
Total:	200	Y	1 month	4 °C
Heparin Na salt (Mat)[b]	1%	N	1 month	4 °C
Heparin Na salt (Fert)[b]	0.10%	N	2 month	−20 °C
Myoinositol (MI)	277	Y	1 month	4 °C
Sodium L-lactate (L-L)	1000	Y	1 month	4 °C
$MgSO_4$ (MS)	50	Y	3 months	4 °C
NaCl	5000	Y	3 months	4 °C
MEM AA sol 100× (NEAAs)	N/A	N/A	1 year	4 °C
Na-pyruvate (P)	50	Y	2 weeks	4 °C
Taurine (T)	10	Y	3 months	−20 °C

[a] 0.005 g phenol red added prior to filtering

[b] Prepared with sterile saline rather than nanopure water

[c] (a) Prepare a 1 M solution of NaOH (0.4 g NaOH in 10 mL water); (b) dissolve the 0.029 g EDTA into 220 μL NaOH solution; (c) add 9.78 mL water to the 220 μL EDTA solution; (d) add the 10 mL EDTA solution to 90 mL water

Table 2
Media formulations

	HCDM-M	IVM	FCDM	HCDM-1	CDM-1	HCDM-2	CDM-2
CDM	78.5	78.5	78.5	78.5	78.5	78.5	78.5
B	5.0	25.0	25.0	5.0	25.0	5.0	25.0
C	2.0	2.0	2.0	2.0	2.0	2.0	2.0
G	0.88	0	0.88	0	0	0	0
F	0	2.0	0	0.5	0.5	2.0	2.0
Glycn	4.90	4.90	4.9	4.9	4.9	4.9	4.9
ALA-GLU	1.0	1.0	1.0	1.0	1.0	1.0	1.0
Hepes (1:1)	20.0	20.0	20.0	20.0	20.0	20.0	20.0
L-L	10.0	10.0	10.0	10.0	10.0	5.0	5.0
P	0.5	0.5	0.5	0.5	0.5	0.2	0.2
MS	0.5	0.5	0.5	0.5	0.5	0.5	0.5
MI	0	2.77	0	0	2.77	0	2.77
NEAAs	1%	1%	1%	1%	1%	1%	1%
EAAs	0	0	0	0	0	2%	2%
T	0	0.1	0	0	0.1	0	0
BSA-FAF	2.5%	5%	5%	2.5%	5%	2.5%	5%
NaCl	22.5	0	14.0	22.5	1.0	25.6	5.0
HP–Mat	20 μg/mL	0	0	0	0	0	0
HP–Fert	0	0	2 μg/mL	0	0	0	0
Cafn	0	0	2.0	0	0	0	0
EDTA	0	0	0	0	0.01	0	0
FCS	0	0	0	0	0	0	5%
Gentamicin	25 μg/mL	25 μg/mL	25 μg/mL	25 μg/mL	25 μg/mL	25 μg/mL	25 μg/mL
Osmolality	275 mOsm	275 mOsm	300 mOsm	275 mOsm	275 mOsm	275 mOsm	275 mOsm
pH	7.3–7.4	7.3–7.4	7.5–7.6	7.3–7.4	7.3–7.4	7.3–7.4	7.3–7.4

See Table 1 for definition of component (first column). All values are mM, unless otherwise indicated. All media are stored at 4 °C for up to 3 weeks

For IVM, on the day of maturation the following hormones are added to the medium: 1 μL/mL of IVM should be added of 15 ng/μL ovine FSH stock, 1 μg/mL bovine LH stock, and 0.1 μg/μL Estradiol 17β; 10 μL/mL of IVM should be added of 50 ng/μL epidermal growth factor stock and 10 mM cysteamine stock (*see* Table 1). For example, if you are making 20 mL of IVM you would add 20 μL of the FSH, LH, and E_2 stocks and 200 μL of EGF and cysteamine stocks

Heparin, gentamicin, and FCS are added after the media are filtered through a 0.22 μm membrane filter

Table 3
Stock solutions for Percoll® gradient preparations

Solution	Component	Sigma #	mM	Filter	Storage time	Storage temp
100× SPAD	$CaCl_2 \cdot 2H_2O$	C-3881	210			
	$MgCl_2 \cdot 6H_2O$	M-2670	40			
				Y	3 months	4 °C
10× SPTL	NaCl	S-9625	800			
	KCl	P-5405	30.8			
	NaH_2PO_4	S-5011	3			
	Hepes free ac.	H-4034	50			
	Hepes Na salt	H-3784	50			
	Na-lactate (60% syrup)	L-1375	328.4			
				Y	3 months	4 °C
1× SPTL	10× SPTL	1.25 mL				
	100× SPAD	0.1 mL				
	H_2O	8.65				
				Y	3 months	4 °C
90% Percoll®	Percoll®	9 mL				
	10× SPTL	1 mL				
	100× SPAD	0.1 mL				
	B (*see* Table 1)	0.1 mL				
				N	2 weeks	4 °C
45% Percoll®	90% Percoll®	3.5 mL				
	1× SPTL	3.5 mL				
				N	2 weeks	4 °C

[a]Note: 45% Percoll® can be made in any volume as long as it is diluted 1:1 with 1× SPTL. Likewise 90% Percoll can be scaled up as needed

well and placed into an incubator at 38.5 °C and 5%CO_2 for a minimum of 3 h.

Preparation of sperm for fertilization (*see* Subheading 3.3) is done via a 45% and 90% Percoll® density gradient centrifugation. Preparation of the solutions for creating these gradient solutions can be found in Table 3.

3 Methods

3.1 Oocyte Collection

Viable oocytes can be collected from abattoir-derived ovaries for several hours after slaughter. These ovaries are held in room temperature saline (~28 °C) in a thermos or small cooler. If identification of individual oocyte donors is critical then a small sandwich bag with ~50 mL of saline can be utilized. If collection conditions are such that cold temperatures are a concern, hand or foot warmers can be placed at the bottom of a cooler and covered with crinkled paper towels to create a buffer between the warmers and the ovaries. Do not lay ovaries, even if contained within a bag, directly on top of warmers.

Oocytes can be aspirated from the ovaries using an 18-gauge needle. Smaller needles risk stripping the oocytes of their cumulus cells, which will compromise maturation. This can be done with a syringe or a vacuum pump set to approximately 50 mmHg. Higher pressures also risk stripping oocytes of cumulus cells. Pressure in the line will increase when one is actively aspirating, so be sure to monitor the pressure during the aspiration process.

Prior to aspirating follicles, we aspirate a few mL of HCDM-M through the aspiration line. This allows you to evaluate the pressure in the line and to provide a stable holding medium into which the oocytes will be aspirated. When collecting from valuable individuals, all follicles on the ovary are collected, even those that fall out of the 3–8 mm diameter range, which is ideal for embryo production. After completion of aspiration, a few more milliliters of HCDM-M is aspirated through the line to ensure all oocytes have been collected into the tube and are not caught in the line.

If follicular aspirates are collected into a tube, the cellular material is allowed to settle and the resulting pellet is collected and placed in a petri dish. The pellet is diluted with HCDM-M warmed to 37 °C to make identification and isolation of oocytes easier while keeping them in a medium buffered for bench top (room air) manipulations. Once identified, oocytes are transferred to a petri dish with HCDM-M, where they are collected, evaluated for quality, and counted prior to being moved to a culture dish for maturation (*see* **Note 1** on oocyte appearance/quality). If there is a lot of debris in the medium with the oocytes, then they are washed through another petri dish with HCDM-M by distributing the oocytes around the dirty dish and moving them to the clean dish with as little medium as possible. This can be done as many times as necessary until there is no debris in the dish with the oocytes.

3.2 Maturation

Once oocytes have been selected for maturation and washed free of debris, they are gathered into groups of 50 and transferred to the culture dishes with 50 oocytes per well containing 1 mL of IVM. Oocytes are matured for 23 h in this medium in an incubator at 38.5 °C and 5% CO_2, 5% O_2, 90% N_2.

3.3 Fertilization

After 23 h of maturation, oocytes are moved to 4-well culture dishes containing 430 μL of FCDM in each well that have been equilibrated for a minimum of 3 h in an incubator at 38.5 °C and 5%CO_2 in air. Oocytes can be moved through a rinse well of FCDM before going into the final well for fertilization or moved to FCDM directly from the maturation medium if they are moved in a minimum volume of no more than 20 μL.

To prepare the sperm, frozen semen is thawed in a water bath that is approximately 35 °C for at least 30 s. The straw is dried off and the contents are expelled into a 15 mL conical tube onto a 45%/90% Percoll® gradient warmed to 37 °C where each layer is 1 mL (*see* **Note 2**). Centrifuge the sperm at 800 × *g* for 10 min. Immediately remove the supernatant leaving the pellet and a small amount of medium (~100 μL) above the pellet so that you do not remove any of the highly motile sperm that will quickly begin to swim out of the pellet. Add 4 mL of wash medium (HCDM-1) and centrifuge the sperm again at 500 × *g* for 5 min. Remove the excess medium leaving a small amount of medium (~100 μL) above the pellet. Determine the concentration of spermatozoa in the pellet (e.g., with a hemocytometer) and dilute to a concentration of 5×10^6 sperm/mL using FCDM that has been equilibrated for a minimum of 3 h in an incubator at 38.5 °C and 5%CO_2 in air. Add 50 μL of the sperm suspension to each well of FCDM containing oocytes for a final concentration of 0.5×10^6 sperm/mL. Allow the oocytes and sperm to coincubate for 18 h at 38.5 °C and 5%CO_2 in air.

3.4 Removal of Cumulus Cells and Culture to the Eight-Cell Stage

After coincubation with sperm, oocytes are stripped of cumulus cells. The preferred method in our lab is mechanical stripping using specialized pipettes (Stripper® pipette, Origio) and tips, though there are other methods that can accomplish this (*see* **Notes 3** and **4**). Prior to removal of cumulus cells, HCDM-1 is warmed to 38 °C. A drop of at least 300 μL of this medium is made in a sterile dish and the zygotes are transferred to the drop. This volume is used in our lab because we rarely strip more than 25 oocytes at a time, so this amount is large enough that we do not have significant evaporation during the stripping process yet small enough that we are efficiently using the media. Stripping can be done in larger volumes and in wells if desired. It is also a good idea to note the motility of the sperm in the well at the time the presumed zygotes are removed. If there is no motility it may be indicative of a low fertilization rate. Typically, 10–20% motility is observed at the end of the 18 h coincubation.

Each zygote is pipetted up and down through a pipette tip that is 135 μm in diameter until all of the cumulus cells are removed. Typically, cells will come off after a few passes through the pipette, but occasionally zygotes retain a small number of cells, particularly if the zona is misshapen. These zygotes are not discarded but may be cultured in a separate well if available. Zygotes are then washed through a series of at least 4 HCDM-1 drops (~70 μL or larger) to remove any remaining spermatozoa and cells in the medium around the zygotes. Once cleaned, a maximum of 50 zygotes are moved to each well of 4-well dishes containing 500 μL of CDM-1 per well that has been equilibrated for at least 3 h at 38.5 °C and 5% CO_2 in air. Zygotes are cultured in these dishes for 56 h at 38.5 °C and 5% CO_2, 5% O_2, 90% N_2.

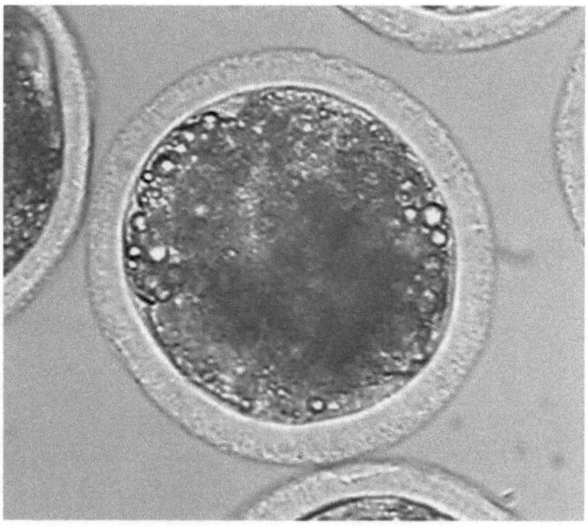

Fig. 1 Bison oocyte with vacuolization

3.5 Culture to the Blastocyst Stage

After culturing in CDM-1 for 56 h, embryos are evaluated for development. Embryos are moved to a small petri dish containing 3 mL HCDM-2 warmed to 38 °C. At this point in development, embryos should have at least eight cells. Each embryo is inspected from all sides to determine how many cells it has and embryos with at least six cells are moved to a 4-well dish containing 500 μL of CDM-2 + 5% FCS per well that has been equilibrated for at least 3 h at 38.5 °C and 5%CO_2 in air. The maximum number of embryos per well is 30. All unfertilized ova and two to five cell embryos are discarded. Embryos are then cultured at 38.5 °C and 5% CO_2, 5% O_2, 90% N_2 for 96 h.

3.6 Embryo Evaluation

At the end of the 96-h culture period in CDM-2, the embryos are approximately 7 days old, that is, fertilization was initiated 7 days prior. Embryos should be at the blastocyst or expanded blastocyst stage of development, though some early blastocysts will be observed. Like their oocytes, bison embryos appear darker than cattle embryos. Some commonly observed morphological characteristics of in vitro-produced bison embryos include high levels of vacuolization (Fig. 1), as well as difficulty in distinguishing morulae postcompaction due to the darkness of the embryo.

4 Notes

1. Bison oocytes have a different appearance than cattle oocytes in that they appear darker (Fig. 2, (a) bison oocytes, (b) cattle oocytes) and many appear to have an uneven distribution of darker cytoplasm that favors a yin-yang conformation (Fig. 3). Preliminary in vitro embryo production experiments in our lab have not demonstrated any differences in the ability of oocytes

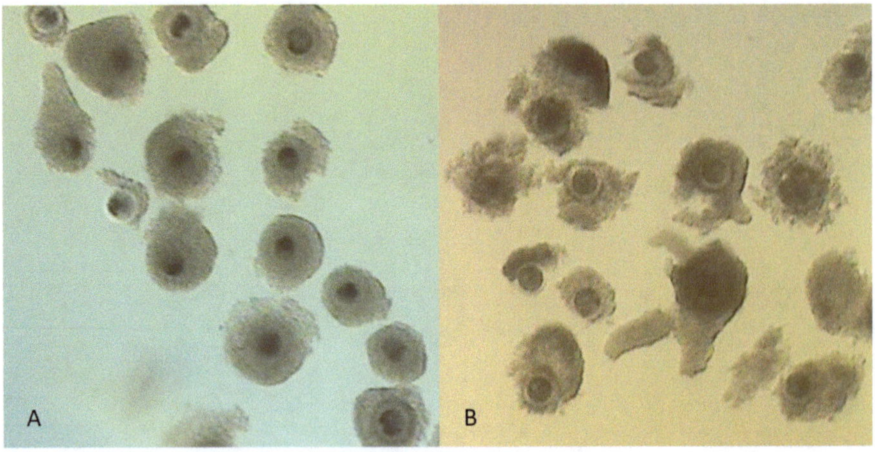

Fig. 2 Comparison of bison oocytes (**a**) and cattle oocytes (**b**)

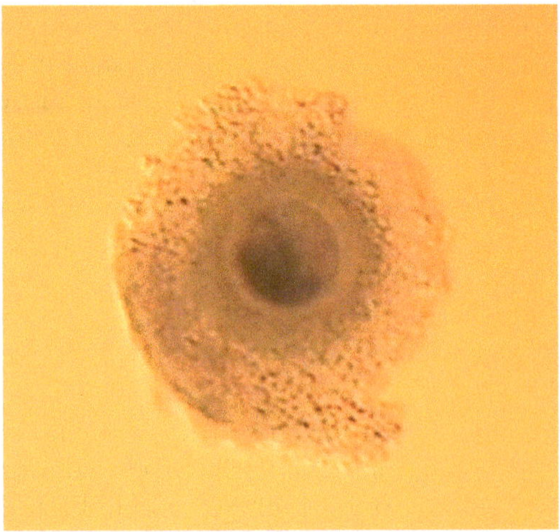

Fig. 3 Uneven cytoplasm of a bison oocyte

with the yin-and-yang appearance to develop into embryos when compared to ones with a homogeneous cytoplasm (unpublished data).

2. Percoll® gradients can be made in different volumes than what is described in Subheading 3.3. One can increase the volume to 2 mL for each layer but that necessitates an increase in centrifugation time to 20 min. Alternatively, a mini-Percoll® prep can be used to clean semen that has a lower sperm number per straw (e.g., sexed semen). This involves a 0.5 mL 45% Percoll® layered on 0.5 mL 90% Percoll® in a microcentrifuge tube. The centrifugal forces are the same for the spins but the timing is shortened to 8 min. If disease or bacterial contamination is of

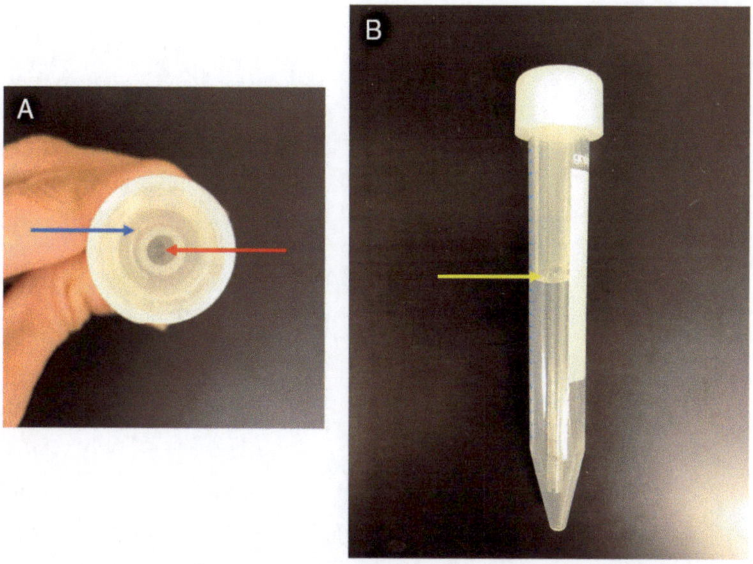

Fig. 4 View of the the Nidacon® ProInsert from the top (**a**) and from the side (**b**). The blue arrow indicates where the gradient and semen are deposited for layering. The red arrow is the central port from which the sperm pellet is retrieved postcentrifugation. The yellow arrow indicates the hole through which the gradient and semen pass during layering

concern, a ProInsert® (Nidacon) can be inserted into a 15 mL conical tube prior to layering the gradient to facilitate accessing the sperm without coming into contact with the seminal fluid after centrifugation (*see* Fig. 4). A general rule of thumb is that the amount of semen layered on the gradient should not exceed the volume of a single layer of Percoll®; for example, if each layer of Percoll® is 2 mL then a maximum of 2 mL of semen should be layered on the gradient.

3. An alternative to stripping presumed zygotes by pipette is vortexing. This involves moving the zygotes from the fertilization wells into a drop of HCDM-1 warmed to 38 °C. Once in a holding medium, they are moved into the bottom of a 0.5 mL microcentrifuge tube in a small volume of fluid. It is best not to fill the tube higher than the tapered portion of the tube (*see* Fig. 5) as this will result in zygotes being caught in the lid making them difficult to retrieve. The zygotes are then vortexed for 1 min at high speed. Immediately after vortexing, fill the tube with HCDM-1. If you hold the tube up to light you will see the zygotes stuck to the side of the tube. This can give you an idea of how much fluid you need to put into the tube to get the zygotes into holding medium. It is critical that you do this as quickly as possible. Set the tube on a warm block for approximately 3 min to allow the oocytes to settle to

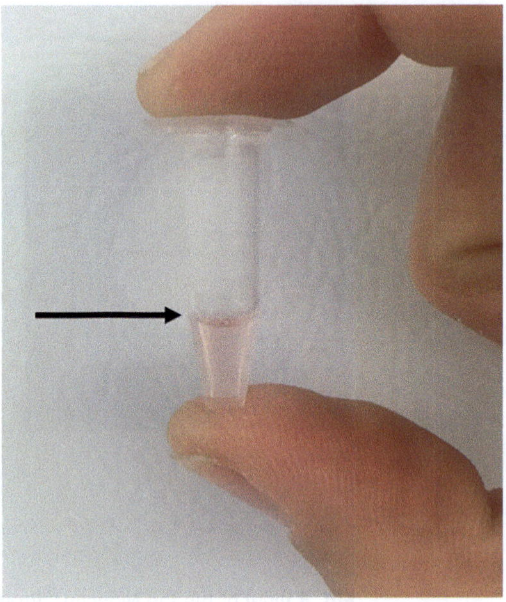

Fig. 5 Microcentrifuge tube with holding medium. Black arrow indicates the end of the tapered portion of the tube above which one should not add liquid prior to vortexing

the bottom of the tube. During this 3 min, make a series of four to eight HCDM-1 drops of approximately 100 μL (volume can vary depending on the amount of debris and number of zygotes you are processing). Pull the pellet out of the bottom of the microcentrifuge tube and add it to one of the HCDM-1 drops. Continue to pull from the microcentrifuge tube until no more zygotes are found. Wash the zygotes through the drops and put them into culture as described in Subheading 3.3.

4. When processing large numbers of bison zygotes, we use the vortex method for removing cumulus cells. However, when processing smaller numbers of oocytes on a per donor basis, we prefer the stripping method as we have found a higher number of broken zonas when vortexing bison zygotes compared to vortexing cattle zygotes.
5. Mitigation of disease transmission can be accomplished by combining these methods with embryo washing techniques outlined by the International Embryo Technologies Society [10]. Embryos can be washed in solutions of trypsin prior to transfer, vitrification, or slow-freeze cryopreservation to ensure that bacterial diseases, like brucellosis, are not transmitted to the recipient bison cow with the embryo. *See* Stringfellow [10] for methodological details.

References

1. Aune K, Jørgensen D and Gates C (2017) *Bison bison*. The IUCN Red list of threatened species 2017: e.T2815A45156541. https://doi.org/10.2305/IUCN.UK.2017-3.RLTS.T2815A45156541.en. Accessed 12 Feb 2018
2. Thundathil J, Whiteside D, Shea B, Ludbrook D, Elkin B, Nishi J (2007) Preliminary assessment of reproductive technologies in wood bison (*Bison bison athabascae*): implications for preserving genetic diversity. Theriogenology 68:93–99
3. Aurini LC, Whiteside DP, Elkin BT, Thundathil JC (2009) Recovery and cryopreservation of epdidymal sperm of plains bison (*Bison bison bison*) as a model for salvaging the genetics of wood bison (*Bison bison athabascae*). Reprod Domest Anim 44:815–822
4. Krishnakumar S, Whiteside DP, Elkin B, Thundathil JC (2015) Effect of reproductive seasonality on gamete quality in the North American bison (*Bison bison bison*). Reprod Domest Anim 50:206–213
5. Cervantes MP, Palomino JM, Anzar M, Mapletoft RJ, Adams GP (2016) In vivo and in vitro maturation of oocytes collected from superstimulated wood bison (*Bison bison athabascae*) during the anovulatory and ovulatory seasons. Anim Reprod Sci 173:87–96
6. Cervantes MP, Palomino JP, Anzar M, Mapletoft RJ, Mastromonaco GF, Adams GP (2017) In vitro embryo production in wood bison (*Bison bison athabascae*) using in vivo matured cumulus-oocyte complexes. Theriogenology 89:122–130
7. Cervantes MP, Palomino JP, Anzar M, Mapletoft RJ, Mastromonaco GF, Adams GP (2017b) In vitro-production of embryos using immature oocytes collected transvaginally from superstimulated wood bison (*Bison bison athabascae*). Theriogenology 92:103–110
8. Barfield JP, Seidel GE Jr (2011) In vitro production of bison embryos. Reprod Fertil Dev 24:197
9. De la Torre-Sanchez JF, Pries K, Seidel GE Jr (2006) Metabolic regulation of *in-vitro*-produced bovine embryos. I. Effects of metabolic regulators at different glucose concentrations with embryos produced by semen from different bulls. Reprod Fertil Dev 18:585–596
10. Stringfellow DA (2009) Recommendations for the sanitary handling of in vivo-derived embryos. In: 4th edition of the IETS manual. International Embryo Technologies Society, Champaign, IL

Chapter 13

In Vitro Embryo Production in Buffalo Species (*Bubalus bubalis*)

Marcello Rubessa, Lucia Boccia, and Serena Di Francesco

Abstract

In the last years there has been a growing attention in the world regarding buffalo (*Bubalus bubalis*) breeding. In tropical countries, climatic and environmental conditions make buffalo an irreplaceable milk producer. Since cattle cannot express their own productive potential in these conditions, buffaloes are best suited to satisfy animal protein demand, as they are easy to maintain, long-lived, and parasite resistant. Here we describe how to produce buffalo embryos from slaughterhouse oocytes, with protocols for media formulation, oocyte selection, semen capacitation, and in the end embryo evaluation.

Key words Buffalo, Oocytes, Embryos, IVM, IVF, IVC, Semen

1 Introduction

Buffalo is a fundamental resource in all developing countries, especially for small-scale farmers. The world population of buffalo (*Bubalus bubalis*) contributing to meat and milk productions is greater than 130 million animals [1]. The overall low reproductive efficiency in buffalo species can be attributed to several factors intrinsic to the species, including delayed puberty, higher age at calving, long postpartum anestrus period, long calving interval, lack of overt sign of heat, and low conception rate. In addition, female buffaloes have few primordial follicles and a high rate of follicular atresia [2]. As these issues consequently determine the low efficiency of superovulation (SO) and embryo transfer (ET) in this species, in the last several years there has been an increasing interest in in vitro embryo production (IVEP) technologies for faster propagation of superior animals and to enhance genetic progress through the maternal contribution [3]. It is now possible to adopt a fully in vitro system, whereby immature oocytes are matured through in vitro maturation (IVM), followed by in vitro fertilization (IVF) with capacitated spermatozoa, and culture (IVC) of newly formed zygotes in suitable media for development of

Jason R. Herrick (ed.), *Comparative Embryo Culture: Methods and Protocols*, Methods in Molecular Biology, vol. 2006, https://doi.org/10.1007/978-1-4939-9566-0_13,

transferable stage embryos. IVEP is a competitive reproductive technology. IVM, IVF, and IVC are also integral parts of a number of other reproductive technologies like cloning, production of transgenic animals, and production of embryonic stem cells [2], which could make a significant contribution to the population of buffalo with superior genetic merit.

2 Materials

2.1 Preparation of Medium and Stock Solutions

1. Prepare all solutions using ultrapure or embryo-quality water and analytical grade reagents.
2. All media are made in graduated cylinders washed with distilled water and sterilized overnight in 122 °C heater (*see* **Note 1**).
3. Prepare all reagents at room temperature, under a flow bench.
4. Filter all the solutions with a disposable syringe and a 0.22 μm filter, unless indicated otherwise.
5. Store all reagents at 4 °C, unless indicated otherwise.

2.2 Hormones

1. Estradiol (E2) Aliquots: This is a 1 mg/ml solution in absolute ethanol. For this purpose, weigh the E2 and dissolve in ethanol in 50 ml tube. Vortex it for 60 s. Do not filter. Place 10 μl in 1.5 ml microcentrifuge tubes and store at −80 °C.
2. Luteinizing hormone (Sheep pituitary, LH) Aliquots: Prepare a 5 μg/ml solution in 0.9% NaCl. Vortex but do not filter. Place 100 μl in 1.5 ml microcentrifuge tubes and store at −80 °C.
3. Follicle-stimulating hormone (Sheep pituitary, FSH) aliquots: Prepare a 0.5 μg/ml solution in 0.9% NaCl. Place 20 μl in 1.5 ml microcentrifuge tubes and store at −80 °C.

2.3 Stock Solutions

1. Stock B (250 mM sodium bicarbonate): Place 2.1 g sodium bicarbonate and 10 mg of phenol red sodium salt in 100 ml of water. The osmolarity should be 435 ± 5 mOsm (*see* **Note 2**). Filter and store at 4 °C for 3 weeks.
2. Stock C (30 mM sodium pyruvate): Dissolve 36 mg in 10 ml of water. The osmolarity should be 60 ± 5 mOsm. Filter and store at 4 °C for 1 week.
3. Stock D Solution: Dissolve 2.52 g of $CaCl_2 \cdot 2H_2O$ (Hygroscopic) in 100 ml of water. The osmolarity: 435 ± 5 mOsm. Filter and keep it in refrigerator for 3 months.
4. Stock G Solution (60 mM Glucose): Dissolve 1.08 g of glucose in 100 ml of water. Osmolarity: 65 ± 5 mOsm. Filter and keep it in refrigerator for 3 months.
5. Stock H (250 mM Hepes): For this purpose, dissolve 3 g of Hepes, free acid, 3.25 g of Hepes sodium salt, and 10 mg of

Phenol Red Sodium in 100 ml of water. Osmolarity: 380–385 mOsm. Filter and keep it in refrigerator for 1 month.

6. Stock L Solution: Prepare solution of 330 mM of sodium lactate. For this purpose, dissolve 4.7 ml of sodium lactate (60%) in 100 ml of water. Osmolarity: 625–660 mOsm. Filter and keep it in refrigerator for 3 months.
7. Stock M Solution: Dissolve 1 g of $MgCl_2 \cdot 6H_2O$ in 100 ml of water. Osmolarity: 120–130 mOsm. Filter and keep it in refrigerator for 3 months.
8. Stock P (73.3 mM of sodium pyruvate): Dissolve 0.08 g of sodium pyruvate in 10 ml water. Vortex for 60 s, or until totally dissolved. Filter and store at 4 °C for 1 week.
9. Gentamycin Sulfate Stock Solution (10 mg/ml): Dissolve 0.1 g gentamycin sulfate in 10 ml of water. Vortex until totally solved. Filter and make aliquots of 30 μl in 1.5 ml microcentrifuge tubes, and store at −20 °C for 6 months.
10. Heparin Stock Solution (10 mg/ml): Dissolve 0.1 g heparin in 10 ml of 0.9% NaCl solution and vortex until totally dissolved. Do not filter. Make 60 μl aliquots in 1.5 ml microcentrifuge tubes and store at −20 °C for 6 months.
11. Sodium Lactate Stock Solution (330 mM): Dilute 4.7 ml sodium lactate solution (D,L-Lactate, 60%) in 100 ml of water. Do filter. Adjust pH to 4.0 using 1 N HCl.
12. Hypotaurine Stock Solution (403.11 mM): Dissolve 0.44 g hypotaurine in 10 ml of water. Vortex until totally dissolved. Do not filter.
13. Penicillamine Stock Solution (80.42 mM): Dissolve 0.12 g penicillamine in 10 ml of water. Vortex until totally dissolved. Do not filter.
14. Epinephrine Stock Solution (10 mM): Dissolve 0.0183 g epinephrine in 10 ml of water in a tube covered with aluminum foil to avoid light exposure. Prepare just before use. It will not completely dissolve, so shake before dispensing into solution.
15. Penicillamine, hypotaurine, and epinephrine (PHE) Stock Solution: Combine 3 ml of Lactate Bisulfate Stock Solution, 1.875 ml of Penicillamine Stock Solution, 187.5 μl of Hypotaurine Stock Solution, and 75 μl Epinephrine Stock Solution. Filter and make 120 μl aliquots in 0.5 ml microcentrifuge tubes. Store at −80 °C for 6 months.
16. Stock IVC Base Solution: In a graduated cylinder, combine 63.0 mg/ml (1078 mM) NaCl, 5.34 mg/ml (71.6 mM) KCl, 1.62 mg/ml (11.9 mM) KH_2PO_4, and 0.25 mg/ml of kanamycin sulfate. The osmolarity of a 1:10 dilution of this solution should be around 216 ± 3 mOsm. Filter and store at 5 °C for 3 months.

2.4 In Vitro Maturation

1. HEPES-Buffered Tissue Culture Medium 199 (H199): Just prior to use, add 10% fetal calf serum (FCS) to Tissue Culture Medium 199 containing 25 mM HEPES.
2. Cystine Stock Solution (0.3 mM): Dissolve 2.5 g cystine in 13.303 ml M199. Vortex and filter. Dilute 1:1 with additional warm M199 [4].
3. Cysteamine Stock Solution (50 μM): Dissolve 5 g cysteamine hydrochloride in 4.4 ml M199. Do not vortex but filter [5, 6]. Dilute 1:1 with M199.
4. IVM Medium: Combine 4.373 ml cystine stock solution, 0.5 ml FCS, 5 μl estradiol stock solution, 10 μl FSH, 50 μl LH, 50 μl cysteamine stock solution, and 12 μl stock P for a final volume of 5 ml IVM medium containing 0.06 mM pyruvate, 0.3 mM cystine, 0.5 μM cysteamine, 1 mg/ml E2, 0.5 mg/ml FSH, 5 mg/ml LH, and 10% (v/v) FCS.

2.5 In Vitro Fertilization (IVF)

1. IVF Base Solution: In a graduated cylinder, combine 6.65 mg/ml (114 mM) NaCl, 0.236 mg/ml (3.17 mM) KCl, 0.042 mg/ml (0.35 mM) NaH_2PO_4, 0.294 mg/ml (2 mM) $CaCl_2{\cdot}2H_2O$, 0.102 mg/ml (0.5 mM) $MgCl_2{\cdot}6H_2O$, 1.9 μl/ml lactic acid (Sodium DL-lactate solution), 2.1 mg/ml (25 mM) $NaHCO_3$. The pH should be adjusted to 7.4 ± 0.05 using 1 M HCl or 1 M NaOH. Osmolarity should be 280 ± 8 mOsm. Filter and make 10 ml aliquots in 15 ml tubes. Store for 6 months at −80 °C.
2. HEPES-TALP Solution (H TALP): On the day of fertilization, combine 10 ml IVF Base Solution, 30 mg fatty acid free bovine serum albumin (BSA, 3.0 mg/ml), 20 μl stock P, and 50 μl of gentamycin stock solution.
3. IVF medium: Combine 5.633 ml of IVF Base Solution, 11.3 μl stock P, 57 μl gentamycin stock solution, 60 μl of heparin stock solution, 240 μl of PHE solution, 33.8 mg of fatty acid free bovine serum albumin. pH should be 7.4 ± 0.02. Filter and place in the incubator (38.5 °C, 5% CO_2, and 20% O_2 and Nitrogen).

2.6 In Vitro Culture

1. Wash medium: Just prior to use, add 5% bovine serum (BS) to Tissue Culture Medium 199 containing 25 mM HEPES.
2. IVC medium: Combine 2 ml of Stock IVC Base Solution, 2 ml of stock B, 0.1 ml of stock C, D, M, L, G, GlutaMAX, 0.2 ml of nonessential amino acids, 0.4 ml essential amino acids, and 160 mg (8 mg/ml) of fatty acid-free BSA in 14.8 ml of water.

2.7 Equipment

1. Heating block.
2. Aspiration pump.

3. Inverted microscope.
4. Centrifuge.
5. Hemocytometer.
6. Microscope.
7. Incubator.
8. Modular incubator chamber.
9. Pipettes, 10, 20, 200, and 1000 μl.

3 Methods

3.1 In Vitro Maturation (IVM)

3.1.1 Preparation of IVM Dishes

1. This procedure has to be carried out at least 4 h before processing the ovaries in the lab.
2. In a laminar hood, place six 40 μl drops of IVM medium in a 35 mm petri dish (not treated polystyrene) and cover the drops with 3 ml of mineral oil (embryo tested) to avoid evaporation.
3. Place the dish in an incubator at 38.5 °C with 5% CO_2.
4. Remaining IVM medium should also be placed in the incubator, with the cap partially open to allow gas exchange.

3.1.2 Ovary Collection and Preparation

1. Buffalo ovaries are collected at the slaughterhouse and transported to the laboratory in a thermos containing 500 ml 0.9% NaCl supplemented with 1000 UI/ml penicillin–streptomycin (30–35 °C). The same solution is used to wash the ovaries once reaching the lab, within 6 h after slaughter [7].
2. Disinfect and clean your working bench with 70% ethanol before and after use.
3. At the lab, empty the thermos containing the ovaries in a glass beaker and discard the saline solution. Wash quickly with 70% ethanol and then wash thoroughly with fresh 0.9% saline (38.5 °C) an additional two times.
4. With scissors and forceps remove extra tissue from the ovaries that could hamper you from seeing the follicles to aspirate on the same ovaries.

3.1.3 Aspiration of Follicles

1. Place 20 ml wash medium in the heating block (37 °C) 1 h before starting. Also, place some empty 15 ml tubes in the heating block, so that they are at the right temperature when starting.
2. Connect an empty 15 ml tube to a vacuum pump (−30 mmHg) using the apparatus described in **Note 3** [8].
3. Dry one ovary at the time on sterile gauze to dry it. Turn on the suction and start by aspirating 1 ml H199 through the system.

Then insert the needle into visible follicles ranging from 1 to 8 mm in diameter.

4. Gently rotate the needle and move forward and back while aspirating. Repeat this procedure for all follicles on the ovary and then repeat with every other ovary.
5. Stop the aspiration before the tube is full (we suggest around 10 ml) and transfer aspiration cap to a new tube. Aspirate another 1 ml wash medium and continue aspiration.
6. When finished, aspirate a little more wash medium to ensure all the oocytes are recovered.

3.1.4 Recovery of Cumulus–Oocyte Complexes (COCs)

1. Transfer the pellet from the bottom of the tube to the center of a 90 mm petri dish using a sterile glass Pasteur pipette. Gently shake the petri dish to distribute the follicular fluid and cells uniformly throughout the dish. Let the cells settle to the bottom of the dish and start searching COCs.
2. Once all COCs are recovered, pick up grade A and B COCs (*see* **Note 4**) and transfer them to a 35 mm petri dish containing 3 ml of wash medium (Fig. 1a).
3. Perform a second washing by moving the oocytes to another 35 mm petri dish containing 3 ml of wash medium.
4. Finally, transfer the COCs to a 35 mm petri dish containing IVM medium for the last wash, while counting them.
5. Transfer 10 μl of IVM medium along with ten (9–11) COCs to each 40 μl droplet of IVM medium to create a final volume of 50 μl of IVM medium.
6. On every dish, write IVM, date, name of operator and the total number of oocytes.

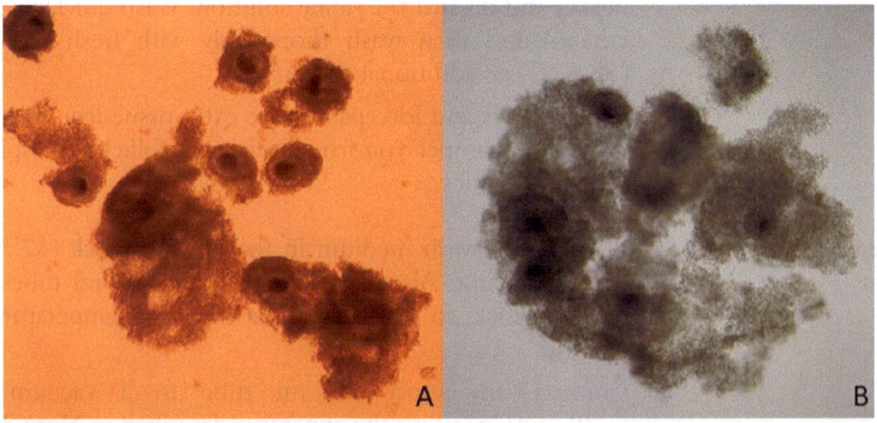

Fig. 1 (**a**) Buffalo oocytes after aspiration of slaughter-derived ovaries. Grade A oocytes with homogeneous cytoplasm and the entire surface surrounded by multiple layers of cumulus cells; Grade B oocytes with homogeneous cytoplasm and with at least 70% of the surface surrounded by multiple layers of cumulus cells; (**b**) buffalo cumulus–oocytes complexes after 20 h of in vitro maturation

7. Place the IVM dishes in the incubator at 38.5 °C for 22 h in a controlled gas atmosphere of 5% CO_2 in humidified air [9, 10].

3.2 In Vitro Fertilization (IVF)

3.2.1 Preparation of IVF Dishes

1. This procedure should be carried out at least 3 h before IVF.
2. Under a laminar hood, place six 45 μl drops IVF medium in each 35 mm petri dish and cover with 3 ml of mineral oil to avoid their evaporation.
3. Place the petri dish in an incubator (38.5 °C with 5% CO_2) to allow equilibration.
4. The remaining IVF medium should be kept in incubator with the cap partially open to allow gas exchange.

3.2.2 Sperm Processing

1. Buffalo semen can be processed with swim up or discontinuous gradient centrifugation, but here we will describe the use of a commercial gradient, BoviPure Sperm 45/80 (Nidacon, Sweden).
2. Put 1 ml each of 45% and 80% BoviPure in the incubator with cap partially open at least 2 h before thawing semen [11].
3. Also put 10 ml HEPES-TALP in a tube with the cap closed in the incubator.
4. Just prior to thawing spermatozoa, place 1 ml 45% BoviPure on top of the 80% BoviPure without mixing the two solutions.
5. Remove a straw of cryopreserved semen and place in a 37 °C water bath for 40 s.
6. Wash the straw with 70% ethanol, dry it, cut the end with scissors, and let the semen drop into a sterile 1.5 ml tube.
7. Put a 5 μl drop of semen on a glass slide and evaluate the percentage of motile spermatozoa. If this percentage is ≤50%, discard the straw and thaw another one.
8. Transfer the semen to the top of the BoviPure gradient without mixing and centrifuge at 1500 rpm (423 × *g*) for 25 min. This process removes the cryo-protectant used to freeze the spermatozoa.
9. Using a sterile pipette, remove the supernatant from the top of the tube and add 2 ml IVF medium (Sperm-TALP) to the sperm pellet. Centrifuge at 1100 rpm (227 × *g*) for 10 min.
10. Using a sterile pipette, remove the supernatant from the top of the tube and discard in biohazard waste container. Add 2 ml IVF medium (Sperm-TALP). Centrifuge at 900 rpm (152 × *g*) for 10 min.
11. After the centrifugation, slowly remove the supernatant, without disturbing the sperm pellet at the bottom of the tube.

12. After centrifugation, remove as much of the supernatant as possible and resuspend the remaining sperm pellet to arrive to a volume of 105 μl. Use IVF medium to dilute the sample.
13. Measure the total volume of the resuspended pellet, knowing that the final desired concentration of buffalo spermatozoa is 2×10^6/ml (*see* **Note 5**).

3.2.3 Preparation of COCs for IVF

1. During the final centrifugation of the sperm, wash the matured COCs. At this time, expansion of cumulus cells can be observed.
2. Pick up COCs from IVM drops with a pipette and wash twice in two different 35 mm dishes containing 3 ml HEPES-TALP and a third dish with 1 ml IVF medium.
3. One should be as quick as possible because the pH of the IVF medium rapidly changes when it is outside the incubator, which is evident by a change of color.
4. As matured oocytes are very difficult to separate, gently move the dish to separate them (Fig. 1b). Then transfer five matured COCs in 10 μl IVF medium to the drops of IVF medium and return the dish to the incubator.
5. At the end of sperm processing, place 5 μl of the sperm suspension in each fertilization droplet containing COCs. 40 μl medium + 10 μl with COCs + 10 μl Sperm = 60 μl.

3.3 In Vitro Culture (IVC)

3.3.1 Preparation of IVC Dishes

1. The day of IVF place 8 ml wash medium in the incubator for each group.
2. In a 35 mm petri dish, prepare six droplets of IVC medium—a central 40 μl drop, surrounded by five 20 μl drops. Cover these drops with 3.5 ml of mineral oil.
3. Put the dishes in a modular incubator chamber and insufflate with a gas mixture containing 7% O_2, 5% CO_2 and 88% N_2 for 5 min (*see* **Note 6**).
4. Prepare similar dishes on day 4 of culture (day of IVF is day 0).

3.3.2 IVC

1. IVC starts 16–20 h from the start of IVF (day 0) [9].
2. Transfer the presumptive zygotes from the IVF drops to a 35 mm petri dish containing 3 ml wash medium.
3. With a 1000 pipette adjusted to 500 μl, repeatedly aspirate the COCs to remove cumulus cells.
4. Transfer zygotes to another dish with 3 ml wash medium and then to the central 40 μl drop of the IVC dish (*see* **Note 7**).
5. Finally transfer ten presumptive zygotes to each of the peripheral 20 μl drops.

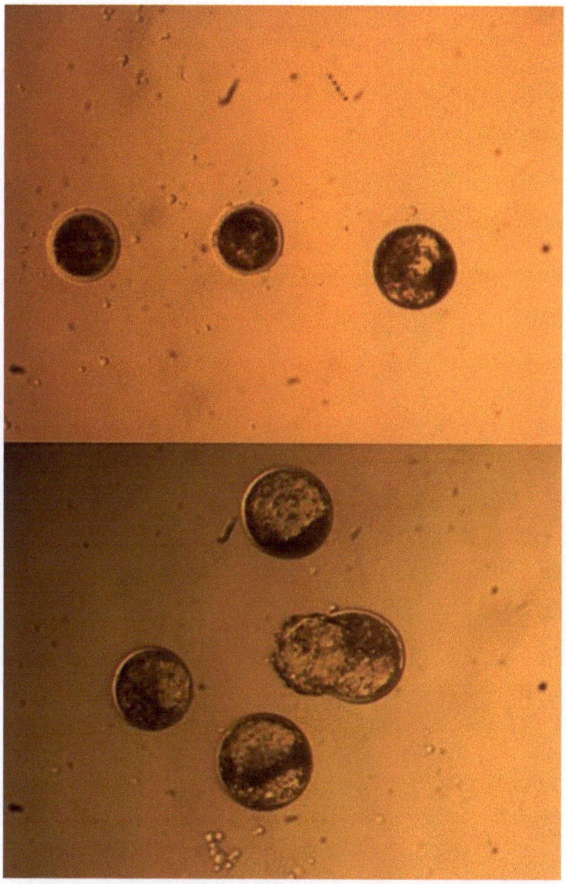

Fig. 2 Buffalo embryos on day 7 of culture. (**a**) Two early blastocysts and one blastocyst (from left to right) and (**b**) A blastocyst, two expanded blastocyst, and a hatching blastocyst

6. Return the dishes to the modular incubator chamber and insufflate with the gas mixture containing 7% O_2, 5% CO_2 and 88% N_2 for 5 min.
7. On day 5, evaluate the proportion of zygotes that cleaved rate and move embryos to fresh drops of IVC medium.
8. Embryos are cultured for an additional 2 days (day 7) and then blastocyst development is evaluated (Fig. 2).

4 Notes

1. Immediately prior to use, glassware for embryos is rinsed twice with distilled water and once with ultrapure or embryo-quality water (in this order).
2. The osmolarity can be reduced adding small amounts of ultrafiltrated water.

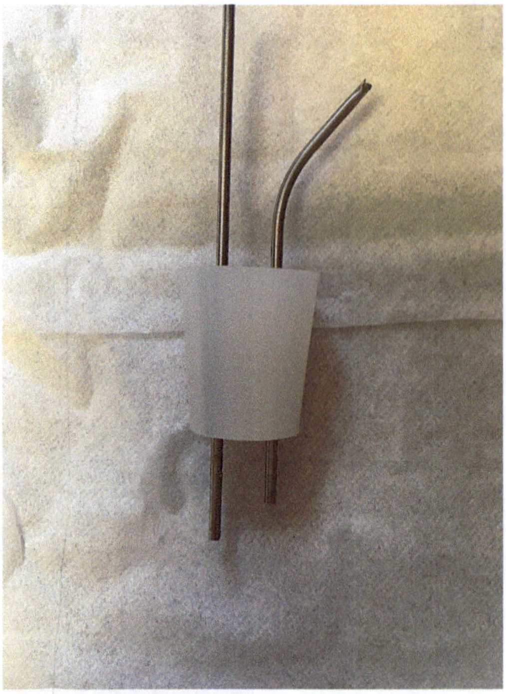

Fig. 3 Silicone cap for 15 ml tube

3. Example of silicone cap for 15 ml tube. It is also possible to use a syringe for follicle aspiration as long as it is free of latex and silicone oil (Fig. 3).
4. Grade A cumulus–oocyte complexes (COCs) contain oocytes with a homogeneous cytoplasm that is completely surrounded by more than five layers of cumulus cells. Grade B COCs have the similar cytoplasm as Grade A COCs, but fewer (3–5) layers of cumulus cells.

   To simplify calculations, it is preferable to dilute the sperm pellet to a final volume of 105 μl. Keep the tube containing processed semen in the heating block. Remove 5 μl of the diluted pellet and dilute 1:40 with distilled water. Put 10 μl on a hemocytometer grid system (e.g., Burker chamber). Count the number of sperm twice and average the two values. The final, desired concentration of buffalo sperm is 2×10^6 per ml or 1×10^5 spermatozoa per IVF drop of 50 μl. The equation that follows could be used to convert the counts in two squares to concentration/ml:

$$\text{Concentration/ml} = (\text{Dilution factor, i.e., } 40) \times (\text{Average count in two squares}) \times (10{,}000).$$

(a) Example: Let us imagine that we count 150 sperm as average of two reads; so now multiply by 40 and by 10,000 your result: 60×10^6 per ml, but because our volume is 100 μl our concentration will be 6×10^6 per 100 μl. Now we have to dilute our sample to attain 1×10^6 per 100 μl; so we add 500 μl of IVF to our 100 μl of sperm (dilution 1:6). Therefore, when we add 10 μl to our 40 μl IVF droplet (final volume 50 μl), we are moving 10,000 sperm (2,000 sperm/μl: i.e. our desired final concentration of 2×10^6 sperm/ml).

5. If you have access to an incubator that can control both CO_2 (5%) and O_2 (7%) concentrations, the modular incubator chamber is not necessary.
6. In this phase of the culture, it is important to move the zygotes using a glass pipette with a smooth tip.
7. The protocol above is an example of in vitro embryo production in buffalo species. It is possible to find in the literature thousands of experiments that try to improve embryo rate and/or embryo quality, usually adding specific protein or sugar in one or all phases of embryo production.

References

1. Singh B, Chauhan MS, Singla SK, Gautam SK, Verma V, Manik RS, Singh AK, Sodhi M, Mukesh M (2009) Reproductive biotechniques in buffaloes (*Bubalus bubalis*): status, prospects and challenges. Reprod Fertil Dev 21 (4):499–510. https://doi.org/10.1071/rd08172
2. Kumar D, Anand T (2012) In vitro embryo production in buffalo: basic concepts. J Buffalo Sci 1:50–54
3. Gasparrini B (2002) In vitro embryo production in buffalo species: state of the art. Theriogenology 57(1):237–256
4. Gasparrini B, Neglia G, Palo RD, Campanile G, Zicarelli L (2000) Effect of cysteamine during in vitro maturation on buffalo embryo development. Theriogenology 54 (9):1537–1542
5. Gasparrini B, Boccia L, Marchandise J, Di Palo R, George F, Donnay I, Zicarelli L (2006) Enrichment of in vitro maturation medium for buffalo (*Bubalus bubalis*) oocytes with thiol compounds: effects of cystine on glutathione synthesis and embryo development. Theriogenology 65(2):275–287. https://doi.org/10.1016/j.theriogenology.2005.05.036
6. Anand T, Kumar D, Chauhan MS, Manik RS, Palta P (2008) Cysteamine supplementation of in vitro maturation medium, in vitro culture medium or both media promotes in vitro development of buffalo (*Bubalus bubalis*) embryos. Reprod Fertil Dev 20(2):253–257
7. Di Francesco S, Boccia L, Di Palo R, Esposito G, Attanasio L, De Rosa A, Gasparrini B (2007) Influence of temperature and time during ovary transportation on in vitro embryo production efficiency in the buffalo species (*Bubalus bubalis*). Ital J Anim Sci 6:755–758
8. Das GK, Jain GC, Solanki VS, Tripathi VN (1996) Efficacy of various collection methods for oocyte retrieval in buffalo. Theriogenology 46(8):1403–1411. https://doi.org/10.1016/S0093-691X(96)00319-6
9. Gasparrini B, De Rosa A, Attanasio L, Boccia L, Di Palo R, Campanile G, Zicarelli L (2008) Influence of the duration of in vitro maturation and gamete co-incubation on the efficiency of in vitro embryo development in Italian Mediterranean buffalo (*Bubalus bubalis*). Anim Reprod Sci 105(3–4):354–364. https://doi.org/10.1016/j.anireprosci.2007.03.022

10. Rubessa M, Di Fenza M, Mariotti E, Di Francesco S, de Dilectis C, Di Palo R, Zicarelli L, Gasparrini B (2008) 217 kinetics of sperm penetration is correlated with in vitro fertility of buffalo (*Bubalus bubalis*) bulls. Reprod Fertil Dev 21(1):206–207. https://doi.org/10.1071/RDv21n1Ab217

11. Gasparrini B, Neglia G, Di Palo R, Vecchio D, Albero G, Esposito L, Campanile G, Zicarelli L (2014) Influence of oocyte donor on in vitro embryo production in buffalo. Anim Reprod Sci 144(3–4):95–101. https://doi.org/10.1016/j.anireprosci.2013.11.010

Chapter 14

In Vitro Culture of Deer Embryos

Debra K. Berg and Ampika Thongphakdee

Abstract

In vitro embryo production of deer species has the potential to increase valuable traits for the agricultural sector, and from a conservation perspective, it is a propagation tool which can improve genetic diversity in small captive populations. In vitro embryo production is a multistep process consisting of oocyte maturation, fertilization, and embryo culture. These techniques provide the backbone for more advanced assisted reproductive technologies such as intracytoplasmic sperm injection (ICSI), somatic cell nuclear transfer (SCNT), a source of embryonic stem cells, and embryos for gene editing. In vitro-produced embryos are a readily available resource for comparative embryology studies and a functional assay to assess oocyte competence and evaluate in vitro embryo requirements during culture. A semidefined fertilization and culture medium system, deer synthetic oviduct fluid (DSOF), has been formulated based on deer oviduct fluid. Red deer calves (*Cervus elaphus*) and Thamin Eld's deer fawn (*Rucervus eldii thamin*) have been produced after the transfer of in vitro embryos (IVF and SCNT) grown in DSOF culture. Here we describe the in vitro method of maturation, fertilization, and embryo culture for deer species.

Key words Semi-defined medium, DSOF, In vitro culture, Deer embryo

1 Introduction

The application of in vitro embryo production for wildlife management has been extensively reviewed to propagate endangered species, preserve genetic diversity, and, in natural populations of red deer, mitigate the effects of inbreeding in small populations [1–3].

The deer family is distributed throughout a wide range of geographical locations from the cold polar regions inhabited by reindeer to the Axis deer living in southern India. These extreme geographical locations affect the beginning and the end of the breeding season as there is a correlation between latitude and the timing of the breeding season [4]. Species which occupy 20° north and 20° south experience little seasonality. In contrast, deer in temperate climes have distinct seasonal breeding patterns, regulated by photoperiod, and typically conceive in the autumn with offspring born in the summer. These endogenous mechanisms controlling seasonality are robust and deer transferred from the

Jason R. Herrick (ed.), *Comparative Embryo Culture: Methods and Protocols*, Methods in Molecular Biology, vol. 2006, https://doi.org/10.1007/978-1-4939-9566-0_14,

northern to the southern hemisphere result in an exact 6-month phase change in reproduction. Tropical species transferred out of the equatorial zone may retain aseasonality, experience a wide breeding season or a complete reverse-phase seasonality (reviewed in [5]). These differences in seasonal breeding combined with the scant reproduction knowledge for some deer species adds an extra layer of difficulty when applying reproductive technologies to a new deer species. As reproductive knowledge increases about a particular deer species, and experience using in vitro technology is obtained, the success rate of in vitro embryo production increases. Blastocyst stage red (*Cervus elaphus scoticus*, [6], wapiti *Cervus elaphus Canadensis* [7], and sika (*Cervus nippon nippon* [8, 9], *Cervus nippon pseudaxis* [10]) deer embryos are produced by different in vitro embryo production systems, and live calves have been born after the subsequent transfer of these blastocysts into synchronized recipients.

Each step of the embryo production process is critical for success. The transportation time and temperature of ovaries or ova to the laboratory is critical for successful oocyte maturation and further embryo development, and this is species dependent [11]. Ovaries or ova recovered from tropical deer species may experience heat stress conditions, whereas oocytes collected from temperate deer species will require systems to prevent chilling. Ovary or ovum collection, which may be remote from the laboratory, will require different transport temperatures dependent upon the storage time. Previous studies have shown that temperature and length of time for ovarian storage affect oocyte nuclear maturation [11, 12]. A short transport time of less than 6 h require the ovaries to be kept in warm, 30 °C, saline (0.9%) or PBS containing antibiotics. Longer storage time requires a lower temperature, and 20–25 °C is the preferred temperature for 6–12 h [11, 12]. Prolonged storage greater than 12 h requires yet lower temperatures between 10 and 15 °C [12].

The timing of in vivo nuclear oocyte maturation [13] and the in vitro maturation (IVM) of immature oocytes are similar, and 20–24 h is recommended [14, 15]. IVM requirements for deer species is based upon domestic ruminant oocyte requirements and is an area that requires further development. Several studies have identified FSH [16] and estradiol [17] as essential requirements for deer IVM. The addition of EGF [16] and insulin, transferrin, and selenium [18] to maturation medium may promote nuclear maturation. Nuclear maturation rates range from 61 to 78% for red deer [8, 14, 15], 75% for white-tailed deer [17], 51–80% for Sika deer [9, 12], and 30–57% for brow-antlered deer or Eld's deer [19]. In addition, nuclear maturation rates of 58–69% were obtained from the postmortem collection (4–16 h after death) of oocytes from ovaries (held at 37 °C D-PBS, within 1 h after collection) for spotted deer, swamp deer, sambar deer, and brow-antlered deer

[18]. The IVM of deer oocytes requires further research, although this is dependent upon an in vitro culture method that can routinely produce blastocyst stage embryos for the assessment of changes in oocyte competence due changes in the IVM medium.

To obtain sperm for IVF, semen is collected from mature stags. Three methods of sperm collection have been described for cervids; natural service into an artificial vagina [20], electroejaculation [21], and postmortem epididymal sperm recovery [22]. Frozen-thawed epididymal and electroejaculated sperm have been used for IVF [11, 23]. Collection should occur during the breeding season to obtain good-quality semen; November to January in white-tailed deer [20], April in red deer [21], November to December in sika deer [9], and February to April in brow-antlered deer [19]. Sheep serum, (20%), either collection at estrus or 1.5 days postestrus appears to be the key capacitating agent for achieving fertilization for various deer species [16, 24]. The optimal sperm concentration of 0.4×10^6 sperm/mL provided high fertilization rate (67%) and reduced polyspermic fertilization in red deer (15%, [24]). Polyspermic fertilization accounted for 22% of sika deer oocytes fertilized [16], but has not been reported in other studies or deer species. The requirement for heparin, the standard bovine capacitating agent, appears to be deer species specific [23]. Wapiti sperm capacitates with heparin using standard bovine fertilization methods [7], Berg unpublished) whereas red deer sperm requires elevated calcium levels (3.5–4.0 mM) in addition to heparin to obtain comparable fertilization rates [23]. It is recommended that IVF medium, sperm capacitation methods, and sperm concentrations are trialed for each new deer species.

In vitro embryo culture (IVC) of deer presumptive zygotes to the blastocyst stage is highly variable, not only among deer species, but within species, and within the same study [10, 19, 23]. Embryonic arrest or 8- to 16-cell in vitro developmental arrest occurred when standard bovine in vitro embryo culture was used [8, 14, 16, 25, 26], and was the impetus to develop deer in vitro embryo culture. Embryos co-cultured with cervine or ovine oviduct epithelial monolayer have successfully overcome the 8–16-cell developmental arrest with calves born after embryo transfer [8, 10, 27]. However, co-culture and/or the addition of fetal calf serum during the in vitro culture of domestic ruminants is associated with faster embryo development, increased lipid like inclusions and less compaction [28]. In addition, the embryos result in calves or lambs with increased birth weight and a longer gestation period when compared with embryos cultured Synthetic Oviduct Fluid (SOF) supplemented with amino acids and bovine serum albumin (SOFaaBSA; [28–31]). Thus far, increased birth weight and longer gestation lengths have not been reported for deer embryos cultured on oviduct monolayers [9, 10]. It may be that deer embryos are not affected by in vitro culture conditions to the same degree as cattle and sheep embryos or alternatively, too few calves/fawns have been born from in vitro co-cultured embryos for large offspring syndrome to be detected. SOF is a semi-defined in vitro culture medium based upon composition of ovine oviduct fluid [32], and, when supplemented with bovine serum albumin (BSA), is the in vitro culture medium used to modulate the effects of embryo culture on pregnancy, perinatal health, and large offspring syndrome, in cattle [30, 33] and sheep [31, 34]. By changing the composition of the medium, embryo requirements can be identified more readily than when confounded by co-culture of other types of cells. Analysis of deer oviduct fluid identified differences in composition compared to published compositions of oviductal fluid from cattle and sheep [23]. Increased Ca^{2+} concentrations throughout the estrous cycle, the appearance of H_2PO_4 on days 6 and 7 after estrus, and a lower glucose concentration was observed [23]. A fertilization and embryo culture medium was formulated, Deer Synthetic Oviduct Fluid (DSOF), using the biphasic embryo culture approach. The fertilization medium, IVF-DSOF increased the capacitating ability of heparin, eliminated the need for the addition of sheep serum and the early culture medium, eDSOF overcame the eight-cell developmental block. Red deer calves, calves derived from somatic cell nuclear transfer and an Eld's deer fawn have been produced using DSOF [6, 19, 35, 36].

Deer IVC consists of two types of culture: the use of oviduct monolayers and a semi-defined system based upon the ionic and carbohydrate composition of red deer oviduct fluid during the first 8 days post estrus. There is a need to refine in vitro culture medium and in vitro embryo production system based upon the specific physiology of deer oocytes and embryos. However, the limited number of deer, abattoir ovaries, and oocytes, and the lower fertilization and embryo development rate make it difficult to design experiments with the statistical power to test in vitro culture components to increase the efficiencies of in vitro embryo production. The rapid development of technologies such as three dimensional cell cultures, coupled with microfluidics would biomimic the secretory milieu of the oviduct which could improve oviduct co-culture to improve sperm capacitation, IVF and early embryo development [37]. Single-cell metabolomics will enable multiple metabolic measures from a single embryo [38]. This technique can be used to test the effect of culture components on oocyte and embryo metabolism to further refine IVM and in vitro culture specifically for each deer species of interest.

Here we describe the in vitro methods of maturation, fertilization and embryo culture for deer species. The fertilization and culture media formulations are based on the analysis of oviduct fluid collected from cannulated red deer oviducts during the first 8 days after estrus onset [23].

2 Materials

Prepare all solutions using "ultrapure" water (18 MΩ-cm) at 25 °C. All chemicals used to prepare the stocks and media are cell culture tested.

2.1 DSOF Stock Preparation

Make up to volume using a volumetric flask. Sterilize stocks using a 0.22 μm filter. We use sterile stocks stored in 25–100 mL flasks at 4 °C for up to 3 months (unless indicated). Stock preparation and concentrations used for IVF-DSOF, HEPES-buffered DSOF, E-DSOF, and L-DSOF are presented in Table 1. Sperm additive concentrations and preparation are listed in Table 2. The formulation and preparation of in vitro maturation stocks and media are listed in Table 3.

Table 1
Stock composition and concentration

Name of stock	Components	Grams	Final volume (mL) of ultrapure water
DSOF stock 10×	NaCl KCl $MgSO_4{\cdot}7H_2O$ Kanamycin sulfate	5.84 0.373 0.17 0.04	100
KCl 10× (10 mM)	KCl	0.07	100
Ca-lactate10× (15 mM)	Ca L-lactate hydrate	0.33	100
Na-lactate10× (330 mM)	D/L sodium lactate 60% syrup	1.18 mL	25
Phosphate10× (1.5 mM)	NaH_2PO_4	0.05	25
Sulfate 10× (10 mM)	$NaSO_4$	0.36	25
Glucose (4 mM)	Glucose	0.02	25
$CaCl_2$ 100× (171 mM)	$CaCl_2{\cdot}2H_2O$	0.63	25
Pyruvate 100× (33 mM)	Pyruvic acid sodium salt Store at −80 °C in 1 mL aliquots	0.09	25
$NaHCO_3$ 10× (250 mM)	$NaHCO_3$ 1 month shelf life at 4 °C	2.101	100
HEPES 10× (250 mM)	Hepes, Na Salt Hepes, free acid 1 month shelf life at 4 °C	3.25 3.00	100
Amino acids	MEM NEAAs (100×) BME AAs (50×) GlutaMAX-1 (100× Gibco 35050) Aliquot into 10 mL lots and store at −20 °C		

Table 2
Sperm stock additives and Percoll gradient

Name of stock	Components	Grams	Final volume (mL) of ultrapure water
Heparin stock	Heparin (180 USP)	0.002	2 in sterile 0.9% NaCl
	Do Not Filter		
	Store at −20 °C in 30 μL aliquots		
Sperm additive	Penicillamine (2 mM)	0.001	10 in 0.9% NaCl
	Hypotaurine stock (1 mM)	0.003	
	Store at −20 °C in 150 μL aliquots		
Sperm pyruvate	Na pyruvate	1.01	10 in 0.9% NaCl
Percoll stock (SPTL, 10×)	NaCl	4.675	100
	KCl	0.238	
	$NaH_2PO_4 \times H_2$	0.040	
	Hepes Na Salt	1.302	
	Hepes free acid	1.192	
	Penicillin	0.060	
	Sodium lactate 60% syrup	3.68 mL	
Percoll stock (SPAD 100×)	$CaCl_2{\cdot}2H_2O$	0.308	10
	$MgCl_2{\cdot}7H_2O$	0.081	
1× SPTL	10× SPTL	2.5	20
	100× SPAD	0.2	
	Ultrapure water	17.3	
90% Percoll (280–300 mOsm)	Percoll	4.5	
	10× SPTL	0.525	
	100× SPAD	0.050	
	Stock B	0.050	
45% Percoll	1× SPTL	2	
	90% Percoll	2	

2.2 Media Preparation Using Stock Solutions

Table 4 lists the volume of stocks to add to prepare culture media whose final composition are provided in Table 5. All media are prepared in a volumetric flask. Approximately half of the volume of ultrapure water is added to the volumetric flask before adding the stocks. After the addition of the stocks, add ultrapure water to final volume. Pour the medium into a beaker, mix thoroughly before measuring the osmolality. Media should be within the listed osmolality range before adding the BSA. Add the BSA and allow the BSA to dissolve in the solution. Gently mix on a stir plate before sterilizing through a 0.22 μm filter.

Table 3
Aspiration and maturation media

Stock	Components	Grams (g) or mL	Final volume or volume of ultrapure water (mL)
TCM 199 stock (2×)[a]	TCM 199 Kanamycin	1 sachet 0.050 g	500
H199 Hepes buffered 199 (270–280 mOsm)	2× TCM 199 stock Stock H Stock B Ultrapure water	50 6 2 42	100
B199 Bicarbonate buffered 199 (270–280 mOsm)	2× TCM 199 stock Stock C Stock B Ultrapure water	50 0.6 10 39.4	100
Cysteamine (10 mM)[b]	2-Mercaptoethylamine–HCl	0.025 g	22 of B199+ FCS
$E_2 17\beta$ (1 mg/mL)[c]	$E_2 17\ \beta$	0.025 g	25 in absolute ethanol
Stock FSH (4 μg/μL)[d]	1 vial of Folltropin (porcine FSH) (400 mg)	10 B199+ 10% FCS	100 of B199+FCS
HCG[f]	Chorionic gonadotrophin 5,000 i.u.)		1 mL M199
Aspiration medium	H199 stock FCS Heparin[e]	90 10 0.185	100
Maturation medium (IVM)	B199 FCS Stock FSH Stock HCG Stock $E_2 17\beta$ Stock Cysteamine	9 1 0.025 0.100 0.010 0.010	10
TCM 199 (2×) for oviduct epithelial cell culture	TCM 199 Gentamycin	1 sachet 0.040 g	500
TCM 199 for oviduct epithelial culture	2× TCM 199 stock Stock C Stock B Ultrapure water	50 0.6 10 39.4	100
Oviduct epithelial cell wash and culture medium	TCM 199 for oviduct epithelial culture FCS	90 10	100

[a]Store filtered at 4 °C. Lasts for 2 weeks
[b]Filter and aliquot 100 μL into Eppendorf tubes. Store at −20 °C
[c]Store at −20 °C
[d]Dissolve contents of vial in 10 mL sterile B199+ 10%FCS. Wash vial with two aliquots of 10 mL B199+ 10% FSC. Make up to final volume of 100 mL. Aliquot into sterile tubes of 1 mL each. Store at −80 °C
[e]Sodium heparin injection 5000 IU/mL
[f]Aliquot 100 μL/tube and store at −80 °C. Final concentration is 50 i.u./mL.

Table 4
DSOF-IVF and culture media formulation using stocks

Stock	H-DSOF volume (mL) of stock for 50 mL	IVF-DSOF volume (mL) of stock for 50 mL	E-DSOF volume (mL) of stock for 20 mL	L-DSOF volume (mL) of stock for 20 mL
DSOF	5	5	2	2
Pyruvate	0.5	0.5	0.2	0.2
KCl	5	5	0	0
MEM NEAAs	0.5	0.5	0.2	0.2
BME AAs	1	1	0.4	0.4
L-alanyl-L-glutamine dipeptide (GlutaMAX)	0.5	0.5	0.2	0.2
Sulfate	0	0	2	2
Na-lactate	0	0	0	0.1
Glucose	0	0	0.5	1
Phosphate	0	0	0	2
$CaCl_2$	0	0.6	0.18	0.18
Ca-lactate	5	5	2	0
HEPES	4	0	0	0
$NaHCO_3$	1	5	2	2
Ultrapure water to final volume	27.5	26.9	10.3	10.3
Fatty acid-free BSA (grams)	0.3	0.05	0.2	0.2
Osmolarity (mOsm)	270–280	270–280	275–285	275–285
Heparin stock[a]		10 μL		
Sperm additive Stock[a]		750 μL		
Sperm pyruvate[a]		50 μL		

Media lasts 3 weeks stored at 4 °C
[a]Add heparin, sperm additives, and pyruvate just before making fertilization drops

3 Methods

3.1 Oocyte Collection from Abattoir/Laparoscopic Aspiration

1. Ovaries collected at slaughter are transported to the lab in normal saline (0.9% NaCl) supplemented with 0.1 mg/mL penicillin and 0.1 mg/mL streptomycin. Transport temperature is dependent upon transport time to the lab: 30 °C for 1–6 h and 20–25 °C for ≥6–12 h.

Table 5
DSOF and SOFaaBSA composition

Stock	H-DSOF (mM)[a]	IVF-DSOF (mM)	E-DSOF (mM)	L-DSOF (mM)	SOFaaBSA Oviduct epithelial cell culture (mM)
NaCl	100	100	100	100	107.7
Pyruvate	0.33	0.33	0.33	0.3	0.3
KCl	6	6	5	5	7.2
Mg-sulfate	0.6	0.6	0.6	0.7	0.7
Glucose	0	0	0.1	0.2	0
KH_2 phosphate	0	0	0	0	1.5
NaH_2 phosphate	0	0	0	1.5	0
$CaCl_2$		2.05	1.5	1.5	1.7
Ca-hemilactate	1.5 3.0	1.5 3.0	1.5 3.0	0 0	0 0
Na-lactate	0	0	0	3.3	3.6
HEPES	20	0	0	0	0
$NaHCO_3$	5	25	25	25	25
MEM NEAAs in 10 mL	0.1	0.1	0.1	0.1	0.1
BME AAs in 10 mL	0.05	0.05	0.05	0.05	0
L-alanyl-L-glutamine dipeptide (GlutaMAX)	2	2	2	2	0
L-Glutamine	0	0	0	0	2
Fatty acid-free BSA (mg/mL)	6	1	10	10	3
Fetal calf serum[a]					10%

[a]Add fetal calf serum for culture of cells. BSA medium only for cell wash

2. Ovaries are washed twice with 37 °C normal saline before aspiration.
3. Follicles of 2–7 mm in diameter are aspirated using a 20-gauge needle connected to a 5 mL syringe with negative pressure or 32 mmHg vacuum pressure. Aspirate is placed into a 15 mL conical tube containing aspiration medium held at 35 °C.
4. Laparoscopic oocyte aspiration: follicles of 2–7 mm in diameter are aspirated using 20-gauge needle with negative pressure

(32–35 mmHg) into a 15 mL conical tube containing aspiration medium and maintained at 38.5 °C.

5. The sediment is aspirated and placed into a 90-mm plastic petri dish containing 10 mL of H199+ 10% FCS.
6. Cumulus-oocyte complexes (COCs) are graded. Only oocytes with a homogenous or slightly mosaic cytoplasm and at least three layers of compact cumulus layers (grade A and B) are selected for IVM.
7. Wash COCs twice in holding medium and once in IVM medium before transferring seven to ten COCs into the IVM drop.

3.2 In Vitro Maturation (IVM)

1. Maturation drops (40 μL) under oil (8 mL) need at least 2 h of equilibration time before adding the oocytes. Two hours is the minimum when using gas and temperature equilibrated oil.
2. IVM drop preparation: pipet 40 μL drops of IVM medium in a 60-mm petri dish (*see* **Note 1**). The IVM drop number will vary depending upon the number of oocytes expected.
3. Overlay the drops with 8 mL of mineral oil stored in a light proof bottle under 5% CO_2 at 38.5 °C. Alternatively equilibrate oil overnight before use.
4. Seven to ten COCs are transferred in a volume of 10 μL and placed into the IVM drop. Final volume of maturation drops is 50 μL.
5. Culture COCs at 38.5 °C in 5% CO_2 for 22–24 h of maturation. Eld's deer oocytes are matured under low oxygen concentrations 5% O_2, 5%CO_2, and 90% N_2.

3.3 Sperm Preparation

1. Thaw frozen semen at 37 °C for 30 s. Semen is microscopically evaluated for motility and concentration. Only semen with superior motility (≥70%) will be selected for IVF.
2. Percoll gradient method: motile sperm from each stag is individually separated on a 45 and 90% Percoll gradient (2 mL of each) centrifuged at 700 × *g* for 20 min (Table 2). Aspirate the sperm pellet, transfer to a new conical 15 mL tube and gradually add 1 mL of H-DOF while mixing gently. Centrifuge at 200 × *g* for 5 min.
3. Swim-up method: the semen sample is centrifuged at 700 × *g* for 5 min and the supernatant discarded. The pellet is resuspended in 1 mL prewarmed H-DSOF and centrifuged at 700 × *g* for 5 min. Discard the supernatant and gently overlay the pellet with 500 μL IVF medium. Close the tube and incubate at 37 °C under 5% CO_2 in air. After 30 min, collect the supernatant, which will contain motile sperm for IVF. Adjust sperm concentration to reach a final sperm concentration of

$1–2 \times 10^6$ sperm/mL in IVF medium. The sperm motility is re-evaluated after dilution.

4. Add 10 μL diluted sperm into the IVF drops.

3.4 In Vitro Fertilization (IVF)

1. IVF is conducted under low O_2 concentrations, 5%O_2, 5% CO_2, and 90% N_2.
2. IVF drop preparation: pipet 12 × 30 μL drops of IVF medium in a 60-mm petri dish and cover with 8 mL equilibrated oil. Equilibrate for a minimum of 2 h at 38.5 °C under 5%O_2, 5% CO_2, and 90% N_2.
3. COCs are removed from the IVM drops and transferred to a 35-mm petri dish containing H-DSOF, washed twice in H-DSOF, and once in IVF medium before placing into the IVF drops.
4. Five oocytes in 10 μL IVF medium are placed in each 30 μL droplet of IVF medium.
5. Add 10 μL of diluted sperm (containing 0.2×10^5 sperm) into the IVF drops (final sperm concentration 0.4×10^6 sperm/mL). Final sperm concentrations will be different for each males and possibly, different species of deer.
6. Return the IVF plates to the incubator at 38.5 °C with 5% CO_2, 5% O_2, and 90% N_2, and culture for 18 h. The day of IVF is Day 0.
7. Eighteen hours after IVF, cumulus cells and dead sperm cells are stripped off the presumptive zygotes before transfer into IVC medium. Mechanical stripping using a fine-bore pipette is used for a small number of presumptive zygotes. For larger numbers of zygotes, we use the vortex method modified from the cattle method. Approximately 150 zygotes are placed into 350 μL of 0.1% hyaluronidase in H-DSOF. The tube containing the zygotes is vortexed for 2–3 min and centrifuged briefly to bring the zygotes to the bottom of the tube. The contents of the tube are placed into a 3 mL of H-DSOF and the zygotes are washed twice in new H-DSOF and placed into in vitro culture.

3.5 In Vitro Culture (IVC)

3.5.1 Embryo Culture in Semidefined Medium

1. Preparation of IVC plates: pipet 5 × 20 μL drops of IVC medium (E-DSOF and L-DSOF composition in Table 2) in a 30-mm petri dish and cover with 3 mL of equilibrated oil. Equilibrate for a minimum of 2 h at 38.5 °C in 5% CO_2, 5% O_2, and 90% N_2.
2. Presumptive zygotes are cultured in vitro in E-DSOF for 4 days.
3. Embryos to be transferred into oviducts of synchronized recipients are cultured for approximately 24 h in E-DSOF. These embryos are visually assessed 32 h post-insemination (hpi) and

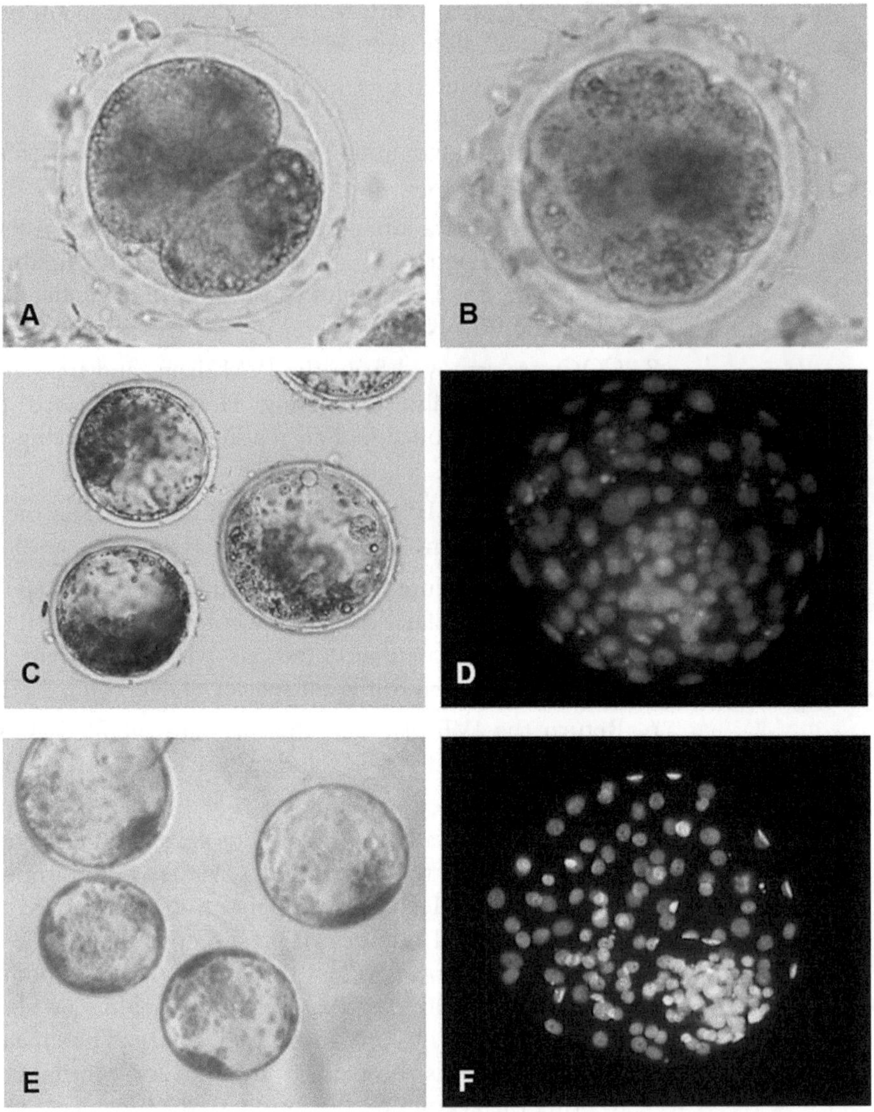

Fig. 1 Thamin Eld's deer embryos culture in DSOF. (**a**) 2-cell, (**b**) 10-cell, (**c**) Blastocyst. (**d**) C blastocyst containing 104 cells stained with Hoechst 33342. (**e**, **f**) Red deer blastocyst culture in DSOF. (**e**) Day 7 expanded blastocysts. (**f**) E blastocyst containing 152 cells stained with Hoechst 33342

again on the morning of transfer, approximately 43 hpi [15, 19] (Fig. 1a, b). Thamin Eld's deer fawn born from oviduct transfer (Fig. 2a).

4. Visual assessment of embryo development is carried out every 24 h for presumptive zygotes cultured to day 7 and/or 8. Once culture conditions have been established, visual assessment of development is limited to the time when embryos are moved from E-DSOF to L-DSOF.

Fig. 2 Eld's deer fawn and three red deer calves from embryos cultured in DSOF. (**a**) Eld's deer fawn produced from in vitro-matured and fertilized oocytes cultured in DSOF. (**b**) Three red deer calves produced by somatic cell nuclear transfer

5. Non-cleaved oocytes are fixed to assess nuclear maturation (*see* **Note 2**).
6. For long-term culture to the blastocyst stage, embryos are transferred into fresh L-DSOF drops on day 4 or 5 and cultured to the blastocyst stage (Day 7 or 8) (Fig. 1).
7. Blastocyst grade 1 and 2 can be frozen and stored for later transfer [36, 39] or fresh transfers into synchronized recipients [35, 40]. Red deer calves born from fresh blastocyst transfer into synchronized red deer recipients (Fig. 2b).

3.5.2 Embryo Culture Using Oviduct Epithelial Cell Coculture

Deer embryos can also be cultured using sheep oviduct coculture with the birth of red deer and sika deer calves [8–10]

1. Presumptive zygotes are washed in SOFaaBSA (3 mg/mL) and cultured for 8 days in SOFaaBSA supplemented with 10% FCS in ovine oviduct epithelial cells monolayer (oOEC) at 38.5 °C in a humidified atmosphere of 5% CO_2 in air.
2. To prepare the oOEC, ovine oviducts ipsilateral to ovary containing ovulations are collected. After several washings of oviduct, the mucosa is expelled by gentle squeezing using a sterile microscope slide on petri dish.
3. Epithelium fragments are washed three times in TCM 199 supplemented with 10% FCS and 40 μg/mL gentamycin and cultured in four-well plates in a final volume of 500 μL.
4. Fragments spontaneously form vesicles after 2–3 days of culture and will attached to the bottom of the plate after a few

days. Within 7–9 days after initiation of culture, monolayers are formed.

5. Medium is renewed every 48 h and replaced 24 h before culture of zygotes by SOFaaBSA supplemented with 10% FCS. Half of the culture medium is renewed every 48 h during in vitro development.

3.6 Embryo Evaluation

1. On Day 2, all noncleaved oocytes will be fixed and stained to determine nuclear status under a microscope equipped with epifluorescence (*see* **Note 2**).
2. For blastocyst cell number evaluation, blastocysts are fixed, stained, and mounted on slides with glycerol, and evaluated for total cell numbers (*see* **Note 3**).

4 Notes

1. Plastic tissue culture plates

   Untreated plastic petri dishes prevent the drops from collapsing when adding the oil. Do not use tissue culture treated plastic petri dishes. The culture drops will spread and flatten.
2. Oocyte staining for nuclear maturation evaluation
   (a) Oocytes are washed in PBS and fixed in ethanol–acetic acid (3:1 v/v) for 48 h at room temperature (20–25 °C) and stained with aceto-orcein (lacmoid, 2.2% glacial acetic acid) for approximately 10 min and rinse excess stain with 45% acetic acid. Aceto-orcein stains the sperm and the sperm tail in the egg cytoplasm. This staining is best for evaluating polyspermic fertilization.
   (b) Stained oocytes are examined using phase-contrast microscopy at 200× and 400× magnification.
   (c) Alternative staining is to fix oocyte with 4% paraformaldehyde for 20 min at RT or overnight at 4 °C, stain using 10 μg/mL Hoechst 33342 (Stock solution 1 mg/mL in MQ water) for 10 min in the dark. Excess stain is removed by washing with PBS and mounted on slides with using antifade product.
   (d) Nuclear status is identified under a microscope equipped with epifluorescence 350 excitation.
   (e) Oocytes are examined for nuclear maturation stages: germinal vesicle (GV), germinal vesicle breakdown (GVBD), metaphase I (MI), anaphase I (AI), telophase I (TI), and metaphase II (MII). Other categories are degeneration and parthenogenic cleavage (in case before IVF).
3. Blastocyst cell staining

(a) Blastocysts are washed in PBS twice and fixed with 4% paraformaldehyde for 20 min at RT overnight at 4 °C.

(b) Blastocysts are stained using 10 μg/mL Hoechst 33342 (Stock solution 1 mg/mL in MQ water) and stained for 20 min at RT in the dark. Excess stain is removed by washing with PBS and mounted on slides with glycerol.

(c) Cell numbers are counted under a microscope equipped with epifluorescence 350 excitation.

References

1. Loskutoff NM, Bartels P, Meintjes M, Godke RA, Schiewe MC (1995) Assisted reproductive technology in nondomestic ungulates: a model approach to preserving and managing genetic diversity. Theriogenology 43(1):3–12
2. Pukazhenthi B, Comizzoli P, Travis AJ, Wildt DE (2005) Applications of emerging technologies to the study and conservation of threatened and endangered species. Reprod Fertil Dev 18(2):77–90
3. Garde JJ, Martínez-Pastor F, Gomendio M, Malo AF, Soler AJ, Fernández-Santos MR, Esteso MC, García AJ, Anel L, Roldán ERS (2006) The application of reproductive technologies to natural populations of red deer. Reprod Domest Anim 41:93–102
4. Loudon ASI, Brinklow BR (1992) Reproduction in deer: adaptations for life in seasonal environments. In: Brown RD (ed) The biology of deer. Springer, New York, NY, pp 261–278
5. Asher GW, Fisher MW, McLeod BJ, Berg DK (1994) Reproductive physiology of cervids: a review. In: Wilson PR (ed) Deer course for veterinarians. Deer Branch of the NZVA, Queenstown, pp 1–21
6. Berg DK, Li C, Asher G, Wells DN, Oback B (2007) Red deer coned from antler stem cells and their differentiated progeny. Biol Reprod 77(3):384–394
7. Pollard JW, Bringans MJ, Buckrell B (1995) *In vitro* production of wapiti and red deer (*Cervus elaphus*) embryos. Theriogenology 43(1):305
8. Locatelli Y, Cognié Y, Vallet JC, Baril G, Verdier M, Poulin N, Legendre X, Mermillod P (2005) Successful use of oviduct epithelial cell coculture for *in vitro* production of viable red deer (*Cervus elaphus*) embryos. Theriogenology 64(8):1729–1739
9. Locatelli Y, Vallet JC, Huyghe FP, Cognie Y, Legendre X, Mermillod P (2006) Laparoscopic ovum pick-up and *in vitro* production of sika deer embryos: effect of season and culture conditions. Theriogenology 66(5):1334–1342
10. Locatelli Y, Hendriks A, Vallet JC, Baril G, Duffard N, Bon N, Ortiz K, Scala C, Maurel MC, Mermillod P, Legendre X (2012) Assessment LOPU-IVF in Japanese sika deer (*Cervus nippon nippon*) and application to Vietnamese sika deer (*Cervus nippon pseudaxis*) a related subspecies threatened with extinction. Theriogenology 78(9):2039–2049
11. García-Álvarez O, Maroto-Morales A, Berlinguer F, Fernández-Santos MR, Esteso MC, Mermillod P, Ortiz JA, Ramon M, Pérez-Guzmán MD, Garde JJ, Soler AJ (2011) Effect of storage temperature during transport of ovaries on *in vitro* embryo production in Iberian red deer (*Cervus elaphus hispanicus*). Theriogenology 75(1):65–72
12. Tulake K, Yanagawa Y, Takahashi Y, Katagiri S, Higaki S, Koyama K, Wang X, Li H (2014) Effects of ovarian storage condition on *in vitro* maturation of Hokkaido sika deer (*Cervus nippon yesoensis*) oocytes. Jpn J Vet Res 62 (4):187–192
13. Berg DK, Thompson JG, Peterson AJ, Asher GW (2008) The temporal relationship between oocyte maturation and early fertilisation events in relation to the pre-ovulatory LH peak and preimplantation embryo development in red deer (*Cervus elaphus*). Anim Reprod Sci 105 (3–4):332–343
14. Fukui Y, McGowan LT, James RW, Asher GW, Tervit HR (1991) Effects of culture duration and time of gonadotropin addition on *in vitro* maturation and fertilization of red deer (*Cervus elaphus*) oocytes. Theriogenology 35 (3):499–512
15. Berg DK, Thompson JG, Asher GW (2002) Development of *in vitro* embryo production systems for red deer (*Cervus elaphus*). 2. The timing of in vitro nuclear maturation. Anim Reprod Sci 70:77–84
16. Comizzoli P, Mermillod P, Cognie Y, Chai N, Legendre X, Mauget R (2001) Successful *in vitro* production of embryos in the red deer

(*Cervus elaphus*) and the sika deer (*Cervus nippon*). Theriogenology 55:649–659

17. Siriaroonrat B, Comizzoli P, Songsasen N, Monfort SL, Wildt DE, Pukazhenthi BS (2010) Oocyte quality and estradiol supplementation affect *in vitro* maturation success in the white-tailed deer (*Odocoileus virginianus*). Theriogenology 73(1):112–119
18. Brahmasani SR, Yelisetti UM, Katari V, Komjeti S, Lakshmikantan U, Pawar RM, Sisinthy S (2013) Developmental ability after parthenogenetic activation of *in vitro* matured oocytes collected postmortem from deers. Small Ruminant Res 113(1):128–135
19. Thongphakdee A, Berg DK, Tharasanit T, Thongtip N, Tipkantha W, Punkong C, Tongthainan D, Noimoon S, Maikeaw U, Kajornklin N, Siriaroonrat B, Comizzoli P, Kamolnorranath S (2017) The impact of ovarian stimulation protocol on oocyte quality, subsequent in vitro embryo development, and pregnancy after transfer to recipients in Eld's deer (*Rucervus eldii thamin*). Theriogenology 91:134–144
20. Plotka ED, Seal US, Schmoller GC, Karns PD, Keenlyne KD (1977) Reproductive steriods in the white-tailed deer (*Odocoileus virginianus borealis*). 1. Seasonal changes in the female. Biol Reprod 16(3):340–343
21. Asher GW, Berg DK, Evans G (2000) Storage of semen and artificial insemination in deer. Anim Reprod Sci 62(1–3):195–211
22. Garde JJ, Ortiz N, Garcia A, Gallego L, Landete-Castillejos T, Lopes A (1998) Postmortem assessment of sperm characteristics of the red deer during the breeding season. Syst Biol Reprod Med 41:195–202
23. Berg DK, Asher GW (2003) New developments in reproductive technologies in deer. Theriogenology 59(1):189–205
24. Berg DK, Thompson JG, Asher GW (2002) Development of *in vitro* embryo production systems for red deer (*Cervus elaphus*). 3. *In vitro* fertilisation using sheep serum as a capacitating agent and the subsequent birth of calves. Anim Reprod Sci 70:85–98
25. Krogenaes A, Ropstad E, Thomassen R, Pedersen O, Forsberg M (1994) *In vitro* maturation and fertilization of oocytes from Norwegian semi-domestic reindeer (*Rangifer tarandus*). Theriogenology 41:371–377
26. Bainbridge DRJ, Catt SL, Evans G, Jabbour HN (1999) Successful in vitro fertilization of *in vivo* matured oocytes aspirated laparoscopicaly from red deer hinds (*Cervus elaphus*). Theriogenology 51:891–898
27. Berg DK, Asher GW, Pugh PA, Tervit HR, Thompson JG (1995) Pregnancies following the transfer of *in vitro* matured and fertilized red deer (*Cervus elaphus*) oocytes. Theriogenology 43(1):166
28. Thompson JG (1997) Comparison between *in vivo*-derived and *in vitro*-produced pre-elongation embryos from domestic ruminants. Reprod Fertil Dev 9(3):341–354
29. Breukelman SP, Reinders JMC, Jonker FH, de Ruigh L, Kaal LMTE, van Wagtendonk-de Leeuw AM, Vos PLAM, Dieleman SJ, Beckers JF, Perényi Z, Taverne MAM (2004) Fetometry and fetal heart rates between Day 35 and 108 in bovine pregnancies resulting from transfer of either MOET, IVP-co-culture or IVP-SOF embryos. Theriogenology 61 (5):867–882
30. van Wagtendonk-de Leeuw AM, Mullaart E, de Roos APW, Merton JS, den Daas JHG, Kemp B, de Ruigh L (2000) Effects of different reproduction techniques: AI, MOET or IVP, on health and welfare of bovine offspring. Theriogenology 53(2):575–597
31. Thompson J, Gardner D, Pugh P, McMillan W, Tervit H (1995) Lamb birth weight is affected by culture system utilized during in vitro pre-elongation development of ovine embryos. Biol Reprod 53(6):1385–1391
32. Tervit HR, Whittingham DG, Rowson LEA (1972) Successful culture *in vitro* of sheep and cattle ova. J Reprod Fertil 30:493–497
33. Farin PW, Crosier AE, Farin CE (2001) Influence of *in vitro* systems on embryo survival and fetal development in cattle. Theriogenology 55:151–171
34. Rooke JA, McEvoy TG, Ashworth CJ, Robinson JJ, Wilmut I, Young LE, Sinclair KD (2007) Ovine fetal development is more sensitive to perturbation by the presence of serum in embryo culture before rather than after compaction. Theriogenology 67(3):639–647
35. Berg DK, Beaumont SE, Berg MC, Asher GW (2004) Red deer (*Cervus elaphus*) calves born from *in vitro* produced blatocysts fertilised and cultured in deer synthetic oviduct fluid. Reprod Fertil Dev 16:201
36. Beaumont SE, Berg MC, Strongman K, Saywell DP, Berg DK (2005) Direct-thaw transcervical transfer of red deer frozen *in vitro* blastocyst can result in pregnancies. Reprod Fertil Dev 17:188
37. Ferraz MAMM, Henning HHW, Stout TAE, Vos PLAM, Gadella BM (2017) Designing 3-dimensional in vitro oviduct culture systems to study mammalian fertilization and embryo

production. Ann Biomed Eng 45 (7):1731–1744

38. Fessenden M (2016) Metabolomics: small molecules, single cells. Nature 540 (7631):153–155

39. Soler JP, Mucci N, Kaiser GG, Aller J, Hunter JW, Dixon TE, Alberio RH (2007) Multiple ovulation and embryo transfer with fresh, frozen and vitrified red deer (*Cervus elaphus*) embryos in Argentina. Anim Reprod Sci 102 (3–4):322–327

40. Berg DK, Thompson JG, Pugh PA, Tervit HR, Asher GW (1995) Successful *in vitro* culture of early cleavage stage embryos recovered from superovulated red deer (*Cervus elaphus*). Theriogenology 44:247–254

Chapter 15

In Vitro Culture of Camelid Embryos

Islam M. Saadeldin, Ayman Abdel-Aziz Swelum, and Abdullah N. Alowaimer

Abstract

The camel is the main meat- and milk-producing animal in the desert environment and is characterized by an induced ovulation pattern, in which ovulation occurs in response to copulation. Little is known about the early embryonic development and placentation in camelid species. Here we describe protocols for the culture of both in vitro-produced and in vivo-retrieved camel embryos. A chemically defined medium enables the development of in vitro-produced embryos from cleavage to the hatching blastocyst stage. In vivo-retrieved embryos will survive in vitro for 23 days postinsemination, reaching a diameter of ~5 mm.

Key words Embryo, In vitro culture, Blastocyst, Defined medium, Camelids

1 Introduction

The Arabian camel (*Camelus dromedarius*) is a unique species that can be a better source of meat and milk in desert areas than cattle, which are severely affected by heat and by scarcity of feed and water [1]. Low reproductive performance of the camel is considered one of the most important factors affecting camel productivity; high fertility of camels is essential both for profitable production and for providing opportunities for selection and genetic improvement [2, 3].

Several factors have been suggested to be associated with reduced camel reproductive efficiency, including seasonality of ovarian activity, long gestation period, and difficulties in detection of the estrous cycle [4, 5]. Additionally, few studies have described follicular growth, oocyte maturation, ovulation induction, fertilization, and subsequent in vitro embryo development [6–9]. Employing in vitro technologies will facilitate the efficient propagation of genetically superior camels. However, refinement of the in vitro

Jason R. Herrick (ed.), *Comparative Embryo Culture: Methods and Protocols*, Methods in Molecular Biology, vol. 2006, https://doi.org/10.1007/978-1-4939-9566-0_15, © Springer Science+Business Media, LLC, part of Springer Nature 2019

embryo production system for oocyte maturation and embryo production is necessary.

Significant losses in camel embryo development occur during in vitro culture (IVC) period, suggesting that the postfertilization environment is critical for development [10–12] due to changes in embryo metabolism [13] and essential transcript expression [14, 15].

Here we describe our recent findings for optimizing the conditions for in vitro culture of camel embryo [12, 16, 17] that would expand the work in assisted reproductive technologies and understanding early embryonic development as well as to optimize the isolation of different embryo-derived stem cell lines in this unique species.

2 Materials

Prepare all media using ultrapure water (prepared by purifying deionized water, to attain a sensitivity of 18 MΩ·cm at 25 °C). Prepare and store all reagents at 4 °C (unless indicated otherwise). Embryo-tested grade mineral oils should be used.

2.1 Preparation of Stock Solutions

1. 0.1 mM Ethylenediaminetetraacetic Acid (EDTA) Stock: Dissolve 3.8 mg of EDTA in 1 ml distilled water. Do not filter. Use the same day as preparation.
2. 20 mM Sodium Pyruvate Stock: Dissolve 22 mg pyruvic acid in 10 ml distilled water. Sterile filter into aluminum foil wrapped 15 ml tube. Store at 4 °C for 2 months.
3. 200 mM L-Glutamine Stock: Dissolve 292 mg L-glutamine in 10 ml saline (0.9% NaCl). Sterile filter into 100 μl aliquots in sterile 0.5 ml microcentrifuge tubes. Store at –20 °C indefinitely.
4. 50 mM Glucose Stock (50 mM): Dissolve 90.1 mg D-glucose in 10 ml distilled water. Sterile filter into 15 ml tube. Store at 4 °C for 2 months.
5. Estradiol Stock Solution (20 μg/ml): Dissolve 1 mg β-estradiol in 1 ml absolute ethanol. Add 49 ml TCM-199 while mixing. Divide aliquots and store at −20 °C indefinitely.
6. EGF Stock Solution: Reconstitute 1 mg EGF in 1 ml phosphate-buffered saline (PBS) containing 0.1% bovine serum albumin (BSA) (w/v) and store in 10 μl aliquots at −20 °C indefinitely.
7. Cysteamine Stock Solution: Prepare 100 mM in ultrapure water. Store 10 μl aliquots at −20 °C indefinitely.

8. Fibroblast Growth Factor 2 Stock Solution (4 μg/ml): Dissolve 100 μg of FGF2 in 25 ml sterile PBS containing 0.1% BSA (w/v). Store 0.5 ml aliquots at −80 °C indefinitely.
9. Ionomycin Stock Solution: Prepare 1 mM in DMSO. Sterile-filter with 0.2 μm syringe filter. Store 5 μl aliquots at −20 °C indefinitely.
10. DMAP Stock Solution: Prepare 100 mM stock solution in PBS store in 8 μl aliquots at −20 °C indefinitely.

2.2 Preparation of Culture Media

1. HEPES-buffered TCM-199: Supplement HEPES-buffered (25 mM HEPES) TCM199 with 2 mM $NaHCO_3$, and 50 μg/ml gentamycin sulfate.
2. TCM199+ FBS: Supplement HEPES-buffered TCM199 with 10% FBS (v/v).
3. *Oocyte in vitro maturation medium*: HEPES-buffered TCM-199 supplemented with 10% (v/v) fetal bovine serum (FBS), 10 μg/ml follicle-stimulating hormone (FSH), 10 μg/ml luteinizing hormone (LH), 1 μg/ml 17β-estradiol, 20 ng/ml epidermal growth factor (EGF), 1 μg/ml insulin, 0.55 μg/ml transferrin, 0.5 ng/ml selenium (ITS), 0.3 μM cysteamine, 1 mM L-glutamine, and 50 μg/ml gentamycin sulfate.
4. *Chemically defined culture medium for in vitro-produced embryos*: Prepare 100 ml according to recipe and notes provided in Table 1. Ingredients added to 100 ml of double distilled H_2O in a volumetric flask. Filter through 0.4 μm filter then 0.2 μm filter into sterile plastic tubes. The fresh medium should be salmon color after preparation and can be stored at 4 °C for 1 week.
5. *Blastocyst medium for in vitro-produced embryos*: Add 0.5 ml FBS to 4.5 ml of the chemically defined medium in sterile tube and mix well. Use it fresh.
6. *Culture medium for in vivo-produced embryos*: Dulbecco's modified Eagle's medium (DMEM) supplemented with 10% FBS, 0.1 mM, β-mercaptoethanol, 1% nonessential amino acids, 2 mM L-glutamine, 10 ng/ml epidermal growth factor, 10 ng/ml fibroblast growth factor (FGF-2), 1 μg/ml insulin, 0.55 μg/ml transferrin, 0.5 ng/ml selenium (ITS), and 1 mg/ml gentamycin.

2.3 Hormones

1. Ovine FSH: Dissolve 18 mg ovine follicle-stimulating hormone (oFSH) in 20 ml saline (0.9% NaCl). Store at 4 °C during the 4 days of treatment course up to 7 days after dissolving. Example for commercial product of oFSH: OVAGEN™ (ICP-bio Ltd., Auckland, New Zealand).

Table 1
Composition of the chemically defined medium used for in vitro culture of camel embryos (to make 100 ml)

Components	Amount	Final conc.
EDTA[a]	100 μl	0.01 mM
NaCl	555.18 mg	95 mM
KCl	18.5 mg	2.5 mM
KH_2PO_4	4.75 mg	0.35 mM
$CaCl_2$-$2H_2O$[b]	25 mg	1.71 mM
$MgCl_2$-$6H_2O$[b]	10 mg	0.2 mM
Glucose	3.6 mg	0.2 mM
$NaHCO_3$	210 mg	25 mM
Myoinositol	48 mg	2.6 mM
PVA	10 mg	0.1 mg/ml
Gentamycin	5 mg	50 μg/ml
EGF	2 μg	20 ng/ml
Na lactate[c]	174 μl	20.49 mM
Na pyruvate	400 μl	0.2 mM
L-glutamine	500 μl	1 mM
ITS 100×	1 ml	1 μg I, 0.55 μg T, 0.5 ng S/ml
BME essential AA 50×	2 ml	1×
MEM nonessential AA 100×	1 ml	1×
Phenol red	100 μl	5 μg/ml

[a]Add first
[b]Dissolve $CaCl_2 \cdot 2H_2O$ and $MgCl_2 \cdot 6H_2O$ in ~15 ml distilled water, then add the total amount to the other ingredients
[c]Sodium DL-lactate solution (Sigma L7900) syrup, suitable for embryo culture, 60% (w/w)

2. Porcine FSH: Dissolve 400 mg porcine follicle-stimulating hormone (pFSH) (1000 IU FSH and 1000 IU LH) in 20 ml saline (0.9% NaCl). Store at 4 °C during the 4 days of treatment course up to 7 days after dissolving. Examples for commercial product of pFSH:
   (a) Folltropin-V: produced by Bioniche Animal Health Canada Inc., Bioniche One Health, distributed by Vétoquinol N.-A. Inc.
   (b) PLUSET: produced by Calier Laboratorios, Calier De, Argentina S.A.
   (c) Stimufol: produced by Reprobiol SPRL, Street of Pel-é-Bois, 20, Belgium.

3 Methods

3.1 In Vitro Culture of In Vitro-Produced Embryos

3.1.1 In Vitro Maturation of Camel Oocytes

1. Transport the camel ovaries from the abattoir to the laboratory in 0.9% NaCl solution at 30–33 °C within 4–6 h.
2. Aspirate the follicular contents from antral follicles (2–8 mm in diameter) using an 18-ga needle attached to a 10 ml disposable syringe.
3. Select the cumulus–oocyte complexes (COCs) with evenly granulated cytoplasm and enclosed by more than three layers of compact cumulus cells.
4. Wash three times with HEPES-buffered tissue culture medium-199 (TCM-199).
5. Incubate 20–25 COCs in 4-well dishes with 500 μl oocyte in vitro maturation medium covered with 500 μl oil at 38.5 °C in an atmosphere of 5% CO_2 in humidified air for 30 h.

3.1.2 Parthenogenetic Activation of Camel Oocytes

1. Remove the cumulus cells from in vitro-matured oocytes by trituration with 0.1% (v/v) hyaluronidase in HEPES-buffered TCM-199.
2. Wash the oocytes twice in HEPES-buffered TCM-199 supplemented with FBS 10%.
3. Incubated oocytes in TCM-199 supplemented with 10% FBS and 5 μM ionomycin for 5 min in a dark chamber.
4. Wash the activated oocytes in TCM-199 supplemented with 10% FBS.
5. Transfer the activated oocytes into 30 μl microdrops of 4 mM 6-dimethylaminopurine (6-DMAP), covered with mineral oil, and incubate for 4 h in a humid atmosphere of 5% CO_2, 5% O_2, and 90% N_2 at 38.5 °C.

3.1.3 Setting Up Microdrops for Embryo Culture

1. Dispense 15 μl drops of defined culture medium for in vitro-produced embryos in an array on the bottom of a noncoated culture dish.
2. Flood the dish with 2 ml paraffin oil.
3. Place the dish in the incubator (38.5 °C in an atmosphere of 5% CO_2, 5% O_2, and 90% N_2) at least 30 min before transferring the embryos to the drops.

3.1.4 Embryo Culture

Embryos can be obtained through parthenogenetic activation, as described above [12, 16], or through in vitro fertilization [18] of in vitro-matured oocytes.

1. Cumulus cells-stripped activated oocytes (Fig. 1a) or fertilized embryos (Fig. 1b) should be selected.

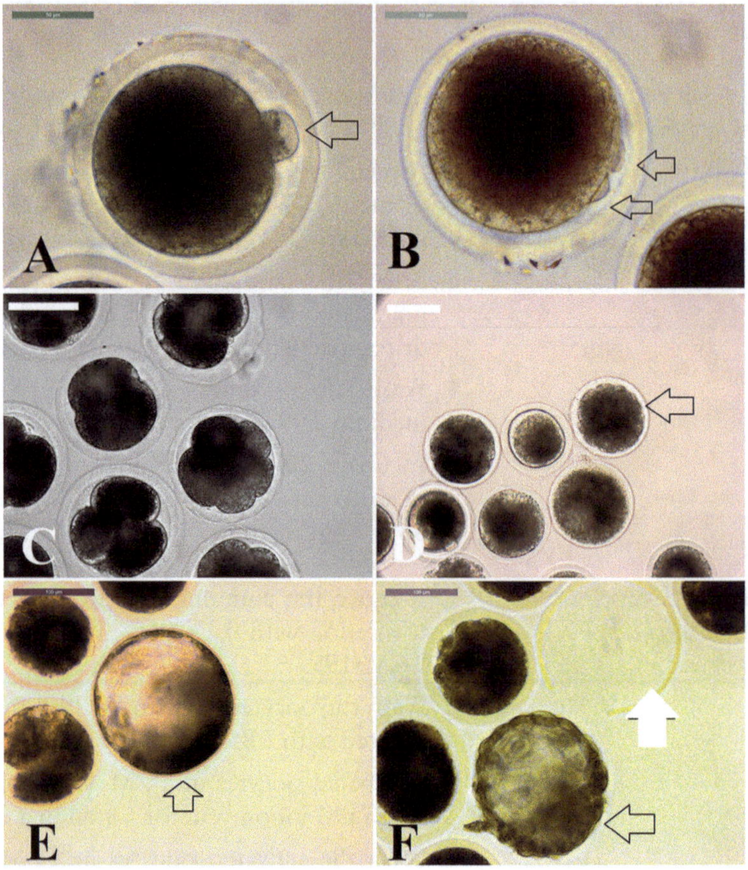

Fig. 1 Microscopic images of the development of mature camel oocytes until the blastocyst stage. (**a**) Mature oocyte with first polar body (arrow). (**b**) Oocyte extruding two polar bodies (arrows). (**c**) Embryos at >2-cell stage. (**d**) Morula (arrow). (**e**) Expanded blastocyst (arrow). (**f**) Hatched blastocyst (arrow) with empty zona pellucida (white arrow)

2. Wash oocytes three times in TCM-199 supplemented with 10% FBS before placement in culture drops.
3. Transfer ten oocytes/embryos (1 embryo per 1.5 μl medium) into the drops and place in the incubator (38.5 °C in an atmosphere of 5% CO_2, 5% O_2, and 90% N_2).
4. Observe the cleavage and subsequent divisions on day-2 until day-5 (Fig. 1c, d).
5. On day-5, transfer the embryos into fresh blastocyst medium, which is the same composition of culture medium with addition of 10% fetal bovine serum (FBS), to enhance the blastocyst formation.
6. Observe the blastocyst expansion on day-7 (Fig. 1e) and hatching on >day-8 (Fig. 1f).

3.2 In Vitro Culture of In Vivo Flushed Embryos

3.2.1 Superovulation

1. Animals aged 7–10 years are preferable to use.
2. Use transrectal linear-array 5 MHz transducer to monitor follicular development.
3. Superovulation treatment is started after elimination of the dominant follicle using 20 μg GnRH agonist (buserelin) or 5000 IU hCG treatment or at the early stage of follicular wave (no follicle >0.2 cm).
4. Use a combination of 3000 IU equine chorionic gonadotropins (eCG) and 18 mg of ovine follicle-stimulating hormone (oFSH) or 400 mg porcine follicle-stimulating hormone (pFSH) (1000 IU FSH and 1000 IU LH).
5. Inject 3000 IU eCG on day 1 of treatment together with the first of the twice daily injections (12 h apart) of oFSH or pFSH, followed by 3 more days of twice daily injections in decreasing doses of oFSH or pFSH (Day 1: 2 × 4 ml, Day 2: 2 × 3 ml, Day 3: 2 × 2 ml, Day 4: 2 × 1 ml) (Fig. 2).
6. Because some females show abstinence or incomplete sexual receptivity despite the presence of mature follicles on the ovaries, mating or AI should be based on follicular development and not on behavioral indicators of estrus. Allow mating or artificial inseminate (AI) when most follicles reach a mature size of 1.2–1.8 cm in diameter [19].
7. Although ovulation occurs in response to mating, inject the females with a single intravenous injection of GnRH analog (20 mg Buserelin) or 3000 IU of hCG at the time of the mating or AI to maximize the ovulation response.
8. Repeat mating or AI 24 h later.

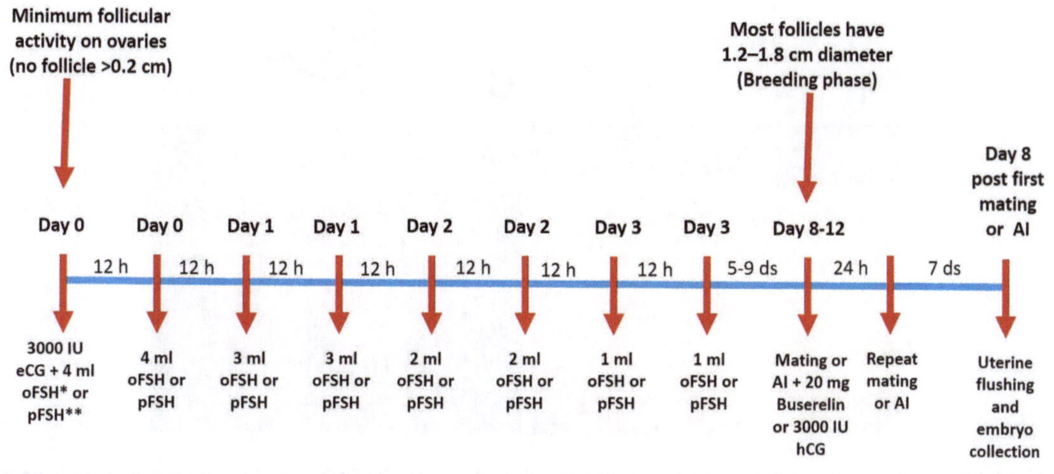

Fig. 2 The timeline for ovarian superovulation, insemination and embryo collection from camels. ∗1 ml oFSH contains 0.9 mg ovine follicle-stimulating hormone. ∗∗1 ml pFSH contains 20 mg porcine follicle-stimulating hormone (50 IU FSH and 50 IU LH)

3.2.2 Embryos Collection

1. Collect embryos on the eighth day after first mating or insemination using a nonsurgical method [17].
2. Keep the females in a restrained sitting position on the ground and sedate with intravenous injection of 0.2 mg/kg xylazine. The rectum, tail, and perineal region should be thoroughly cleaned and/or disinfected.
3. Introduce a Foley catheter (20 gauge) guided through the vagina.
4. Dilate the cervix gently using fingers and pass the catheter through it.
5. Once the catheter is through the cervix, inflate the cuff with 35 ml of saline and pull it back against the internal cervical *os* to seal it.
6. Flush the two uterine horns repeatedly with a total of 1000 ml of flushing medium.
7. Filter the collected flushing medium through an embryo filter until only 20 ml of the medium remains.
8. Examine the flushing medium under a stereomicroscope for the presence of embryos.
9. Evaluate the condition of each embryo according to developmental stage and morphological characteristics (Fig. 3a).

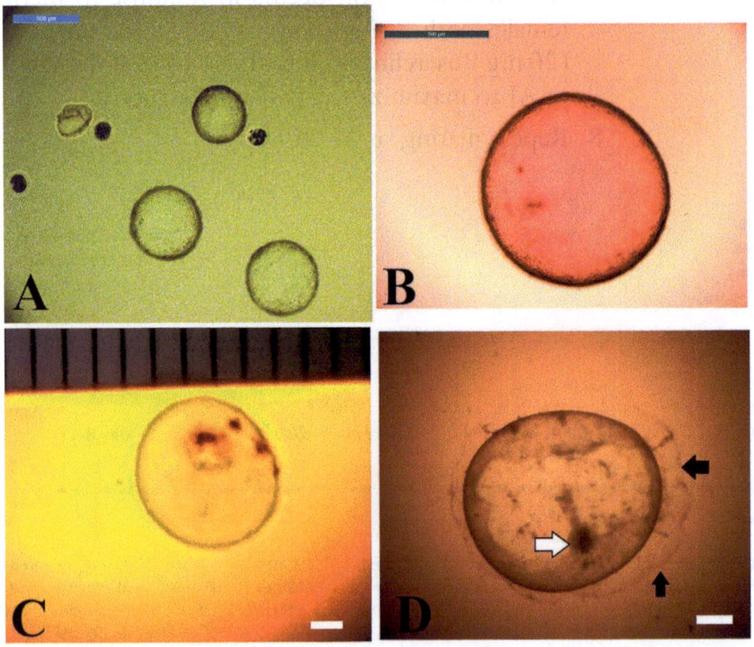

Fig. 3 Photomicrographs of in vivo flushed embryos cultured in vitro. (**a**) Embryos at the day of flushing (eighth dpi). (**b**) At third day of culture. (**c**) At 11th day of culture. (**d**) At 13th day of culture showing grown inner cell mass derived cells (white arrow) and trophoblast outgrowths (black arrows). (**a**) and (**b**) were visualized by a stereomicroscope (scale bar = 500 μm), while (**c**) and (**d**) were visualized with a mobile camera phone (scale bar = 1 mm)

3.2.3 Embryo Culture

1. Wash the blastocysts three times in medium comprised of TCM199 supplemented with 10% FBS and 50 μg/ml gentamycin.
2. Add 2 ml culture medium into 35 mm polyesterene-coated tissue culture dishes and incubate (38.5 °C in an atmosphere of 5% CO_2).
3. Place individual blastocyst into the culture dish and incubate at 38.5 °C in an atmosphere of 5% CO_2.
4. Carefully replace some of the medium with freshly prepared medium every 3 days.
5. Embryos grow with the rate of 0.42 ± 0.85 mm/day (Fig. 3b) and attach to the culture dish around day-5 of culture and remain growing till trophoblast outgrowths on day-12 of culture (i.e., the 20th day postinsemination dpi) (Fig. 3c).
6. Along with the growth of trophectoderm layer, the vesicular shape of the embryos deflates and trophectoderm cell layer (characterized by cuboidal morphology with big nucleus) become easily differentiated from the inner cell mass derived cell growths which characterized by dark aggregation of cell masses (Fig. 3d).

References

1. Schmidt-Nielsen K (1959) The physiology of the camel. Sci Am 201:140–151
2. Gordon I (2004) Old and new world camelids. In Reproductive technologies in farm animals. CABI Publishing, CAB International, Wallingford, Oxfordshire, pp 7–8
3. Skidmore JA, Morton KM, Billah M (2010) Unique strategies to control reproduction in camels. Soc Reprod Fertil Suppl 67:467–474
4. Tibary A, Anoussi A (1997) Theriogenology in camelidea: anatomy, physiology, pathology and artificial breeding, 1st edn. Acts Editions, Morocco, Rabat, pp 423–442
5. Novoa C (1970) Reproduction in Camelidae. Reproduction 22:3–20
6. Sghiri A, Driancourt MA (1999) Seasonal effects on fertility and ovarian follicular growth and maturation in camels (*Camelus dromedarius*). Anim Reprod Sci 55:223–237
7. Abdoon ASS (2001) Factors affecting follicular population, oocyte yield and quality in camels (*Camelus dromedarius*) ovary with special reference to maturation time in vitro. Anim Reprod Sci 66:71–79
8. Torner H, Heleil B, Alm H, Ghoneim IM, Srsen V, Kanitz W, Tuchscherer A, Fattouh EM (2003) Changes in cumulus-oocyte complexes of pregnant and non-pregnant camels (*Camelus dromedarius*) during maturation in vitro. Theriogenology 60:977–987
9. Monaco D, Padalino B, Lacalandra GM (2015) Distinctive features of female reproductive physiology and artificial insemination in the dromedary camel species. Emir J Food Agric 27:328–337
10. Saadeldin IM, Kim B, Lee B, Jang G (2011) Effect of different culture media on the temporal gene expression in the bovine developing embryos. Theriogenology 75:995–1004
11. Lonergan P, Fair T, Corcoran D, Evans AC (2006) Effect of culture environment on gene expression and developmental characteristics in IVF-derived embryos. Theriogenology 65:137–152
12. Saadeldin IM, Swelum AA-A, Yaqoob SH, Alowaimer AN (2017) Morphometric assessment of in vitro matured dromedary camel oocytes determines the developmental competence after parthenogenetic activation. Theriogenology 95:141–148
13. Absalón-Medina VA, Butler WR, Gilbert RO (2014) Preimplantation embryo metabolism and culture systems: experience from domestic animals and clinical implications. J Assist Reprod Genet 31:393–409

14. Wrenzycki C, Herrmann D, Carnwath JW, Niemann H (1999) Alterations in the relative abundance of gene transcripts in preimplantation bovine embryos cultured in medium supplemented with either serum or PVA. Mol Reprod Dev 53:8–18
15. Wrenzycki C (2001) Effects of culture system and protein supplementation on mRNA expression in pre-implantation bovine embryos. Hum Reprod 16:893–901
16. Yaqoob SH, Saadeldin IM, Swelum AA-A, Alowaimer AN (2017) Optimizing camel (*Camelus dromedarius*) oocytes in vitro maturation and early embryo culture after parthenogenetic activation. Small Rumin Res 153:81–86
17. Saadeldin IM, Swelum AA-A, Elsafadi M, Moumen AF, Alzahrani FA, Mahmood A, Alfayez M, Alowaimer AN (2017) Isolation and characterization of the trophectoderm from the Arabian camel (*Camelus dromedarius*). Placenta 57:113–122
18. Wani NA (2009) In vitro embryo production in camel (*Camelus dromedarius*) from in vitro matured oocytes fertilized with epididymal spermatozoa stored at 4°C. Anim Reprod Sci 111:69–79
19. Swelum AA, Saadeldin IM, Moumen A, Ba-Awadh H, Alowaimer AN (2018) Efficient follicular wave synchronization using a progesterone-releasing intravaginal device (PRIDΔ) in Camelus dromedarius. Theriogenology 118:203–211

Chapter 16

In Vitro Culture of Embryos from Horses

Elaine M. Carnevale, Lisa J. Maclellan, and Jo Anne E. Stokes

Abstract

Establishment of optimal methods for equine embryo culture has been slow when compared to some domestic species. In part, this delay was caused by the failure of standard in vitro fertilization techniques in horses. However, the development of intracytoplasmic sperm injection (ICSI) for the assisted fertilization of equine oocytes has resulted in a renewed interest in establishing optimal methods for embryo culture. Currently, ICSI-produced equine embryos are cultured using media designed for other species or other cell cultures and, typically, with the addition of serum. Although systems specifically for horse embryo culture still have not been established, ICSI-produced embryos are developmentally competent and capable of producing live offspring.

Key words Equine, Embryo, Culture, ICSI, Cleavage, Blastocyst, Glucose, Media

1 Introduction

Equine embryos remain in the oviduct for 5–6 days before entering the uterus as morulae or early blastocysts [1, 2]. With blastocoel development, the embryo rapidly expands and produces a glycoprotein layer between the trophoblast and zona pellucida, called the capsule [3, 4]. The specific function of the capsule can be debated; however, this acellular membrane potentially helps the embryo to retain its integrity during pregnancy when dynamic uterine contractions move the conceptus throughout the uterus [3]. After entry into the uterus, the equine embryo does not "hatch" from the zona pellucida, but instead it thins and dissipates as the embryo continues to grow [1]. In contrast to embryos collected from mares' uteri, equine embryos produced in vitro do not form a functional capsule, the zona pellucida fails to completely expand, fewer cells are present, and the inner cell mass is not well defined; differences in in vivo and in vitro equine embryos have been described [5]. However, with transfer into a mare's uterus, embryos that are produced in vitro and are still within the zona

Jason R. Herrick (ed.), *Comparative Embryo Culture: Methods and Protocols*, Methods in Molecular Biology, vol. 2006, https://doi.org/10.1007/978-1-4939-9566-0_16, © Springer Science+Business Media, LLC, part of Springer Nature 2019

pellucida appear to function similar to their in vivo counterparts, with formation of a capsule and loss of the zona pellucida.

After embryo transfer, the developmental competency of embryos produced in vitro appears to be reduced when compared to embryos collected from mares' uteri. However, variation in oocyte quality for in vitro embryo production is probably a major contributor. The mare will typically select only one follicle for ovulation, with a pattern of one or two follicle waves during the interovulatory interval [6]. Oocytes for ICSI can be collected from large, follicular-phase follicles after induction of follicle/oocyte maturation or from all follicles on the ovaries [7–9]. Consequently, oocytes can be collected in various stages of competency or atresia. Variability in oocyte developmental competence and quality can ultimately impact embryo viability, resulting in rates of embryo loss that are often higher after transfer of embryos produced in vitro versus in vivo. In addition, equine ICSI is often performed when oocyte or semen parameters are not optimal; that is, gametes from subfertile/old mares and/or compromised sperm (e.g., semen quality, freezing or refreezing procedures) can be incorporated into an ICSI program, resulting in reduced oocyte activation and more embryonic defects and losses.

1.1 Equine Embryo Culture

No current system for equine embryo culture results in the in vitro production of embryos that are identical in morphology and function as embryos recovered from the mares' uteri. Consequently, culture methods continue to be established and improved for equine embryos, although current culture systems do produce embryos that will result in offspring. Many of these systems use media developed for embryos of other species or for more general cell culture. The culture systems for equine embryos often include a serum source, primarily fetal bovine serum or newborn calf serum, to optimize results. The use of serum in horse media does not appear to have a detrimental effect, and problems with large offspring syndrome are not observed, as described with early bovine and ovine culture systems [10]. In general, comparing culture systems among facilities is difficult because of differences in many confounding factors including donor mare populations, semen sources, types of oocytes, storage or shipment of oocytes, ICSI techniques, and variations in defining cleavage and embryo stages. However, direct comparisons among culture components and culture systems from laboratories producing viable embryos and live offspring will be reviewed.

1.1.1 Base Media

Probably the most predominant culture medium used for equine embryo culture has been Dulbecco's Modified Eagle Medium/Nutrient Mixture F-12 (DMEM/F12), which is a complex culture medium designed to accommodate the growth of various cell types. The base medium that is typically used is buffered with sodium bicarbonate (14 mM), contains a high concentration of glucose (17 mM), and is often used with HEPES (15 mM), a zwitterionic buffer that helps to stabilize pH. The base medium of DMEM/F12 has been used in various laboratories with reports of blastocyst formation and foal production. Other media have been used with variable success. The commercial G1.3/G2.3 system (Vitrolife) resulted in low equine blastocyst development, when compared to DMEM/F12 [11]. However, other components of the G-series™ media, such as G-MOPS, have been incorporated into equine ICSI programs [8]. Modified SOF medium has been used to culture equine embryos to Day 5 or later [12, 13], and commercial media designed for the human embryos have been incorporated into equine ICSI programs, including Global® (personal communications, E. Carnevale, R. Foss, and C. Herrera) and Sage™ [14].

1.1.2 Glucose

One of the more controversial media components for equine embryo culture has been glucose. Because many of the early culture systems used DMEM/F12 with a high concentration of glucose (17 mM), it was postulated that high concentrations of the simple sugar, well above normal equine blood glucose concentrations (4.2–7.3 mM) [15], could be important for embryo development. This was supported when sperm-injected oocytes were cultured in SOF with 0, 1.5 or 19 mM glucose, and the embryo development rate was significantly higher with 19 mM glucose than with 0 or 1.5 mM glucose [13]. The amount of glucose used by equine embryos in early developmental stages in the oviduct has not been established; however, glucose consumption by embryos (morulae and blastocysts) that were collected from mares' uteri have been assessed [16]. Embryo uptake of glucose was not significantly different in morulae and early blastocysts (159.7 ± 22 and 261.7 ± 53.2 pmol/embryo/h), but a substantial increase was noted in expanded blastocysts (643.5 ± 64.3 pmol/embryo/h). Pyruvate uptake remained consistent across embryo development stages (150 pmol/embryo/h) [16]. The increased uptake of glucose by the blastocyst is consistent with other species [17].

In contrast to the high glucose concentrations in DMEM/F12, most commercial culture media designed for embryos of other species, including human embryos, have glucose concentrations of $\leq$1 mM for the initial stages of culture [18, 19]. The exposure of early cleavage-stage embryos from other species to high concentrations of glucose is considered nonphysiological and detrimental to subsequent developmental potential [17]. After partial zona resection and exposure of the developing equine embryos to different glucose combinations, Azuma et al. [20] concluded that the best embryo development occurred when glucose was supplemented at 0.5 mM from Days 1 to 4 and 5.5 mM from Days 5 to 8. The addition of glucose to culture media for ICSI-produced equine embryos was investigated using a commercial culture medium with minimum glucose (approximately 0.2 mM glucose, Global® from LifeGlobal Group LLC) [18, 19] using 1 μl of medium per embryo and up to five embryos in a 5-μl drop [21]. Fetal bovine serum (10% with 5.5 mM glucose in the serum) was added to the medium. Medium was refreshed or changed at Day 5 (Day 0 = Day of ICSI). The authors of this study did not observe significant differences in blastocyst development rates over a wide range of glucose concentrations (<1 to ~20 mM glucose), but incorporated the use of Global® medium (<1 mM glucose) in subsequent work with the addition of 20 mM glucose when medium was changed on Day 5 [22]. Although equine embryos seem resistant to high glucose concentrations, embryo uptake of glucose and current culture systems suggest that the equine embryo does not require such high glucose concentrations in the medium, although the total available glucose needs to be appropriate for embryos' needs at compaction and blastocoel expansion.

1.1.3 Current Culture Systems

Because an optimal medium has not been established for equine embryos, various culture systems have been used. Historically, DMEM/F12 was the primary medium used for equine embryo culture, and some laboratories will still use the medium for the entire culture period [23]. However, many laboratories are incorporating other media, especially for the initial stages of embryo development. A combination of Global® with DMEM/F12 has been successfully used for the entire duration of equine embryo culture [8, 24]. More specific embryo culture media with lower concentrations of glucose in the early stages of development have also been used, with the addition of DMEM/F12 as the embryo continues to grow. For instance, an SOF medium is used until Day 5, at which time embryos are transferred into DMEM/F12 [9], or embryos are cultured in Global® until Day 5 and then transferred into a mixture of Global® and DMEM/F12 (Foss, personal communication). Global® is also being used until Day 5, with the addition of 20 mM glucose to the medium for culture after Day 5 [23]. In our laboratory, we have historically used DEMEM/F12 throughout or after Days 3 or 4 of embryo culture. However, we have recently switched to using Global® for the initial stage of embryo culture. To increase glucose availability for morula and blastocyst development, embryos are moved to Global® for Fertilization at the approximate time of compaction, usually on Day 4. Global® for Fertilization has the same base formulation as Global® but with a higher concentration of glucose, approximately 2.5 mM glucose (D. Rieger, personal communication). Because we typically culture single embryos in 30-μl drops and transfer embryos into recipients when they form an early blastocyst, the medium should supply sufficient glucose. If embryos were cultured in smaller drops, as with the 1-μl medium/embryo system previously discussed, higher concentrations of glucose would be needed to supply sufficient glucose for embryo requirements, as the glucose content (concentration × volume) of a drop is lower. In equine culture systems, some protein source is used, which is most commonly 10% fetal bovine serum or newborn calf serum. Other protein sources have been tried, in whole or part, although any advantage from using them has not been proven. In general, culture temperatures range from 37.9 to 38.2 °C in the aforementioned laboratories. Embryos are cultured in mixed gas incubators with a range of 5–7% CO_2, depending on medium and altitude, 5% O_2, and nitrogen. The final pH of medium is usually indicated to be about 7.3. Embryos are cultured in 4-well dishes or various sizes of medium per drop under oil (1 μl/embryo to 30 μl per drop).

2 Materials

2.1 Supplies

1. Petri dishes (35 × 10 mm) or four-well dishes of tissue culture grade.
2. Global® and Global® for Fertilization (LifeGlobal Group, without human serum albumin) with the addition of 10% fetal bovine serum. For Global® at our altitude (~1500 m above sea level), incubators are set at 7% CO_2, 5% O_2, and N_2 with the final pH of the medium at approximately 7.3.
3. DMEM/F12 containing 15 mM HEPES with the addition of 10% fetal bovine serum. Culture with DMEM/F12 was typically performed at 5% CO_2, 5% O_2, and N_2. Commercial DMEM/F12 can range in pH depending on the source and lot. Consequently, the pH for each lot should be determined prior to culture and adjusted as needed.
4. Fetal bovine serum or newborn calf serum is routinely added for equine embryo cultures.
5. Other additives are not consistently used in culture systems. If antibiotics are needed, gentamicin (25 μg/ml) is typically added.
6. Cell culture-tested mineral or paraffin oil.

2.2 Equipment

1. Laminar flow hood for preparation of culture dishes.
2. Thermocouple.
3. Gas analyzer or method to confirm incubator CO_2 and O_2 concentrations.
4. Stereomicroscope with heated stage 38.2 °C.
5. Mixed gas incubator.

3 Methods

1. Thermocouple is used to confirm incubator temperatures and differences between shelves. Variability can exist between unit readouts and inner temperature or between incubator shelves [25], and this should be recorded and temperatures modified as needed.
2. Concentrations of CO_2 and O_2 are checked at regular intervals using a gas analyzer. Medium is equilibrated (approximately 1 day) in tubes to confirm pH at regular intervals and prior to use of new lots of medium or serum. Incubator water pans are filled using sterilized, ultrapure water.
3. Culture dishes are prepared 12–18 h prior to use, allowing for equilibration in a mixed-gas and humidified incubator at 38 °C. Drops of 30 μl of medium are placed in a cell culture petri dish (35 × 10 mm) and covered with 3 ml of oil (paraffin or mineral oil) allowing for extra drops of medium for washing the embryo(s). We will culture 1–3 embryos per drop, with only potential zygotes or embryos from the same mare and stallion in each well-labelled dish.
4. Following maturation and removal of cumulus cells, oocytes are washed in a MOPS-buffered medium with BSA and placed into drops of the same medium to undergo ICSI.
5. After ICSI, the sperm-injected oocytes are rinsed through a drop of medium and moved to a final drop of Global or DMEM/F12 for culture, being careful to move a minimum amount of medium between drops and avoiding air bubbles. Dishes remain out of the incubator for a minimal interval of time, and embryo movements are performed on a heated stereomicroscope stage.
6. The number of embryos per incubator is limited as much as possible, to minimize opening incubators.
7. At Day 1 (Day 0 = day of ICSI), cleavage can be assessed for most potential zygotes (Fig. 1a, *see* **Note 1**). If cell division has not occurred, indications of pending cleavage should be determined. These include membrane fragmentation or indentation. Occasionally, a second polar body can be observed.
8. If cleavage has not occurred on Day 1, embryos are reassessed on Day 2. Determination of cleavage is important in our program, as most embryos are transferred into recipient mares on our facility. We synchronize a recipient for an embryo based on the day of cleavage, with administration of ovulation-inducing drugs to the recipient on the day of cleavage. If recipients are not needed, embryos checks can be delayed until Days 3–5 or

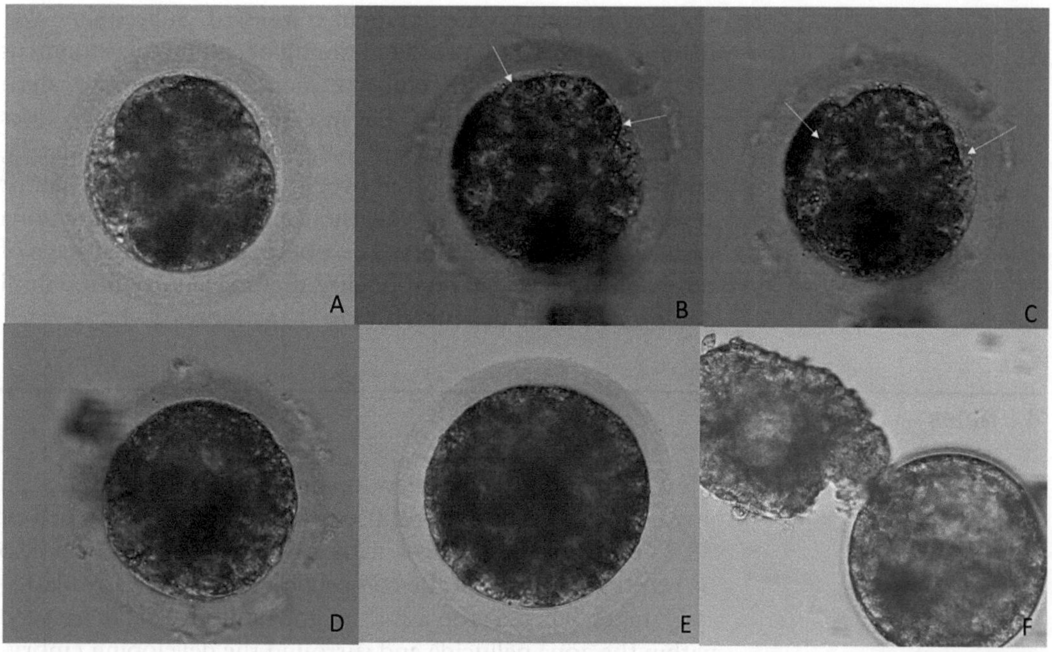

Fig. 1 Images (**a–d**) represent sequential stages of development for an individual equine embryo. Cleavage is observed on Day 1 (**a**), with ICSI performed on Day 0. In the morning of Day 6 (**b**), a small compact morula (arrows) is observed with a large amount of surrounding cellular debris; the morula could be easily be mistaken for cellular debris. Later on Day 6 (**c**), the morula has increased in size and is more distinct. Although extruded cellular debris is often observed within the zona pellucida of equine embryos, this amount of debris and the irregular outer zona pellucida are not standard in our system. On Day 7 (**d**), a late morula/early blastocyst is observed, and the embryo is transferred and produces a pregnancy. The Day-7 early blastocyst (**e**) is of good quality; the dark appearance is normal for early equine embryos. The zona pellucidae of equine embryos remaining in culture after blastocyst formation do not thin appropriately or form a capsule, and embryonic cells will start to extrude from the zona pellucida (**f**)

when the medium is changed or refreshed. However, in our laboratory, assessment of sperm-injected oocytes after Day 2 is not considered an accurate indicator of cleavage rate, as aged oocytes can fragment during longer incubation intervals. Embryos with delayed cleavage ($\leq$2 cells on Day 2) seldom develop into embryos or viable pregnancies in our system.

9. Developing embryos can be assessed and moved to a new culture drop, depending on medium and number of embryos per drop. Embryos initially cultured in Global® are assessed at Day 4, rinsed, and moved into an equilibrated, 30-μl drop of Global® for Fertilization with the addition of 10% fetal bovine serum. Embryo(s) cultured in DMEM/F12 in 30-μl drops can remain in the same drop for the entire culture duration, be changed to a new drop on Day 3–5, or have some of the medium replaced on Day 3–5.

10. On Day 6, embryo development is assessed. Subsequent evaluations are based on predicted timing of embryo development and ultimate use of the embryo, such as fresh transfer, shipment or cryopreservation. Because the equine embryo does not form a capsule or properly hatch from the zona pellucida, embryos are preferentially removed from culture for transfer or cryopreservation prior to extrusion of cells through the zona pellucida (Fig. 1f). If embryos do not reach the desired stage of development or are of poor quality, they are left in culture until determined to be degenerate (Fig. 1, *see* **Notes 2–4**).

4 Notes

1. Cleavage and embryo development is assessed using an inverted microscope at 200× with a warm stage set at 38.2 °C. Imaging embryos with a stereomicroscope can also be helpful, especially at the compacting morula or early blastocyst stages when cytoplasmic fragmentation can be imaged within the zona pellucida and surround the developing embryonic mass. Embryos are held outside of the incubator for a minimal interval of time during evaluations.
2. The assessment of equine embryo development can be difficult, particularly at the time of early morula formation (Fig. 1b, c). The cells are dark and contain high lipid concentrations. It is not unusual for embryos to have extruded debris/cells during culture, and the early morula can be "lost" within the surrounding debris and/or look like a single, large cell.
3. Most embryos develop into a morula or blastocyst at 6–8 days after ICSI (Fig. 1d, e). However, some embryos will develop slowly and can form a blastocyst at Day 10 or 11. These late-developing embryos can be viable and result in pregnancies. Good indicators of a viable embryo, even if development is delayed, is rapid development from a morula to blastocyst or rapid blastocyst expansion.
4. Embryo morphology and developmental timing (such as intervals to cleavage, morula to blastocyst, and blastocyst expansion) are important for assessing developmental potential. Ideally in our laboratory, cleavage to the 2-cell stage occurs within 24 h, and a compact morula is observed within 5 days after ICSI. When a morula is observed, it will ideally form into an early blastocyst or blastocyst within 2 days, and the late morula/early blastocyst should have a well-defined blastocoel within 1 or 2 days.

References

1. Betteridge KJ, Eaglesome MD, Mitchell D, Flood PF, Beriault R (1982) Development of horse embryos up to twenty two days after ovulation: observations on fresh specimens. J Anat 135:191–209
2. Freeman DA, Weber JA, Geary RT, Woods GL (1991) Time of embryo transport through the mare oviduct. Theriogenology 36:823–830
3. Betteridge KJ (1989) The structure and function of the equine capsule in relation to embryo manipulation and transfer. Equine Vet J 21 (S8):92–100
4. Oriol JG, Betteridge KJ, Clarke AJ, Sharom FJ (1993) Mucin-like glycoproteins in the equine embryonic capsule. Mol Reprod Dev 34:255–265
5. Tremoleda JL, Stout TAE, Lagutina I, Lazzari G, Bevers MM, Colenbrander B, Galli C (2003) Effects of in vitro production on horse embryo morphology, cytoskeletal characteristics, and blastocyst capsule formation. Biol Reprod 69:189501906
6. Ginther OJ (2017) Systemic and intrafollicular components of follicle selection in mares. Domest Anim Endocrinol 59:116–133
7. Carnevale EM (2016) Advances in collection, transport and maturation of equine oocytes for assisted reproductive techniques. Vet Clin North Am Equine Pract 32:379–399
8. Foss R, Ortis H, Hinrichs K (2013) Effect of potential oocyte transport protocols on blastocyst rates after intracytoplasmic sperm injection in the horse. Equine Vet J 45(Suppl 45):39–43
9. Hinrichs K (2010) In vitro production of equine embryos: state of the art. Reprod Domest Anim 45(Suppl 2):3–8
10. Young LE, Sinclair KD, Wilmut I (1998) Large offspring syndrome in cattle and sheep. Rev Reprod 3:155–163
11. Choi YH, Roasa LM, Love CC, Varner DD, Brinsko SP, Hinrichs K (2004) Blastocyst formation rates *in vivo* and *in vitro* of *in vitro*-matured equine oocytes fertilized by intracytoplasmic sperm injection. Biol Reprod 70:1231–1238
12. Colleoni S, Lagutina I, Lazzari G, Rodriguez-Martinez H, Galli C, Morrell JM (2011) New methods for selecting stallion spermatozoa for assisted reproduction. J Equine Vet Sci 31:536–541
13. Herrera C, Revora M, Vivani L, Miragaya MH, Losinno L, Quintans C, Pasqualini RS (2008) Effect of high glucose concentrations during equine *in vitro* culture of equine embryos. Proceedings of the 7th international symposium on equine embryo transfer and technology. R&W Communications, Suffolk, pp 52–53
14. Leisinger CA, Markle ML, Paccamonti DL, Cramer E, Pinto CRF (2016) Production of equine embryos in vitro using conventional intracytoplasmic sperm injection and a complete human embryo culture system. J Equine Vet Sci 4:78
15. Hollis AR, Boston RC, Corley KTT (2007) Blood glucose in horses with acute abdominal disease. J Vet Intern Med 21:1099–1103
16. Lane M, O'Donovan MK, Squires EL, Seidel GE Jr, Gardner DK (2001) Assessment of metabolism of equine morulae and blastocysts. Mol Reprod Dev 59:33–37
17. Hu K, Yu Y (2017) Metabolite availability as a window to view the early embryo microenvironment in vivo. Mol Reprod Dev 84:1027–2038. https://doi.org/10.1002/mrd.22868
18. Morbeck DE, Krisher RL, Herrick JR, Baumann NA, Matern D, Moyer T (2014) Composition of commercial media used for human embryo culture. Fertil Steril 102:759–766
19. Morbeck DE, Bauman NA, Oglesbee D (2017) Composition of single-step media used for human embryo culture. Fertil Steril 107:1055–1060
20. Azuma T, Choi YH, Hochi S, Oguri N (1995) Effect of glucose in the culture medium on development of horse oocytes matured and microfertilized *in vitro*. Reprod Fertil Dev 7:1067–1071
21. Choi YH, Ross P, Valez IC, Macias-Garcia B, Riera FL (2015) Cell linage allocation in equine blastocysts produced in vitro under varying glucose conditions. Reproduction 150:31–41
22. Choi YH, Gibbons JR, Canesin HS, Hinrichs K (2016) Effect of medium variants (zinc supplementation during oocyte maturation, perifertilization pH, and embryo culture protein source) on equine embryo development after intracytoplasmic sperm injection. Theriogenology 86:1782–1788
23. Rader K, Choi YH, Hinrichs K (2016) Intracytoplasmic sperm injection, embryo culture, and transfer of in vitro-produced blastocysts. Vet Clin Equine 32:401–413
24. Maserati M, Mutto A (2016) In vitro production of equine embryos and cloning: today's status. J Equine Vet Sci 41:42–50
25. Herrera C, Jeannerat E, Wyck S, Bittner L, Van den Bergh M, Janett F, Burger D, Bollwein H (2016) Successful in vitro production of mammalian embryos: a strict quality management approach. J Equine Vet Sci 41:74

Chapter 17

In Vitro Culture of Embryos from Domestic Cats

Jason R. Herrick

Abstract

A variety of protocols are available for the production of feline blastocysts from oocytes that are matured and fertilized in vitro. However, most of these protocols utilize media that were developed for other cell types or the embryos of other species. The protocol outlined in this chapter is based on a series of studies that led to the development of feline optimized culture medium (FOCM), a feline-specific, sequential medium system. The initial embryo culture medium was subsequently modified for both IVM and IVF. Using a small number of stock solutions, a series of media can be easily prepared that allows for immature cat oocytes to be matured and fertilized in vitro, with the resulting embryos developing to the blastocyst stage without the use of serum.

Key words Cat, Feline, Carnivore, Oocyte, Embryo, Blastocyst

1 Introduction

The development of assisted reproductive technologies (ART) for cats has largely been driven by the desire to use ART for the propagation of endangered cat species in managed zoo populations [1, 2]. The sustainability of these populations, and therefore their role in conservation efforts for these species, is dependent on successful offspring production between specific pairs of animals. While many pairs do produce offspring, others fail to reproduce due to behavioral incompatibility, physical limitations, and/or compromised fertility. In addition, managed populations are spread across numerous zoos throughout the world, which requires animals to be transported long distances to accommodate breeding recommendations. This can be stressful to the animals, require special permits, and be expensive for the institutions involved. Combining ART with gamete and/or embryo cryopreservation would allow offspring to be produced from animals at different institutions without transporting live animals. In addition, long-term storage of cryopreserved samples would allow animals to reproduce well beyond their natural life span, which could help maintain

Jason R. Herrick (ed.), *Comparative Embryo Culture: Methods and Protocols*, Methods in Molecular Biology, vol. 2006,
https://doi.org/10.1007/978-1-4939-9566-0_17, © Springer Science+Business Media, LLC, part of Springer Nature 2019

representation of rare founder lines and reduce adaptation to captivity. Many of the benefits of ART for endangered cat specie also have applications for managing research colonies of domestic cats with rare genetic mutations [3]. However, the potential of ART to manage populations of cats can only be realized if normal, viable embryos can be routinely and efficiently produced. Although offspring have been produced by in vitro fertilization (IVF) and embryo transfer (ET), as well as other ARTs, in domestic and nondomestic cats, protocols are far from routine or efficient.

Successful in vitro fertilization in the cat was first reported in 1970 by Hamner et al. using spermatozoa that were capacitated in utero [4]. A few years later Bowen [5] demonstrated that cat spermatozoa from the ductus deferens were capable of fertilizing oocytes following capacitation in vitro when cultured in a simple salt solution (BSW) [6] or a complex tissue culture medium (Ham's F-10). This study illustrates the relative ease of inducing capacitation of feline spermatozoa in simple, bicarbonate-buffered media containing Ca^{+2} and a protein source. The use of heparin, caffeine, or other cAMP-modulating agents in IVF media is not necessary in cats, as it is for other species [7–9]. Both of the studies by Hamner et al. and Bowen utilized in vivo-matured oocytes collected from gonadotropin-treated females. The ability of the resulting embryos to develop in vivo following ET was demonstrated in 1988 by Goodrowe et al. [10] with the birth of domestic cat kittens. Successful production of offspring following IVF of in vivo-matured oocytes and ET has also been reported in tigers [11], ocelots [12], sand cats [13], servals, caracals, black-footed cats, and fishing cats [14], indicating the broad application of protocols developed in the domestic cat to other cat species.

Successful embryo production from in vitro-matured oocytes was first reported in 1989 by Johnston et al. [15], with production of kittens reported by Pope et al. [16]. The ability to use ovaries from local veterinary clinics as a source of viable oocytes, in a similar manner to ovaries of livestock species collected from slaughterhouses, made it possible for many laboratories throughout the world to begin working with feline gametes and developing ART. Oocyte quality is resilient to both the temperature and duration of storage, with kittens being produced from oocytes collected from ovaries that were stored at 4 °C for 24 h [17, 18]. The primary limiting factor to in vitro maturation appears to be seasonality, with reduced oocyte quality in the fall and winter [19, 20]. Since the initial quality of the cumulus–oocyte complex is perhaps the most important determinant of embryo quality, an improved understanding of follicle and oocyte biology in cats will be a critical component of developing efficient feline ART. Significant contributions in these areas have been made by multiple research groups [21–29], but our understanding of the acquisition of developmental competence by the feline oocyte is still limited compared to rodents and livestock.

Following in vitro fertilization of in vitro-matured oocytes, successful development of blastocyst stage embryos has been reported by a variety of research groups using many different culture systems. In fact, blastocysts were produced in the very first attempt at embryo culture following IVF [5]. To date, cat blastocysts have been produced in simple salt solutions [5, 30–32], complex cell culture media [19, 30, 33, 34], and embryo culture media developed for livestock, mice, or humans [20, 35–37]. Despite the number of laboratories working with feline embryos, relatively few attempts have been made to develop feline-specific culture conditions [32, 38–41]. Effects of gas atmosphere, temperature, and protein supplementation have been evaluated by numerous groups [30, 42, 43]. However, the effects of pH, osmolarity, ions, carbohydrates, amino acids, and growth factor have received little, if any attention, despite the pronounced effects of these factors on the development of embryos from other species.

The lack of interest in studies of culture medium composition are likely related to the relative ease of producing acceptable quantities of blastocysts (>50% of cleaved embryos). However, studies of embryo quality indicate that cultured feline embryos are compromised. Cell numbers of in vitro produced blastocyst are less than half of those found in age-matched embryos that have developed in vivo [41, 44]. Similarly, the process of hatching seems to be altered in cultured blastocysts. The zona pellucida appears to restrict the growth of the feline embryo, as enzymatic removal of the zona significantly increases cell numbers [45]. In vitro hatching usually involves escape from a small hole in the zona around day 7 or 8 post-insemination, compared to the slow dissolution of a large portion of the zona that happens closer to day 10 or 11 post-fertilization in vivo [14, 46, 47]. Unfortunately, we know very little about in vivo-derived feline embryos [41, 44, 48, 49], making it difficult to accurately evaluate the normality of cultured feline embryos. Additional studies of the mechanisms driving development of the feline embryo, including further characterization of in vivo embryos, and how these mechanisms are influenced by the embryo's environment will be instrumental in the development of feline-specific culture conditions and, ultimately, the production of normal, "in vivo-like" cat embryos.

The protocol outlined in this chapter is based on a series of studies that led to the development of feline optimized culture medium (FOCM), a feline-specific, sequential medium system [41, 50]. The initial embryo culture medium was subsequently modified for both IVM and IVF [51]. Using a small number of stock solutions, a series of media can be easily prepared that allows immature cat oocytes to be matured and fertilized in vitro, with the resulting embryos developing to the blastocyst stage without the use of serum.

2 Materials

2.1 Equipment

1. Dissecting microscope, preferably with heated stage.
2. Incubators capable of controlling CO_2 and O_2 concentrations.
3. Sterile instruments (scissors, forceps, #11 scalpel blades, etc.) for dissection.
4. 35 and 60 mm sterile, polystyrene petri dishes.
5. Organ culture or 4-well dishes.
6. Large, square dishes with an etched grid.

2.2 Glassware

Glassware for media preparation is exclusively used for this purpose. Glassware is thoroughly rinsed immediately after use and washed by additional rinses (14×) with culture-grade water (18.2 MΩ, <5 ppb total organic carbon, and endotoxin- and pyrogen-free). After drying, glassware is covered with aluminum foil and sterilized with dry heat (125 °C for 12 h).

2.3 Commercially Available Media Supplements

1. Alanyl-glutamine (200 mM).
2. Minimum essential medium (MEM) nonessential amino acids (NEAA, 100×).
3. MEM essential amino acids (EAA), without glutamine (EAA, 50×).
4. MEM vitamins (100×).
5. Gentamicin sulfate (50 mg/mL).
6. Insulin (recombinant human, 1000 mg/L), transferrin (recombinant human, 550 mg/L), and selenium (0.67 mg/L) (ITS).

2.4 Stock Solutions for Preparation of Culture Media

All stock solutions are prepared with culture-grade water and cell-culture or pharmaceutical grade chemicals (Tables 1 and 2). Whenever possible, chemicals that have been tested by the supplier and confirmed to have low concentrations of endotoxins, heavy metals, DNases, RNases, and proteases are used.

When the volume of the stock solution is ≤50 mL (Table 2), the required amount of each chemical is placed in a sterile, polystyrene culture tube or flask and the corresponding volume of culture-grade water is added. For larger volumes (≥100 mL; Table 1), a beaker is filled to approximately half of the final volume with culture-grade water and chemicals are dissolved with the aid of a stir bar (dedicated to medium preparation). Once all chemicals have completely dissolved, the contents of the beaker are poured into an appropriate volumetric flask. The beaker is rinsed with 10–20 mL of culture-grade water, which is then poured into the volumetric flask. This process is repeated at least three or four times until the final volume is reached.

Table 1
Composition of monthly stock solutions for the preparation of various formulations of feline optimized culture medium (FOCM)

Stock[a]	MW	g/100 mL	mM
FOCM base			
NaCl	58.44	5.844	1000
KCl	74.55	0.5964	80
KH_2PO_4	136.10	0.1361	10
$CaCl_2$ $2H_2O$[b]	147.00	0.2940	20
$MgSO_4$ $7H_2O$	246.50	1.2325	50
Glucose	180.20	1.802	100
L-Lactic acid[c]	90.08	0.9008	100
Taurine	125.10	0.6260	50

[a]Stock solutions are filtered (0.22 μm) immediately after preparation and stored at 5 °C for 1 month.
[b]*See* **Note 1**
[c]MP Biomedicals 190228, *see* **Note 2**

Table 2
Composition of weekly stock solutions for the preparation of various formulations of feline optimized culture medium (FOCM)

Stock[a]	MW	g/10 mL	mM
Na-bicarbonate	84.01	0.2100	250
Na-pyruvate	110.00	0.0220	20
Citric acid	192.12	0.0961	50
Myoinositol	180.16	0.0135	7.5
Cysteine–HCl·H_2O	175.64	0.1054	60
Cysteamine–HCl[b]	113.61	0.0568	50

[a]Stock solutions are stored at 5 °C for 1 week
[b]Sometimes listed by suppliers as 2-mercaptoethylamine–HCl, *see* **Note 3**

Additional stock solutions are prepared as follows:

1. Gentamicin Stock (5 mg/mL): 400 μL of the purchased solution is diluted with 3.6 mL of culture-grade water and stored at 5 °C until expiration of the original, undiluted product.
2. ITS Stock (100 μg I, 55 μg T, and 67 ng S/mL): 400 μL of the purchased solution is diluted with 3.6 mL of culture-grade water and stored at 5 °C until expiration of the original, undiluted product.
3. 1.0 M NaOH: 2.0 g in 50 mL culture-grade water. Filter (0.22 μm) and store at 5 °C for 3 months.

4. Albumin Stock (100 mg/mL): Place 1 mL of culture-grade water in a sterile, polystyrene tube and add 1.0 g albumin (bovine serum albumin or recombinant human albumin, *see* **Note 4**). Add 4 mL of culture-grade water "on top" of the albumin and allow it to dissolve at room temperature. Add an additional 5 mL of culture-grade water to bring the total volume to 10 mL. Mix well, but do not shake or vortex, and filter (0.22 μm). Store at 5 °C for up to 3 months.
5. Heparin Stock (4000 U/mL): Weigh 0.25 to 0.3 g of heparin, record the exact amount in milligrams, and transfer to a sterile polystyrene tube. Determine the lot-specific activity of the product (U/mg) provided by the supplier. Multiply the weight in milligrams by the stated activity to determine the total number of units and divide this amount by 4000 to determine the volume of HEPES-buffered FOCM (FOCMH; Table 3) needed. Mix well, but do not shake or vortex, and filter (0.22 μm). Place ~550 μL aliquots into sterile microcentrifuge tubes and store at −80 °C indefinitely.
6. Fatty acid-free (FAF) bovine serum albumin (BSA) Diluent (1.0 mg/mL): Dissolve 0.0100 g fatty acid-free BSA in 10 mL of culture-grade water. Mix well, but do not shake or vortex, and filter (0.22 μm). Store at 5 °C and use within 1 week of preparation.
7. Equine Chorionic Gonadotropin (eCG, 0.5 IU/μL): Add 10 mL FAF BSA Diluent to a vial containing 5000 IU lyophilized eCG. Mix well, but do not shake or vortex. Aliquot 25–50 μL into sterile 0.5 microcentrifuge tubes and store at −80 °C for 1 year.
8. Epidermal Growth Factor (recombinant human EGF, 5 ng/μ L): Add 10 mL of FAF BSA Diluent to a vial containing 200 μg EGF (20 ng/μL). Mix well, but do not shake or vortex. Aliquot 1.1 mL into sterile microcentrifuge tubes, label as EGF Stock 1, and store at −80 °C for 1 year. Dilute 1 mL of EGF Stock 1 with 3 mL of FAF BSA Diluent (5 ng/μL). Mix well, but do not shake or vortex. Aliquot ~60 μL into sterile microcentrifuge tubes and store at −80 °C for 1 year.
9. Saline for Ovary Collections: Add 9.0 g NaCl and 1.0 mL of purchased, undiluted (50 mg/mL) gentamicin solution to 1 L of culture grade water. Mix well and filter (0.22 μm). Aliquot 20–25 mL into sterile, polypropylene tubes (50 mL) and store at 5 °C for 6 months.

2.5 Culture Media

All culture media are prepared from stock solutions as described in Table 3. This approach facilitates altering the composition of the medium for various experiments. Individual components can be added or deleted from the medium, or added in greater or lesser

Table 3
Preparation of feline optimized culture medium (FOCM) for in vitro maturation (IVM), in vitro fertilization (IVF), in vitro culture (FOCM1, day 1–3 and FOCM2, day 3–7), and HEPES-buffered FOCM (FOCMH) from stock solutions

	μL of stock solutions per 10 mL of medium				
Component (final concentration)	FOCM IVM	FOCM IVF	FOCM1	FOCM2	FOCMH[a]
H_2O	4536	6346	6406	5526	7230
FOCM Base[b]	1000	1000	1000	1000	1000
$MgSO_4$ (1.0 mM)	200	200	200	200	200
Bicarbonate (mM)	1000 (25)	1000 (25)	1000 (25)	1000 (25)	200 (5.0)
HEPES (20 mM)					0.0477 g
Glucose (mM)	600 (6.0)	150 (1.5)	150 (1.5)	150 (1.5)	300 (3.0)
Pyruvate (0.1 mM)	50	50	50	50	50
L-Lactate (6.0 mM)	600	600	600	600	600
Citrate (0.5 mM)	100	100	100	100	
Ala-Gln (1.0 mM)	50	50	50	50	50
Taurine (0.1 mM)	20	20	20	20	20
NEAA (1× MEM)	100	100	100	100	100
EAA (× MEM)	100 (0.5)	50 (0.25)	50 (0.25)	200 (1.0)	
Cysteine (0.6 mM)	100				
Cysteamine (0.1 mM)	20				
Vitamins (1× MEM)	100			100	
Myoinositol (0.075 mM)				100	
eCG (1.0 IU/mL)	20				
EGF (25 ng/mL)	50				
ITS (μg I, μg T, ng S/mL)	1000 (10, 5.5, 6.7)			500 (5, 2.75, 3.4)	
Albumin (mg/mL)	200 (2.0)	200 (2.0)	200 (2.0)	200 (2.0)	0.04 g (4.0)
Gentamicin (μg/mL)	100 (50)	100 (50)	40 (20)	40 (20)	100 (50)
1.0 M NaOH	54	34	34	64	150

[a]Typically 100–200 mL of HEPES-buffered FOCM (FOCMH) is prepared at one time and stored for 4–6 weeks at 5 °C
[b]Final concentrations of salts in FOCM Base are as follows: 100.0 mM NaCl, 8.0 mM KCl, 1.0 mM KH_2PO_4, 2.0 mM $CaCl_2{\cdot}2H_2O$

amounts, simply by altering the volume of water to maintain the final volume (10 mL in Table 3). Using a slightly different salt solution (NaCl, KCl, KH_2PO_4, and $CaCl_2{\cdot}2H_2O$) as the base medium, this same set of stocks can be used to prepare media for the culture of murine and bovine embryos [52, 53].

The desired volume of culture-grade water is added to a sterile polystyrene tube. Stock solutions are then added individually with sterile pipette tips. After each stock solution is expelled into the water, the tip is rinsed by aspirating medium into the tip and then expelling it back out. Albumin is always added last, either from a stock solution (Table 3) or by adding lyophilized albumin to the medium. Media can be stored at 5 °C for one week, but are typically prepared fresh for each experimental replicate. If FOCM IVM is to be stored, do not add eCG or EGF until the day of use.

2.6 Preparation of Culture Drops and Medium Equilibration

The bicarbonate-buffered media (FOCM IVM, IVF, IVC1, and IVC2) require equilibration with CO_2 before use. Using the volumes of NaOH indicated in Table 3 and equilibrating media in incubators set to 6.0% CO_2, the final pH of all media should be 7.25 ± 0.05. The volume of NaOH may need to be adjusted depending on individual incubator settings. Similarly, laboratories at higher elevations may need to increase the concentrations of CO_2 and O_2 to compensate for reduced atmospheric pressure.

Culture dishes are prepared and placed into the incubator at least 4 h before use to ensure the medium is at the appropriate temperature and pH when oocytes or embryos are placed into culture. Ovary processing is usually done in the afternoon, with subsequent steps being performed at 24 h intervals. Therefore, media are usually prepared before 10:00 am and equilibrated until use after 3:00 p.m. Sterile, polystyrene culture dishes are filled with 4.0 (35 mm dishes) or 8.0 (60 mm dishes) mL of equilibrated mineral/paraffin oil. Drops are made by placing the desired amount of medium underneath the oil. The pipette tip containing the medium is placed into the oil and the medium is expelled when the end of tip touches the bottom of the dish. Alternatively, small amounts of medium (≤half of the final drop volume) can be placed in the dish and then quickly covered with oil. The remaining volume of medium is then added to each drop. When dishes are prepared with warm medium, in a laminar flow hood, and/or in a dry environment, the first method will minimize evaporation of the drops, which can increase the osmolarity of the medium.

2.7 Determination of Blastocyst Cell Number

1. 16% Paraformaldehyde.
2. PBS: Ca^{+2}/Mg^{+2}-free phosphate-buffered saline.
3. PBS-PVP: PBS with 1.0 mg/mL (0.1%) PVP.
4. Storage Buffer: PBS with 5.0 mg/mL (0.5%) BSA.
5. Wash Buffer: PBS-PVP with 0.1% Triton X-100 (TX100).
6. Permeabilization Buffer: PBS-PVP with 1.0% TX100.
7. NucBlue Fixed Cell Stain (Thermo Fisher Scientific, Waltham, MA, USA) containing DAPI.

8. ProLong Gold Antifade Reagent (Thermo Fisher Scientific, Waltham, MA, USA).
9. Poly-L-lysine-coated slides (Thermo Fisher Scientific, Waltham, MA, USA).

3 Methods

3.1 Collection of Ovaries

1. The best sources of ovaries are usually high-volume, low-cost clinics that participate in trap–neuter–return programs for feral cats. These clinics usually get more adult females and perform larger numbers of surgeries each day than the average pet hospital. Age is the only factor used for ovary selection. Ovaries should only be collected from females that are at least 6 months old. The stage of the estrous cycle and/or gestation can alter the number of oocytes recovered, but all ovaries are processed.
2. Each clinic is provided with tubes of saline that they can keep in their refrigerator. On the day of surgery, these tubes can be kept in the surgery area at room temperature. Surgeons are asked to remove the ovaries from the rest of the reproductive tract and place them in saline as soon as possible after excision. Most of the time they place the distal end of the uterine horn, with the attached ovary and oviduct, into the tube of saline. Depending on how much excess tissue is removed by the surgeon, 10–15 pairs of ovaries can be placed into a single tube.
3. Tubes full of ovaries should be maintained at 5 °C until processing (*see* **Note 5**).

3.2 Collection of Cumulus–Oocyte Complexes (COCs) and In Vitro Maturation

1. On the day of ovary collection (Day −1), FOCM IVM should be prepared in the morning and dishes placed into the incubator (6% CO_2 in air, 38.7 °C). An organ culture dish containing 750 μL of medium and 750 μL of oil is used for washing. In vitro maturation drops are 50 μL. Fifty milliliters of FOCMH should be supplemented with 500 μL of the heparin stock (40 U/mL final) and kept at 5 °C. Ten milliliters of FOCMH without heparin should be placed in a warming oven (38.7 °C, without CO_2).
2. Five to six ovaries are placed into the lid of a 60 mm petri dish and any excess tissue is removed using sterile scissors and forceps. The ovaries are then transferred into a 60 mm dish containing ~8 to 10 mL of FOCMH with heparin (5 °C).
3. Ovaries are sliced longitudinally with a #11 scalpel blade to facilitate visualization of individual follicles. Each half of the ovary is then held in place with forceps and repeatedly sliced with a #11 scalpel blade until all visible follicles have been ruptured.

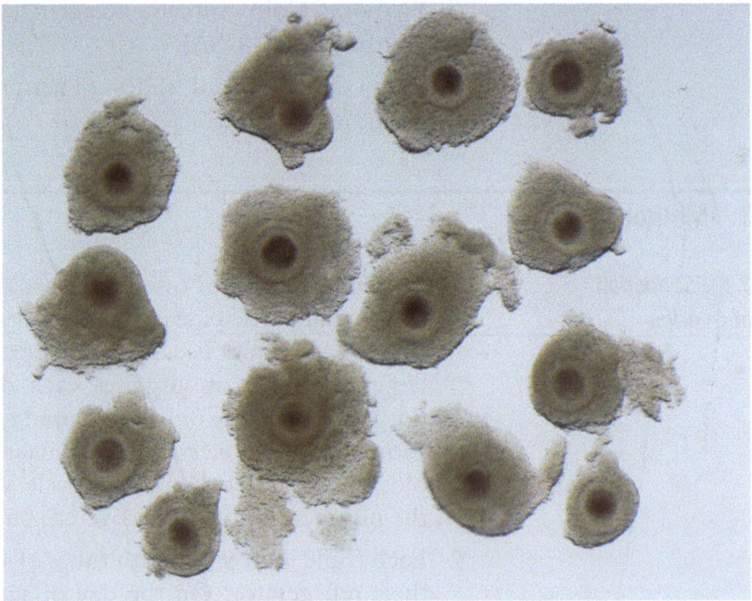

Fig. 1 Immature feline cumulus–oocyte complexes (COC) selected for maturation based on the darkness of the oocyte cytoplasm and amount of surrounding cumulus cells

4. The contents of the dish are poured into a square, grid dish on a slide warmer and allowed to warm to 38.7 °C while the rest of the ovaries are processed. Make sure to keep the dish covered to minimize evaporation and the associated changes in medium osmolarity.
5. Media from two groups of ovaries can be poured into the same grid dish, such that each grid dish will contain ~20 mL of FOCMH and the cellular debris from 10 to 12 ovaries.
6. After all ovaries have been processed, the grid dishes are searched for COCs.
7. COCs are transferred to a 35 mm dish with ~3 mL warm FOCMH (no heparin) for further grading and selection.
8. Cat oocytes have a very high content of intracellular lipid. Oocytes from excellent and good quality COCs should have a homogeneous, dark (brown to black) cytoplasm and be surrounded by multiple layers of cumulus cells [16, 54] (Fig. 1).
9. Oocytes with pale and/or irregular cytoplasm should be avoided. Some preantral follicles look similar to COCs, but the oocytes are usually slightly smaller and/or paler and the cumulus cells are smoother around the edges than a COC [55]. These follicles should also be avoided.
10. Oocyte quality is dependent on many factors, including age and stage of the estrous cycle, so recovery can be highly variable

between batches of ovaries. On average 2–3 excellent quality COCs are recovered per ovary during the breeding season (February to August in the Northern Hemisphere).

11. COCs are washed through another 35 mm dish containing ~3 mL FOCMH (no heparin) and then through the organ culture dish of FOCM IVM.
12. COCs are cultured in groups of 5 in 50 μL drops of FOCM IVM for 24 h in an atmosphere of 6% CO_2 in air at 38.7 °C.

3.3 *In Vitro* Fertilization

1. In the morning following oocyte collection (Day 0), FOCM IVF (*see* **Note 6**) should be prepared and dishes placed into the incubator (6% CO_2 in air, 38.7 °C). An organ culture dish containing 750 μL of IVF medium and 750 μL of oil is used for washing. In vitro fertilization drops are 45 μL. Two milliliters of FOCMH should be placed in a warming oven (38.7 °C, without CO_2).
2. About 1 h before IVF (~23 h post-maturation), straws of cryopreserved, ejaculated sperm (~50 μL per 0.25 mL straw, ~25 to 50 × 10^6 motile sperm/mL) are thawed in air for 10 s and then placed into a 37–39 °C water bath for 30 s.
3. Straws are wiped dry and the contents of the straw are expelled into a sterile, microcentrifuge tube.
4. The entire content of the straw is diluted slowly with 150 μL FOCMH, resulting in a 1:4 dilution (one part sperm: three parts medium). Freshly ejaculated sperm diluted 1:1 with FOCMH can also be used.
5. Samples are centrifuged for 15 min at 200 × *g*.
6. While the sample is in the centrifuge, wash COCs through the organ culture dish containing FOCM IVF and place groups of five COCs into 45 μL drops of FOCM IVF. Return the dish to the incubator until insemination.
7. The supernatant is removed and the pellet containing spermatozoa is resuspended in 25–50 μL FOCMH.
8. Aliquots are used to evaluate motility and sperm concentration (hemocytometer).
9. Multiply the concentration of spermatozoa by the proportion of spermatozoa that are motile to determine the concentration of motile sperm. The sample is then diluted to 5 × 10^6 motile sperm/mL.
10. Add 5 μL of the sperm suspension to the 45 μL drops of FOCM IVF containing COCs. This results in a final concentration of 0.5 × 10^6 motile sperm/mL in the 50 μL drop.
11. COCs and spermatozoa are coincubated for 20–24 h in 6% CO_2 in air at 38.7 °C.

3.4 Embryo Culture

1. The day after IVF (Day 1), FOCM1 is prepared and dishes are placed into the incubator (6% CO_2 and 5% O_2 at 38.7 °C). An organ culture dish containing 750 μL of FOCM1 medium and 750 μL of oil is used for washing. Embryos will also be washed through two 50 μL drops of FOCM1 and then cultured in 20 μL drops of FOCM1. Each dish should contain two wash drops and the appropriate number of culture drops.
2. Cleavage can begin by 18 h post-insemination. Most embryos have reached the two-cell stage and some embryos have already reached the four-cell stage by 24 h post-insemination [35].
3. Loosely bound spermatozoa and cumulus cells are removed using pulled Pasteur pipettes. Pipettes used to remove COCs from the IVF drops have a larger diameter for removal of the outer layers of cumulus cells. After COCs are transferred to the organ culture dish of FOCM1, smaller diameter pipettes are used to remove any remaining cumulus cells.
4. Two- to four-cell embryos, presumptive zygotes, and lysed/degenerated oocytes are counted. Although most embryos that will cleave will have done so by 24 h post-insemination, some zygotes will cleave later. Two- and four-cell embryos are pooled, washed through both wash drops of FOCM1, and then placed in groups of five (±1) into the culture drops of FOCM1. Zygotes are then pooled, washed, and cultured in separate drops, so late developing embryos can be identified. If different treatments are to be utilized for the first step of culture, embryos and zygotes should be separately allocated to treatments to ensure that each treatment receives a similar number of high (early cleaving)- and low (late cleaving)-quality embryos [35]. Lysed or degenerate embryos (<10% of the total) are typically discarded. However, if the experiment is evaluating IVM or IVF conditions the number of discarded oocytes should be recorded, as it may be affected by treatment.
5. Embryos are cultured (6% CO_2 and 5% O_2 at 38.7 °C) for 48 h in FOCM1.
6. On the morning of Day 3, FOCM2 is prepared and dishes are placed into the incubator (6% CO_2 and 5% O_2 at 38.7 °C). Embryos will be washed through two 50 μL drops of FOCM2 and then cultured in 20 μL drops of FOCM2. Each dish should contain two wash drops and the appropriate number of culture drops.
7. In the afternoon of Day 3 (72 h post-insemination), embryonic development is evaluated, including the total number of cleaved embryos (≥2-cell). At this time, embryos should have ≥16 cells. Embryos with ≥8 cells are transferred to FOCM2

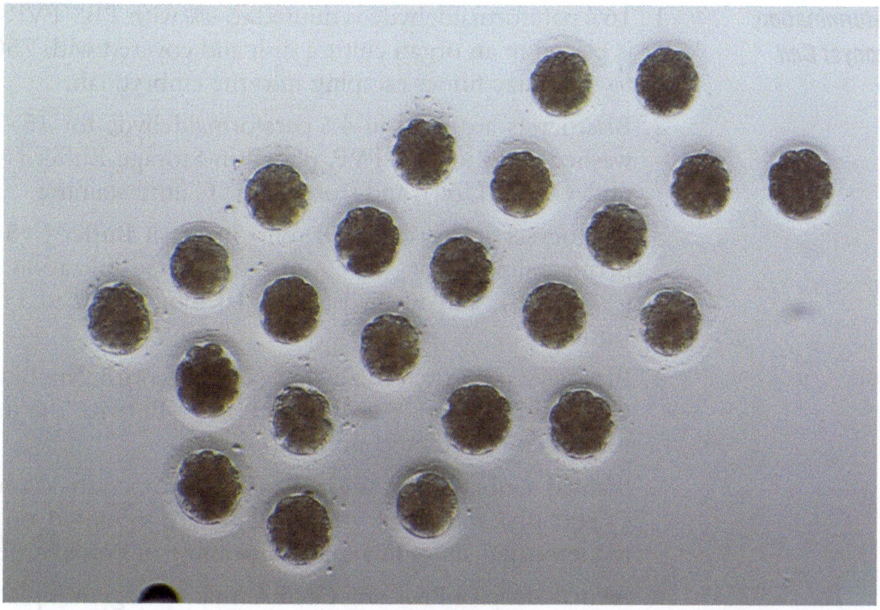

Fig. 2 Feline embryos on day 3 prior to the move to FOCM2

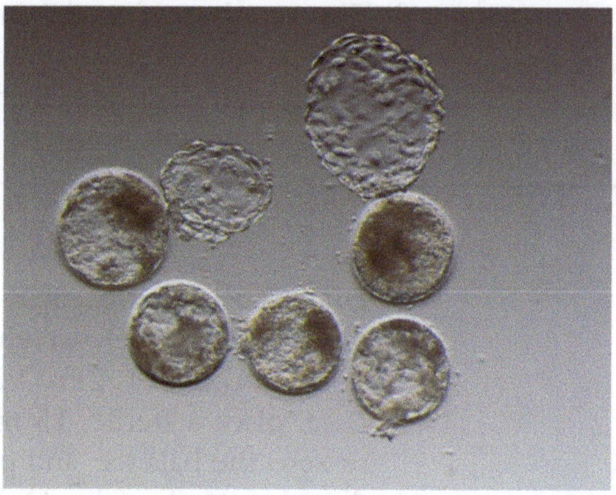

Fig. 3 Expanded and hatching feline blastocysts on day 7 of culture

for additional culture (Fig. 2). Embryos with <8 cells are discarded.

8. Embryos are washed through the two wash drops of FOCM2 and then groups of five (± 1) are placed in culture drops of FOCM2.
9. Embryos are cultured (6% CO_2 and 5% O_2 at 38.7 °C) in FOCM2 for an additional 96 h to Day 7 in FOCM2.
10. Blastocyst formation and hatching are evaluated in the afternoon of Day 7 (168 h post-insemination) (Fig. 3).

3.5 Determination of Blastocyst Cell Number

1. 16% paraformaldehyde is diluted to 4% with PBS-PVP. 750 μL is placed in an organ culture dish and covered with 750 μL oil to minimize fumes escaping into the embryo lab.
2. Blastocysts are fixed in 4% paraformaldehyde for 15–20 min, washed once in PBS-PVP, placed in Storage Buffer (~500 μL under 500 μL oil), and stored at 5 °C until staining.
3. Blastocysts are washed (10 min) in Wash Buffer (≥500 μL), permeabilized for 60 min in Permeabilization Buffer (≥500 μL), and washed again (10 min; ≥500 μL) in Wash Buffer.
4. Blastocysts are then incubated (20 min) with NucBlue Fixed Cell Stain (2 drops per 500 μL PBS-PVP; twice the manufacturer recommended concentration).
5. Stained embryos are washed (10 min) once in Wash Buffer (≥500 μL) and small groups (2–4) are mounted on poly-L-lysine-coated slides in a minimal volume of Wash Buffer.
6. A small drop of ProLong Gold Antifade Reagent is placed over the embryos and then a cover slip is applied.
7. Slides are kept flat until the mounting medium has dried and stored in the dark until evaluated.
8. The number of nuclei in the blastocyst is counted using a fluorescent microscope and can be facilitated with software with a manual count function.

4 Notes

1. Due to the hygroscopic nature of $CaCl_2 \cdot 2H_2O$, it can be difficult to precisely weigh this chemical. Alternatively, a stock solution of $CaCl_2 \cdot 2H_2O$ is prepared. For 100 mL of FOCM Base, place a sterile, 15 mL polystyrene tube on the balance, re-zero the balance, and place 0.3000–0.3500 g $CaCl_2 \cdot 2H_2O$ directly into the tube. Quickly record the weight and calculate how much culture-grade H_2O to add to make 0.2940 g/ 10 mL or 0.0294 g/mL. Vortex and add 10 mL (=0.2940 g) to the beaker containing other reagents.
2. Due to the hygroscopic nature of L-lactic acid (MP Biomedicals 190228), it can be difficult to precisely weigh this chemical. Place half of the desired volume of culture-grade H_2O in to a beaker with a stir bar. Add 0.8–1.1 g of L-lactate directly to the beaker. Quickly record the weight and calculate how much culture-grade H_2O to add to the beaker to obtain the desired concentration (0.0090 g/mL, 100 mM).
3. Due to the hygroscopic nature of cysteamine–HCl (sometimes listed as 2-mercaptoethylamine–HCl), it can be difficult to

precisely weigh this chemical. Place a sterile, 15 mL polystyrene tube directly on the balance, rezero the balance, and place 0.0284–0.0682 g cysteamine directly into the tube. Quickly record the weight and calculate how much culture-grade H_2O to add to the tube to obtain the desired concentration (0.0057 g/mL, 50 mM).

4. Albumins with and without lipids will support blastocyst development, but reduced fatty-acid or fatty acid-free preparations are preferable. All preparations should have low endotoxin levels (≤1 EU/mg).
5. Unlike most species, the developmental competence of cat oocytes is better if the ovaries are kept cold (5 °C) [56]. Blastocysts, and even offspring, can be produced from oocytes recovered from ovaries stored at 5 °C for up to 24 h, but developmental competence is compromised compared to shorter (<12 h) storage times [17, 18].
6. For experiments in which cumulus cells are removed prior to IVF, the concentration of $MgSO_4$ in FOCM IVF should be reduced to 0.2 mM, and albumin should be replaced by 5% (vol/vol) fetal calf serum [57].

References

1. Swanson WF (2006) Application of assisted reproduction for population management in felids: the potential and reality for conservation of small cats. Theriogenology 66(1):49–58. https://doi.org/10.1016/j.theriogenology.2006.03.024
2. Wildt DE, Roth TL (1997) Assisted reproduction for managing and conserving threatened felids. Int Zoo Yearb 35:164–172
3. Lyons LA (2015) DNA mutations of the cat: the good, the bad and the ugly. J Feline Med Surg 17(3):203–219. https://doi.org/10.1177/1098612X15571878
4. Hamner CE, Jennings LL, Sojka NJ (1970) Cat (*Felis catus* L.) spermatozoa require capacitation. J Reprod Fertil 23(3):477–480
5. Bowen RA (1977) Fertilization in vitro of feline ova by spermatozoa from the ductus deferens. Biol Reprod 17(1):144–147
6. Biggers JDW, Whiten WK, Whittingham DG (1971) The culture of mouse embryos in vitro. In: Daniel JC Jr (ed) Methods in mammalian embryology. Freeman, Dallas, TX, pp 87–116
7. Bavister BD, Boatman DE, Leibfried L, Loose M, Vernon MW (1983) Fertilization and cleavage of rhesus monkey oocytes in vitro. Biol Reprod 28(4):983–999
8. Wang WH, Niwa K, Okuda K (1991) In-vitro penetration of pig oocytes matured in culture by frozen-thawed ejaculated spermatozoa. J Reprod Fertil 93(2):491–496
9. Parrish JJ (2014) Bovine in vitro fertilization: in vitro oocyte maturation and sperm capacitation with heparin. Theriogenology 81 (1):67–73. https://doi.org/10.1016/j.theriogenology.2013.08.005
10. Goodrowe KL, Wall RJ, O'Brien SJ, Schmidt PM, Wildt DE (1988) Developmental competence of domestic cat follicular oocytes after fertilization in vitro. Biol Reprod 39 (2):355–372
11. Donoghue AM, Johnston LA, Seal US, Armstrong DL, Tilson RL, Wolf P, Petrini K, Simmons LG, Gross T, Wildt DE (1990) In vitro fertilization and embryo development in vitro and in vivo in the tiger (*Panthera tigris*). Biol Reprod 43(5):733–744
12. Swanson WF (2003) Research in nondomestic species: experiences in reproductive physiology research for conservation of endangered felids. ILAR J 44(4):307–316
13. Herrick JR, Mehrdadfar F, Campbell M, Levens G, Leiske K, Swanson WF (2010) Birth of sand cat (*Felis margarita*) kittens

following in vitro fertilization and embryo transfer. Biol Reprod 83:6–7

14. Pope CE (2014) Aspects of in vivo oocyte production, blastocyst development, and embryo transfer in the cat. Theriogenology 81(1):126–137. https://doi.org/10.1016/j.theriogenology.2013.09.006
15. Johnston LA, O'Brien SJ, Wildt DE (1989) In vitro maturation and fertilization of domestic cat follicular oocytes. Gamete Res 24(3):343–356. https://doi.org/10.1002/mrd.1120240310
16. Pope CE, McRae MA, Plair BL, Keller GL, Dresser BL (1997) In vitro and in vivo development of embryos produced by in vitro maturation and in vitro fertilization of cat oocytes. J Reprod Fertil Suppl 51:69–82
17. Wolfe BA, Wildt DE (1996) Development to blastocysts of domestic cat oocytes matured and fertilized in vitro after prolonged cold storage. J Reprod Fertil 106(1):135–141
18. Pope CE, Gomez MC, King AL, Harris RF, Dresser BL (2003) Embryos produced in vitro after recovery of oocytes from cat ovaries stored at 4°C for 24 to 28 hours retain the competence to develop into live kittens after transfer to recipients. Theriogenology 59(1):308
19. Spindler RE, Wildt DE (1999) Circannual variations in intraovarian oocyte but not epididymal sperm quality in the domestic Cat. Biol Reprod 61(1):188–194
20. Freistedt P, Stojkovic M, Wolf E (2001) Efficient in vitro production of cat embryos in modified synthetic oviduct fluid medium: effects of season and ovarian status. Biol Reprod 65(1):9–13
21. Goritz F, Jewgenow K, Meyer HH (1996) Epidermal growth factor and epidermal growth factor receptor in the ovary of the domestic cat (*Felis catus*). J Reprod Fertil 106(1):117–124
22. Spindler RE, Pukazhenthi BS, Wildt DE (2000) Oocyte metabolism predicts the development of cat embryos to blastocyst in vitro. Mol Reprod Dev 56(2):163–171. https://doi.org/10.1002/(SICI)1098-2795(200006)56:2<163::AID-MRD7>3.0.CO;2-3
23. Bogliolo L, Leoni G, Ledda S, Zedda MT, Bonelli P, Madau L, Santucciu C, Naitana S, Pau S (2004) M-phase promoting factor (MPF) and mitogen activated protein kinases (MAPK) activities of domestic cat oocytes matured in vitro and in vivo. Cloning Stem Cells 6(1):15–23. https://doi.org/10.1089/15362300460743790
24. Bristol SK, Woodruff TK (2004) Follicle-restricted compartmentalization of transforming growth factor beta superfamily ligands in the feline ovary. Biol Reprod 70(3):846–859. https://doi.org/10.1095/biolreprod.103.021857
25. Saint-Dizier M, Malandain E, Thoumire S, Remy B, Chastant-Maillard S (2007) Expression of follicle-stimulating hormone and luteinizing hormone receptors during follicular growth in the domestic cat ovary. Mol Reprod Dev 74(8):989–996.
26. Comizzoli P, Pukazhenthi BS, Wildt DE (2011) The competence of germinal vesicle oocytes is unrelated to nuclear chromatin configuration and strictly depends on cytoplasmic quantity and quality in the cat model. Hum Reprod 26(8):2165–2177. https://doi.org/10.1093/humrep/der176
27. Rojo JL, Linari M, Musse MP, Peluffo MC (2015) *Felis catus* ovary as a model to study follicle biology in vitro. J Assist Reprod Genet 32(7):1105–1111. https://doi.org/10.1007/s10815-015-0511-5
28. Lee PC, Wildt DE, Comizzoli P (2018) Proteomic analysis of germinal vesicles in the domestic cat model reveals candidate nuclear proteins involved in oocyte competence acquisition. Mol Hum Reprod 24(1):14–26. https://doi.org/10.1093/molehr/gax059
29. Lee PC, Wildt DE, Comizzoli P (2015) Nucleolar translocation of histone deacetylase 2 is involved in regulation of transcriptional silencing in the cat germinal vesicle. Biol Reprod 93(2):33. https://doi.org/10.1095/biolreprod.115.129106
30. Johnston LA, Donoghue AM, O'Brien SJ, Wildt DE (1991) Culture medium and protein supplementation influence in vitro fertilization and embryo development in the domestic cat. J Exp Zool 257(3):350–359. https://doi.org/10.1002/jez.1402570308
31. Pope CE, Keller GL, Dresser BL (1993) In vitro fertilization in domestic and non-domestic cats including sequences of early nuclear events, development in vitro, cryopreservation and successful intra- and interspecies embryo transfer. J Reprod Fertil Suppl 47:189–201
32. Swanson WF, Roth TL, Godke RA (1996) Persistence of the developmental block of in vitro fertilized domestic cat embryos to temporal variations in culture conditions. Mol Reprod Dev 43(3):298–305. https://doi.org/10.1002/(SICI)1098-2795(199603)43:3<298::AID-MRD3>3.0.CO;2-L
33. Verstegen JP, Onclin K, Silva LD, Donnay I, Mettens P, Ectors F (1993) Superovulation and embryo culture in vitro following treatment

with ultra-pure follicle-stimulating hormone in cats. J Reprod Fertil Suppl 47:209–218

34. Kanda M, Oikawa H, Nakao H, Tsutsui T (1995) Early embryonic development in vitro and embryo transfer in the cat. J Vet Med Sci 57(4):641–646
35. Spindler RE, Crichton EG, Agca Y, Loskutoff N, Critser J, Gardner DK, Wildt DE (2006) Improved felid embryo development by group culture is maintained with heterospecific companions. Theriogenology 66 (1):82–92. https://doi.org/10.1016/j.theriogenology.2006.03.021
36. Merlo B, Iacono E, Regazzini M, Zambelli D (2008) Cat blastocysts produced in vitro from oocytes vitrified using the cryoloop technique and cryopreserved electroejaculated semen. Theriogenology 70(1):126–130. https://doi.org/10.1016/j.theriogenology.2008.03.009
37. Nestle E, Graves-Herring J, Keefer C, Comizzoli P (2012) Source of protein supplementation during in vitro culture does not affect the quality of resulting blastocysts in the domestic cat. Reprod Domest Anim 47(Suppl 6):152–155. https://doi.org/10.1111/rda.12047
38. Kanda M, Miyazaki T, Kanda M, Nakao H, Tsutsui T (1998) Development of in vitro fertilized feline embryos in a modified Earle's balanced salt solution: influence of protein supplements and culture dishes on fertilization success and blastocyst formation. J Vet Med Sci 60(4):423–431
39. Pope CE, Gomez MC, Dresser BL (2006) In vitro production and transfer of cat embryos in the 21st century. Theriogenology 66 (1):59–71. https://doi.org/10.1016/j.theriogenology.2006.03.014
40. Murakami M, Otoi T, Karja NW, Ooka A, Suzuki T (2002) Effects of serum-free culture media on in vitro development of domestic cat embryos following in vitro maturation and fertilization. Reprod Domest Anim 37 (6):352–356
41. Herrick JR, Bond JB, Magarey GM, Bateman HL, Krisher RL, Dunford SA, Swanson WF (2007) Toward a feline-optimized culture medium: impact of ions, carbohydrates, essential amino acids, vitamins, and serum on development and metabolism of in vitro fertilization-derived feline embryos relative to embryos grown in vivo. Biol Reprod 76 (5):858–870. https://doi.org/10.1095/biolreprod.106.058065
42. Johnston LA, Donoghue AM, O'Brien SJ, Wildt DE (1991) Influence of temperature and gas atmosphere on in-vitro fertilization and embryo development in domestic cats. J Reprod Fertil 92(2):377–382
43. Karja NW, Otoi T, Murakami M, Yuge M, Fahrudin M, Suzuki T (2002) Effect of protein supplementation on development to the hatching and hatched blastocyst stages of cat IVF embryos. Reprod Fertil Dev 14(5–6):291–296
44. Yin XJ, Lee HS, Jeon JT, Kong IK (2007) Effect of culture medium and prostaglandin I (2) (PGI(2)) analogue on in vitro development of parthenogenetically activated cat oocytes. J Reprod Dev 53(5):1053–1058
45. Freistedt P, Stojkovic P, Wolf E, Stojkovic M (2001) Energy status of nonmatured and in vitro-matured domestic cat oocytes and of different stages of in vitro-produced embryos: enzymatic removal of the zona pellucida increases adenosine triphosphate content and total cell number of blastocysts. Biol Reprod 65 (3):793–798
46. Denker HW, Eng LA, Mootz U, Hamner CE (1978) Studies on the early development and implantation in the cat: 1. Cleavage and blastocyst formation. Anat Anz 144(5):457–468
47. Roth TL, Swanson WF, Wildt DE (1994) Developmental competence of domestic cat embryos fertilized in vivo versus in vitro. Biol Reprod 51(3):441–451
48. Swanson WF, Roth TL, Wildt DE (1994) In vivo embryogenesis, embryo migration, and embryonic mortality in the domestic cat. Biol Reprod 51(3):452–464
49. Filliers M, Goossens K, Van Soom A, Merlo B, Pope CE, de Rooster H, Smits K, Vandaele L, Peelman LJ (2012) Gene expression profiling of pluripotency and differentiation-related markers in cat oocytes and preimplantation embryos. Reprod Fertil Dev 24(5):691–703. https://doi.org/10.1071/RD11068
50. Herrick JR, Lyons SM, Greene-Ermisch AF, Broeckling CD, Schoolcraft WB, Krisher RL (2018) A carnivore embryo's perspective on essential amino acids and ammonium in culture medium: effects on the development of feline embryos. Biol Reprod 99(5):1070–1081. https://doi.org/10.1093/biolre/ioy122
51. Herrick JR (2014) Reversible meiotic arrest in feline oocytes. Reprod Fertil Dev 26 (2):258–267. https://doi.org/10.1071/RD12341
52. Herrick JR, Lyons SM, Greene AF, Broeckling CD, Schoolcraft WB, Krisher RL (2016) Direct and osmolarity-dependent effects of glycine on preimplantation bovine embryos. PLoS One 11(7):e0159581. https://doi.org/10.1371/journal.pone.0159581

53. Herrick JR, Paik T, Strauss KJ, Schoolcraft WB, Krisher RL (2016) Building a better mouse embryo assay: effects of mouse strain and in vitro maturation on sensitivity to contaminants of the culture environment. J Assist Reprod Genet 33(2):237–245. https://doi.org/10.1007/s10815-015-0623-y
54. Wood TC, Wildt DE (1997) Effect of the quality of the cumulus-oocyte complex in the domestic cat on the ability of oocytes to mature, fertilize and develop into blastocysts in vitro. J Reprod Fertil 110(2):355–360
55. Reynaud K, Gicquel C, Thoumire S, Chebrout M, Ficheux C, Bestandji M, Chastant-Maillard S (2009) Folliculogenesis and morphometry of oocyte and follicle growth in the feline ovary. Reprod Domest Anim 44(2):174–179. https://doi.org/10.1111/j.1439-0531.2007.01012.x
56. Naoi H, Otoi T, Shimamura T, Karja NW, Agung B, Shimizu R, Taniguchi M, Nagai T (2007) Developmental competence of cat oocytes from ovaries stored at various temperature for 24 h. J Reprod Dev 53(2):271–277
57. Herrick JR, Strauss KJ, Schneiderman A, Rawlins M, Stevens J, Schoolcraft WB, Krisher RL (2015) The beneficial effects of reduced magnesium during the oocyte-to-embryo transition are conserved in mice, domestic cats and humans. Reprod Fertil Dev 27(2):323–331. https://doi.org/10.1071/RD13268

Chapter 18

The Domestic Dog Embryo: In Vitro Fertilization, Culture, and Transfer

J. B. Nagashima, A. J. Travis, and N. Songsasen

Abstract

Advances in embryo technologies in the domestic dog have made significant strides in the past decade. This progress has been spurred by interests in taking advantage of the dog as a biomedical research model for human and companion animal medicine, developing assisted reproductive technologies to manage genetic diversity in endangered canids maintained ex situ, and improving breeding in rare or working breeds of dogs. Here, we focus on recent advancements and techniques for collection of in vivo-matured oocytes, in vitro fertilization (IVF), in vitro culture of early (≤8-cell) and advanced stage (≥16-cell) embryos, and embryo transfer.

Key words Domestic dog, In vitro fertilization, Oocyte, Embryo culture, Embryo cleavage kinetics

1 Introduction

Understanding of reproductive biology in the domestic dog is surprisingly limited. This includes the regulators of ovarian follicle development, oocyte maturation, anestrus termination, and embryo development. In turn, this lack of understanding has restricted our ability to develop assisted reproductive technologies (ART) in canids and advance the domestic dog as a research model. Over 410 genetic diseases have been identified in domestic dogs with homologies to human conditions [1, 2], making it an attractive model for human biomedical research. Toward these ends, transgenic puppies have been produced via somatic cell nuclear transfer [3–5], and CRISPR/Cas9 [6]. Still, rates of successful transgenesis remain low, and the process is dependent on the use of zygotes, which limits broad utilization. The domestic dog is also the ideal model to develop ART for the conservation of its endangered cousins. The International Union for Conservation of Nature has designated 5 of the 35 extant wild canid species as either endangered or critically endangered, and two have already gone extinct [7]. As result, there is growing need for the creation of ex

Jason R. Herrick (ed.), *Comparative Embryo Culture: Methods and Protocols*, Methods in Molecular Biology, vol. 2006, https://doi.org/10.1007/978-1-4939-9566-0_18,

situ insurance populations. However, maintenance of genetic diversity within a small group of animals over many generations is a critical challenge. Natural breeding requires the animals to be at the same geographic location, which means that achieving specific matches involves the often-stressful and expensive transport of valuable animals to distant locations. Improved ART such as in vitro fertilization introduces the ability to transfer gametes or embryos to distant locations, rather than the animals themselves, and still produce genetically valuable, live offspring.

Unfortunately, ART have been historically challenging to develop in domestic dogs due to their unique reproductive biology. Domestic dogs experience a reproductive cycle consisting of roughly 1-week periods of proestrus and estrus, followed by a 2-month obligatory diestrus [8]. Diestrus is followed by a variably long (2–10 month) anestrus [9], resulting in ovulation only 1–2 times annually. The variable duration and final termination of anestrus are not well understood, despite decades of work investigating the traditional reproductive hormones (gonadotropins, estradiol, progesterone) that drive estrus. Many attempts have been made to develop estrus induction protocols for the dog, with varying results (*see* Review [10]). For example, supplementation of exogenous follicle-stimulating hormone (FSH) has been used to stimulate follicle growth [11]. Dopamine agonists and gonadotropin-releasing hormone agonists have both been utilized to induce estrus by stimulating gonadotropin release [12, 13]. The latter two treatment types have successfully induced estrus in most studies, but effectiveness is highly dependent on the stage of reproductive cycle at the time of hormone treatment [14, 15] and, in the case of GnRH agonists, long-term treatment resulted in estrus suppression [16], and contraception [17]. Thus, while many different estrus induction techniques have been attempted in the dog, few have yielded consistent results.

After the onset of estrous behavior and two days following the surge of LH, a germinal vesicle stage oocyte is ovulated into the bitch's oviduct, which is embedded in the fatty bursa encompassing the ovary. Dog oocytes require an additional 48–72 h in the oviduct to reach metaphase II (MII) and become fertilizable [18]. The evolutionary advantage for this timing, which differs between canids and other mammals, is unclear. In addition, the mechanism underlying this delayed oocyte maturation is not yet understood, though it has been postulated to involve the prolonged close association of cumulus cells with ovulated oocytes [18]. Nevertheless, attempts to recapitulate this maturation process in vitro have generally yielded low rates of MII development. Several excellent reviews exist on dog in vitro oocyte maturation (IVM) efforts [19–21], but the overall conclusion is that attempts to mature eggs in vitro have generally yielded low rates of MII oocytes. Additionally, a comprehensive review of domestic dog embryo biotechnologies, from in vivo embryo development to cloning is also already available [22]. Therefore, in this chapter, we concentrate on recent advancements in in vitro embryo production, with an emphasis on culture medium composition and early embryo cleavage kinetics in vitro, followed by detailed techniques for the collection of in vivo-matured oocytes, in vitro fertilization (IVF), in vitro culture of early and late stage embryos, and embryo transfer.

1.1 Techniques for Collection of In Vivo-Matured Oocytes

As a result of the unique characteristics of the domestic dog's reproductive biology described above, obtaining mature oocytes is a particular challenge in this species. Ideally, the vast stores of immature oocytes present in the ovary at any stage of the reproductive cycle could be utilized for ART. Detailed methods exist on oocyte collection from ovaries [23]; however, the success of various in vitro oocyte maturation protocols have proved variable [19–21]. Thus far, the only reliable source of MII dog oocytes is via collection of in vivo-matured ova. This strategy requires precise timing of the luteinizing hormone (LH) surge and/or ovulation. This is best accomplished via serum progesterone [24, 25] and/or identifying the LH surge [26–28], with monitoring of changes in vaginal cytology being used in a supportive, confirmatory fashion [29]. Recovery of MII oocytes from the oviduct can be accomplished in situ, via inserting a catheter into the distal oviduct (isthmus) then retrograde flushing toward a needle or another catheter inserted into the infundibulum [30–32]. This in situ technique has been successfully applied to collect mature oocytes for use in somatic cell nuclear transfer, and zygotes for autologous transfer after modification via CRISPR/Cas9 [33]. In a less technically challenging method, in vivo-matured oocytes can also be flushed from isolated oviducts following specifically timed ovariohysterectomy. Day 6 post-LH surge (i.e., in vivo-matured) dog oocytes collected in this manner have recently produced live young following IVF, demonstrating their fertilization competence on this day of collection [28].

1.2 In Vitro Fertilization and Embryo Culture in the Dog

While the production of IVF puppies was primarily attributed to use of in vivo-matured oocytes and modifications in sperm capacitation medium [28], the extent to which fertilization and embryo culture media contributed to the success is unclear. In vivo-matured ova recovered on Day 6 post-LH surge had previously been used in IVF in the dog [34]. Hori and Tsutsui timed ovulation based on plasma progesterone levels over 2 ng/ml, based on previous work in their laboratory correlating progesterone with LH and ultrasonographic imaging of the ovary [27]. These collected oocytes were coincubated in a modified Krebs–Ringer bicarbonate solution (*see* TYH-BSA and mTYH-BSA in Table 1) with sperm capacitated in TYH for 20 h [34]. Two-cell embryos were observed 48–108 h following initiation of coculture, with cleavage to the 4-cell stage

Table 1
Details of media preparations used in IVF studies with in vivo-matured oocytes

Ingredient (mM)	TYH-BSA	mTYH-BSA	mSOF	mNCSU	cNCSU
NaCl	119.37	119.37	107.70	108.73	108.73
KCl	4.78	4.78	7.16	4.78	4.78
$CaCl_2{\cdot}2H_2O$	1.71	1.71	1.71	1.67	1.67
KH_2PO_4	1.19	–	1.19	1.19	1.19
$MgSO_4{\cdot}7H_2O$	1.19	1.19	–	–	–
$MgCl_2$	–	–	0.49	1.19	1.19
$NaHCO_3$	25.07	25.07	25.07	25.07	25.07
Glucose	5.56	–	1.50	0.55	0.55
BSA	4 mg/ml	4 mg/ml	32 mg/ml	4 mg/ml	4 mg/ml
Penicillin G	75 μg/ml	75 μg/ml	100 U/ml	100 U/ml	100 U/ml
Streptomycin sulfate	50 μg/ml	50 μg/ml	50 ug/ml	100 U/ml	100 U/ml
Methyl-β-cyclodextrin	0.74	0.74	–	–	–
Polyvinylalcohol	22.70	22.70	–	–	–
Minimum essential medium (100×)	–	–	–	–	10 μl/ml
Basal medium eagle (50×)	–	–	–	–	20 μl/ml
Na pyruvate	1.0	1.0	0.33	0.33	0.33
Sodium lactate	–	–	3.3		
Taurine	–	–	–	7.00	7.00
Hypotaurine	–	–	–	5.00	5.00
Glutamine				1.00	1.00

after an additional 12–24 h. Across all treatment groups, ten oocytes cleaved out of 94 oocytes fertilized (Table 2). The authors attributed the low rate of cleavage to suboptimal fertilization and embryo culture medium.

In our study of IVF that yielded live births, in vivo-matured oocytes were collected on Day 6 following the LH surge, which was determined based on serum progesterone levels of 1.5–2.5 ng/ml and a positive LH snap test. Serum progesterone levels of ~4.5 ng/ml confirmed the day of ovulation, 2 days after the LH surge [28]. Ova were coincubated with sperm for 14 h in a "complete" NCSU-23 medium (cNCSU, *see* Table 1), with both essential and nonessential amino acids, as well as 6 μg/ml progesterone. Sperm had been collected and preincubated for 2.5–3.5 h under capacitating conditions in modified canine capacitation medium (containing 25 mM HEPES and 1.0 mM $MgCl_2$). No cleavage was observed

Table 2
Comparison of oocyte collection day, and IVF and embryo culture medium in vivo-matured oocyte IVF studies

Study	Day of oocyte collection	IVF medium	Culture medium	# Oocytes	# Embryos produced (%)
Hori and Tsutui [34]	6	TYH-BSA	TYH-BSA	43	5 (11.6)
	6	mTYH-BSA	mTYH-BSA	38	5 (21.1)
	6	TYH-BSA	TYH-FCS	13	0 (0)
Nagashima et al. [28] and unpublished data	5	M199	mSOF	13	0 (0)
	5	M199	mNCSU	15	0 (0)
	5	cNCSU	cNCSU	37	11 (29.9)
	6	M199	mSOF	3	2 (67)
	6	M199	mNCSU	2	2 (100)
	6	cNCSU	cNCSU	146	115 (78.8)

24 h after initiation of coincubation, but 115 oocytes cleaved out of 146 fertilized (Table 2), with the majority developing to the 4-cell stage by 48 h postcoincubation.

In the same study, Day 5 oocytes were similarly collected and fertilized, yielding 11 embryos of 37 oocytes (Table 2), although six of these did not cleave until 72 h postcoincubation. Our laboratory has also previously evaluated Tissue Culture Medium 199 (M199) + 20% fetal bovine serum (FBS) for IVF, followed by culture in a modified synthetic oviductal fluid (mSOF) or NCSU-23 (mNCSU). Both mSOF and mNCSU lack the full complement of amino acids used in cNCSU (*see* Table 1). This strategy produced four embryos from five Day 6 oocytes (2/3 and 2/2 embryo development in mSOF and mNCSU, respectively, Table 2), three of which developed to the 8-cell stage in culture.

It should be noted that at least part of the difference in embryo production rates between the Nagashima et al., and Hori and Tsutsui studies may be due to a difference in the designation of Day 0. Although it is difficult to compare assays performed using different methods in different laboratories, one can make the observation that serum progesterone values in our study on the designated day of ovulation were typically ~4.5 ng/ml [28], whereas plasma progesterone values higher than 2 ng/ml were used to identify ovulation in the Hori and Tsutsui study [34].

When assessing these studies in the context of the broader literature on embryo production and development, it is apparent that the presence of phosphate and/or glucose does not hinder early embryo (<8-cell) development in the dog, as in the hamster [35]. Also, although essential and nonessential amino acids were both present in the media producing live young, it cannot be said conclusively that their presence benefited embryo production and early development, as in the cow [36], because the relevant comparisons were not made. Fertilization was accomplished in the Hori and Tsutsui study (as well as those of others [37–39]) in the absence of essential and nonessential amino acids. In our laboratory, IVF embryos cultured in mSOF developed to the 8-cell stage as readily as those cultured in the presence of the full complement of amino acids. Additional investigation is needed to determine the influence of amino acid supplementation on developmental competence of canine embryos generated by IVF.

1.3 Kinetics of Dog Embryo Cleavage Events and In Vitro Culture

Time-lapse imaging has advanced our ability to understand the relationship between embryo cleavage kinetics and developmental capacity in vitro. In human IVF, shorter times of cleavage to the 8-cell stage are associated with increased development to the blastocyst stage, improved blastocyst quality [40], and better implantation rates [41, 42]. Alterations in ovarian stimulation protocols [43] and culture medium [44] do not appear to significantly influence cleavage kinetics of human embryos. However, developmentally delayed embryos are typically aneuploid [45], which may explain the deficiencies observed at the blastocyst and implantation stages.

To our knowledge, time-lapse imaging of domestic dog embryo development has not been completed. Still, it is likely that cleavage rates which align to in vivo development rates are indicative of better embryo quality. Early embryo development in vivo has been well described in the dog [22]. Briefly, 2-cell stage embryos are typically observed between 144 and 216 h post-LH surge, 4-cell between 160 and 192 h, and 8-cell stage embryos between 160 and 336 h after the LH surge [18, 46] (Fig. 1). The high temporal variability is likely due in part to

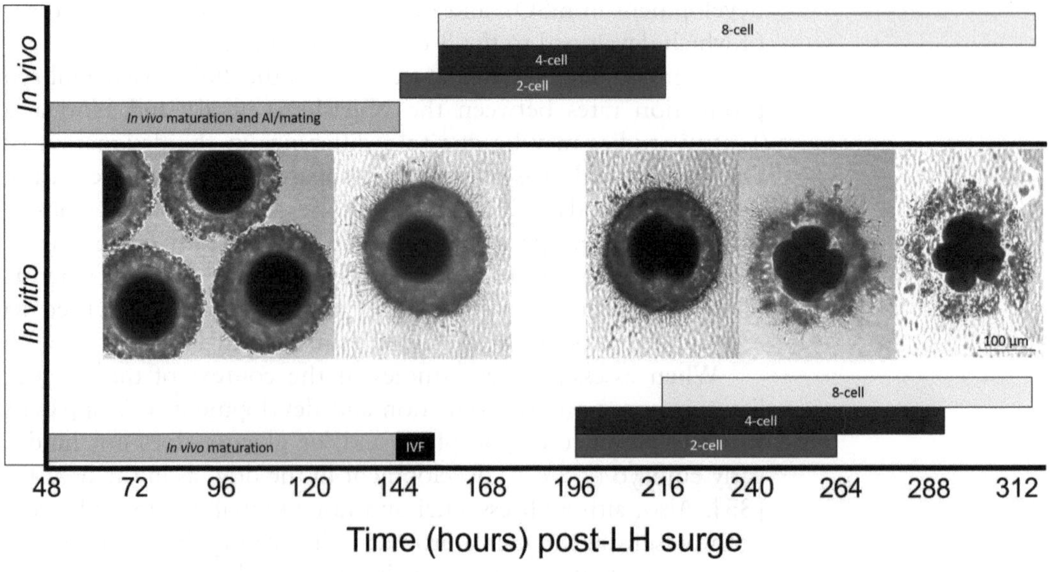

Fig. 1 Kinetics of early embryo development in vivo (top, from Reynaud et al. [46]) and in vitro (bottom, from Nagashima et al. [28])

asynchrony of ovulation after the LH surge (which can range up to 36–50 h [46]), in time from ovulation to complete nuclear maturation (48–60 h [18, 47]), and in fertilization time point following the former two events.

Of the oocytes which developed into embryos in our IVF study, >92% did so within 48 h of coincubation with sperm (~192 h or Day 8 post-LH surge) [28]. At this time point, the vast majority (81%) of embryos were at the 4-cell stage. No cleavage was observed at the 24-h post coincubation time point in our study; therefore, the first two cleavage events both took place between hours 24 and 48 following the onset of coincubation in the fertilization droplet. By the subsequent check (72 h post coincubation), embryos were typically at the 6- or 8-cell stage. Despite using different sources of oocytes, embryo cleavage kinetics reported from dog IVF studies utilizing in vitro-matured ova previously reached the same consensus, with initial embryo cleavage occurring between 24 and 48 h postcoincubation with sperm, and development to 4- to 8-cell stage by the 72-h mark [37–39]. Therefore, it appears that the kinetics of embryo development in the dog are fairly conserved across gamete sources, media preparations, and protocol variations.

In spite of this apparent consistency, we also noted that a subset of embryos (5.4% of oocytes coincubated with sperm) experienced "delayed" cleavage, wherein development to the 2-cell stage did not occur until sometime around 72–120 h postcoincubation, or 216–264 h (Day 9–11) post-LH surge [28]. Frequency of delayed cleaving embryos was higher in oocytes collected on Day 5 post-LH surge, and with one of the individual males used as a sperm donor. A similar phenomenon was described by Hori and Tsutsui, with cleavage to the 2-cell stage occurring up to 108 h after initial coculture. As a comprehensive evaluation of the developmental competence of these "delayed" embryos has not been completed, it is still uncertain if delay is indicative of poor embryo quality, as in the human. Domestic dog ova are known to have a uniquely long period of fertility, with live young having been produced after intrauterine artificial insemination up to 10 days post-LH surge [48]. Therefore, it is also possible that these "delayed" embryos could produce normal pregnancies.

1.4 Embryonic Genome Activation and Advanced Stage Embryo Culture

In the Nagashima et al. IVF study, 15 of the 21 embryos cultured for 96 h (i.e., to 240 h or Day 10 post-LH surge) developed to ≥8-cell stage, but development halted between the 8- and 16-cell stage in vitro. This block coincides with the time of genome activation in the canid embryo [49]. Results suggest that, although culture conditions described are sufficient for short term culture of early stage domestic dog embryos, they are not sufficient to overcome this developmental block during long-term in vitro culture. Nevertheless, embryo development in vitro to the 16-cell stage and beyond has been observed, albeit infrequently, following in vitro oocyte maturation and IVF in the dog. Otoi et al. produced four embryos that were at the 16-cell stage or beyond from 30 embryos produced using an M199-based medium [50]. IVF and culture of in vitro-matured dog oocytes in M199-based media have produced morulae and blastocyst stage embryos in other studies as well [51, 52]. The formulation of M199 is much more complex compared to even the modified NCSU used for the fertilization and early embryo culture of the IVF-derived embryos that produced live young, and so may provide embryos with better support during embryonic genome activation. Further studies are needed to identify component(s) that may be responsible for this effect.

In vivo-derived wild canid embryos have also been cultured to advanced stages in vitro. Post-embryonic genome activation-stage silver fox embryos have been maintained in vitro using M199 [53]. Renton et al. reported a 1-cell stage embryo collected 5 days after ovulation developing to the morula stage after 120 h in vitro [54]. The embryo was also cultured in M199 supplemented with 20% FBS. Further, our laboratory has had some success in culturing compact morula-stage embryos in a similar medium to the blastocyst stage (Fig. 2), indicating this base medium is supportive to the development of advanced stage embryos.

Interestingly, in the latter two experiments, as well as in the elegant work describing preimplantation development of later stage dog embryos by Abe et al., in vivo-derived dog blastocysts were observed to initially shrink within their zona pellucidae around the

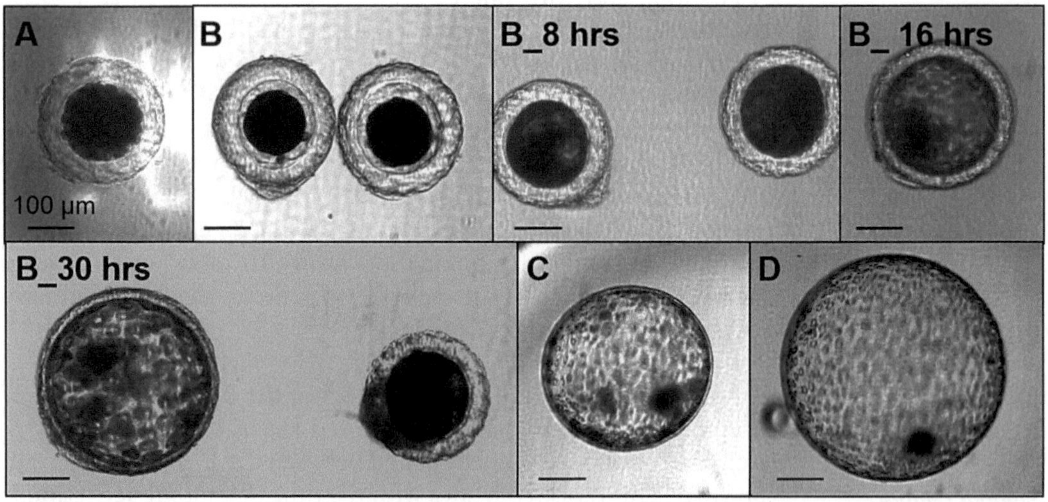

Fig. 2 Advanced stage embryo development. Images of (**a**) a freshly collected morula and (**b**) compact morulae. Panels **b**—8, 16 and 30 h show the compact morulae from (**b**) cultured for the referenced time, with both initially developing to the early blastocyst stage after 8 h, but only one expanding further after 30 h and the other degenerating. (**c**, **d**) show expanding blastocysts collected Days 14 and 15 post-LH surge, respectively

time of collection [54–56]. In all cases, the blastocoel cavity recovered after a short period of in vitro culture. Renton et al. noted that recovery could be accomplished quickly utilizing a 1:1 Ham's F10-distilled water medium, suggesting the rapid change in cavity volume is related to a change in the osmolarity of their environment. In our work, blastocoel reexpansion was typically achieved after ~2 h incubation in M199 with 20% FBS. Further, we have previously physically collapsed cavities of expanded blastocysts via microinjection ($n = 8$), then transferred to a recipient uterus and produced two live births (Data not shown). Human blastocyst studies have deliberately collapsed this cavity to reduce opportunity for ice crystal formation during cryopreservation [57, 58], and also produced successful pregnancies. Although not ideal, these data indicate blastocoel cavity collapse in vitro does not prohibit successful birth in humans or dogs following transfer.

1.5 Embryo Transfer

Currently, there are three key methods to transfer embryos which have been documented as resulting in live births. Choice of method depends in part on the stage of embryos being transferred, as transfer location must be physiologically appropriate for the embryo. (Details of embryo stages or locations have been well described in the dog [55, 59].) The first two are surgical transfers, to either the oviduct or the uterine horn. The oviduct is preferred for early stage embryos (<8-cell), whereas transfer into the uterine horn is technically more straightforward but appears better suited for later stage embryos (≥8-cell stage). Both strategies are currently performed via laparotomy, but have the potential for completion via laparoscopy to reduce invasiveness. The third method, nonsurgical embryo transfer, is a minimally invasive standing procedure, involving passage of a cystoscope-equipped catheter through the cervix into the uterine body and horns [59]. As deposition of embryos is into the uterine horn, this again is preferably used with ≥8-cell stage embryos.

1.6 Future Directions

Although we have focused on the recent advances in canine embryo technologies, we would be remiss not to also highlight the areas in need of additional study. The development of in vitro fertilization in this species opens the door to reevaluate in vitro oocyte maturation protocols with a proven-method end point. Establishment of protocols that will result in the production of in vitro-matured oocytes will, in turn, allow for more studies to understand the mechanisms regulating embryo development and embryo genome activation. Therefore, developing a robust IVM protocol for oocytes collected from diestrus and anestrus dogs is the highest priority. Following this, high rates of success in terms of live birth after oviductal embryo transfer should quickly facilitate translation of these advances to broader applications, such as producing viable biomedical research models for human medicine [33], or applying IVF to the conservation of endangered canid species.

2 Materials

All media should be prepared using ultrapure water (18 mΩ with low VOC) and cell culture grade reagents. Unless otherwise indicated, store all reagents at room temperature. All plastic materials in contact with gametes (i.e., culture tubes, petri dishes) should be polystyrene.

2.1 Oocyte Collection

1. PB1 [60]: 136.87 mM NaCl, 2.68 mM KCl, 1.47 mM KH_2PO_4, 0.49 mM $MgCl_2{\cdot}6H_2O$, 8.09 Na_2HPO_4, 0.33 Na pyruvate, 5.56 mM glucose, 3.00 g/l BSA, 100 U/ml penicillin, and streptomycin sulfate, and 0.90 mM $CaCl_2{\cdot}2H_2O$ (*see* **Note 1**). Adjust pH to 7.55–7.6, then make up to 1 l with ultrapure water. Store at 4 °C up to 2 weeks.
2. Warming pad.
3. #10 blades and scalpel blade holder.
4. Sterile forceps.
5. Sterile scissors.
6. Luer-lock syringe (5–20 cc).
7. Sterile curved fine-tip forceps.
8. 23 G winged infusion set (*see* **Note 2**), or straight needle.
9. 100 × 15 mm petri dishes.
10. 60 x 15 mm petri dishes.
11. 35 x 10 mm petri dishes.

2.2 IVF and Early Embryo Culture

1. mCCM [28, 61]: 83.29 mM NaCl, 4.78 mM KCl, 1.71 mM $CaCl_2$ (*see* **Note 1**), 1.19 mM KH_2PO_4, 0.25 mM Na pyruvate, 2.78 mM glucose, 100 IU/ml penicillin G, and streptomycin sulfate, and 1.0 mM $MgCl_2{\cdot}6H_2O$ in ultrapure water. Store at 4 °C for up to 2 weeks. Osmolality should be around 300 mOsm. Add 37.61 mM $NaHCO_3$, 21.66 mM Na lactate 60% syrup (*see* **Note 3**), and 2 g/ml bovine serum albumin (BSA) on the day of IVF. Adjust pH to 7.8. Warm in incubator at 38.5 °C and 5% CO_2.
2. cNCSU: *see* Table 2. Prepare cNCSU with ultrapure water, pH to 7.55–7.6 and store at 4 °C. Add 1× BME and 1× minimum essential medium (MEM) nonessential amino acids solutions (also stored at 4 °C) on the day of IVF.
3. Polyvinyl alcohol-coated tubes [62] (*see* **Note 4**).
4. Mineral oil, cell culture grade.
5. 35 × 10 mm non-tissue culture treated petri dishes.
6. Mouth pipette.

7. Hypoxia chamber.
8. 5% CO_2, 5% O_2, 90% N_2 gas.
9. 10 mM progesterone in DMSO (optional).

2.3 Late Stage Embryo Culture

1. Blastocyst Culture Medium: M199, 10% v/v FBS, and 100 U/ml penicillin and streptomycin sulfate (pH 7.5, Osm 340).
2. Mineral oil, cell culture grade.
3. Hypoxia chamber.
4. 5% CO_2, 5% O_2, 90% N_2 gas.

2.4 Oviductal Embryo Transfer (Surgical)

1. Transfer medium: M199, 20% w/v FBS (*see* **Note 5**).
2. Recipient bitch at synchronized stage of estrous cycle.
3. Sterile forceps.
4. Sterile surgical scissors.
5. 3.5 FR, 4.5″ (11.4 cm) tomcat catheter.
6. Luer-lock syringe (1–3 cc).
7. Dissecting scope with warming plate.

2.5 Uterine Horn Embryo Transfer (Surgical)

1. Transfer medium (*see* Subheading 2.4, **item 1**).
2. Recipient bitch at synchronized stage of estrous cycle.
3. #11 scalpel blade and holder, sterile.
4. Autoclaved wide-orifice 200 μl pipette tips.
5. 200 μl pipette.
6. Dissecting scope with warming plate.
7. Suture (optional).

3 Methods

3.1 In Vivo-Matured Oocyte Retrieval

1. Monitor onset of donor bitch's estrous cycle via behavioral and physical signs of proestrus. Use vaginal cytology, serum progesterone [24, 25], and/or LH surge, to determine the date of the LH surge (counted as Day 0). Specifically time ovariohysterectomy for Day 6 post-LH surge.
2. Warm 500 ml PB1 overnight in incubator at 38.5 °C prior to collection.
3. Transfer excised tract from Day 6 post-LH surge bitch immediately after removal into at least 300 ml warmed PB1 in a large beaker. Transport to laboratory.
4. In tissue culture hood using forceps and scissors, remove entire bursal compartment around an ovary, including a small portion of the uterine horn just distal to the uterotubal junction.

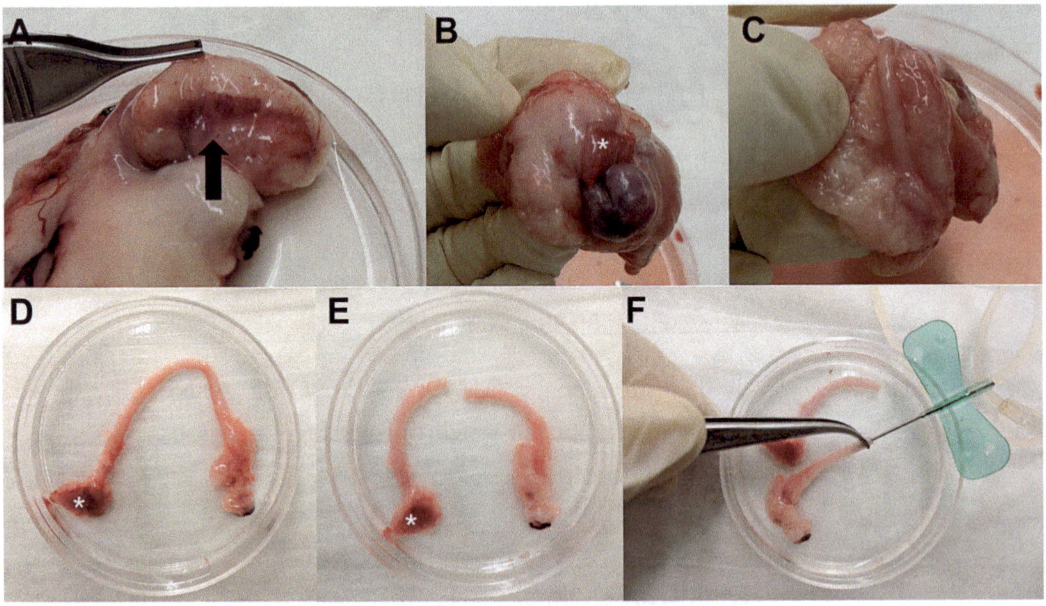

Fig. 3 Isolation and flushing of the dog oviduct. (**a**) Excised bursa-encompassed ovary and cranial uterine horn, with arrow indicating bursal window. (**b**) Opened and inverted bursa with a white asterisk indicating infundibulum. (**c**, **d**) Isolation of oviduct within bursal fat pad. (**e**) Dissected oviduct for ease of cannulation (**f**) and flushing to collect oocytes or early embryos

Transfer to 60 × 100 mm petri dish on warming pad or warmed surface, keeping moist with additional PB1 (approx. 5 ml, but will depend on the size or breed of the dog).

5. Identify bursal window then create an opening in this region large enough to invert the bursal sac without cutting into the oviduct running through (Fig. 3a, b, *see* **Note 6**).
6. Record # of corpora lutea on each ovary.
7. Isolate oviduct from bursa using #10 scalpel blade (Fig. 3c, *see* **Note 7**).
8. Make a transverse section of each oviduct with a scalpel blade (Fig. 3d, e, *see* **Note 8**) and transfer each isolated oviduct section to an individual 60 mm petri dish with ~3 ml fresh warmed PB1.
9. For each oviduct half, cannulate from the end cut in Subheading 3.1, **Step 7** with 23 g needle infusion set attached to a syringe filled with warmed PB1. This can be done using curved fine-tip forceps to sheath the oviduct over the needle.
10. Flush (Fig. 3f) oviduct by holding cannulated oviduct firmly over needle end with forceps to prevent backflow while slowly depressing syringe with PB1. Flush each side with at least 3 ml of PB1, into the petri dish, then transfer oviductal tissue to a fresh dish with PB1 and set aside on a warming plate.

11. Observe dish with flushed medium under a dissecting microscope. Oocytes will typically settle to the base of the dish. Using either a mouth pipette or pipette set to small volume (<10 μl), collect oocytes (*see* **Note 9**) into a new 35 mm petri dish with warmed cNCSU to wash. Maintain dish in incubator at 38.5 °C and 5% CO_2 until all oocytes have been collected.
12. Repeat flushing of oviducts set aside in Subheading 3.1, **Step 9** if the number of oocytes recovered is significantly less than the number of CLs on the corresponding ovary.
13. Maintain room lights on low or off once oocytes have been recovered (*see* **Note 10**).

3.2 Sperm Preparation (See Note 11)

1. If possible, set temperature-controlled centrifuge to 37 °C.
2. Prepare CCM with BSA, bicarbonate, and sodium lactate, pH to 7.8, sterile filter and maintain at 38.5 °C and 5% CO_2.
3. Collect first and second fraction of semen via manual stimulation into sterile tube or collection bag (*see* **Note 12**).
4. Transfer semen to PVA-coated tube.
5. Centrifuge at 100 × *g* for 1 min to remove debris (primarily epithelial cells).
6. Transfer supernatant to fresh PVA-coated tube diluted 1:1 with warmed mCCM.
7. Centrifuge at 400 × *g* × 5 min to pellet the sperm. Remove supernatant and discard, minimally disrupting pellet.
8. Resuspend sperm pellet in 1 ml mCCM.
9. Using a hemocytometer and a small sample of the washed sperm in water (typically 1:200 dilution), assess sperm concentration.
10. Incubate washed sperm at a concentration of 7.5×10^6 sperm/ml in mCCM in PVA-coated tubes at 38.5 °C and 5% CO_2 for at least 2 h to capacitate (*see* **Note 13**).

3.3 Fertilization

1. Prepare 100 μl droplets of cNCSU overlaid with mineral oil by first pipetting 5 μl onto the base of a non tissue-culture treated 35 mm petri dish, then overlaying with ~4.5 ml cell culture grade mineral oil, then pipetting the remaining 95 μl cNCSU into the 5 μl droplet under the oil (*see* **Note 14**). Equilibrate for at least 1 h.
2. After 2 h capacitation, count sperm again (*see* **Note 15**), then adjust volume of cNCSU droplet for a final concentration of 1.0×10^6 sperm/ml in a 100 μl droplet following the addition of capacitated sperm (*see* **Note 16**).
3. Transfer mature oocytes by mouth pipette (minimal medium) into the cNCSU fertilization droplet with capacitated sperm.

4. Coincubate gametes for 14 h at 38.5 °C and 5% CO_2, 5%O_2, and 90% N_2 in a humidified hypoxia chamber (*see* **Note 17**).
5. Prepare individual culture dishes for each oocyte collected, with a 50 μl droplet of cNCSU under mineral oil as in Subheading 3.3, **Step 1**.
6. After 14 h coincubation (*see* **Note 18**), transfer presumptive zygotes into a dish with 5 ml warmed cNCSU and wash by gentle pipetting to remove loosely attached sperm and/or cumulus cells.
7. Transfer each presumptive zygote in minimal medium to a preequilibrated culture dish prepared in Subheading 3.3, **Step 5**, then return all dishes to the hypoxia chamber for culture at 38 °C and 5% CO_2, 5%O_2, and 90% N_2 (*see* **Note 19**).
8. Evaluate embryo cleavage at 48 h post-IVF.

3.4 Embryo Culture: Early Embryos (1- to 8-Cell Stage)

1. Prepare 50 μl droplets of cNCSU overlaid with mineral oil, as in Subheading 3.3, **Step 1**. Allow the medium to equilibrate for at least 1 h at 38.5 °C and 5% CO_2, 5% O_2, and 90% N_2 in a humidified hypoxia chamber.
2. Culture each embryo individually in the preequilibrated dishes at 38.5 °C and 5% CO_2, 5% O_2, and 90% N_2 in a humidified hypoxia chamber.
3. Exchange half the medium (25 μl) with fresh every 48 h.

3.5 Embryo Culture: Advanced Embryos (Compact Morulae to Blastocysts) (See Note 20)

1. Prepare 25 μl droplets of Blastocyst Culture Medium overlaid with mineral oil, as in Subheading 3.3, **Step 1**.
2. Transfer blastocysts using wide-orifice 200 μl or smaller pipette tips (*see* **Note 21**) and culture each embryo individually in the preequilibrated dishes at 38.5 °C and 5% CO_2, 5%O_2, and 90% N_2 in a humidified hypoxia chamber.
3. Exchange half the medium (12.5 μl) with fresh every 24 h.

3.6 Oviductal Embryo Transfer

1. Monitor estrus as in Subheading 3.1, **Step 1** to time synchronization of recipient bitch with the stage of the embryos.
2. Exteriorize both left and right ovarian bursa of recipient via laparotomy (*see* **Note 22**) and manually dissect the suspensory ligament as far from the ovarian bursa as possible while minimizing manipulation of the area.
3. Using forceps and scissors, make a slit through the bursal window wide enough to access the infundibulum. This will vary depending on the size of the animal and maneuverability possible based on field exteriorized via laparotomy.
4. Perform a "dry run" cannulation of the infundibulum with a 3.5 FR, 4.5″ tomcat catheter to locate lumen.

5. Draw up embryos with ~20 μl medium into the tomcat catheter connected to a 1 cc syringe (*see* **Note 23**). First, draw up air, then embryos in medium, then additional medium with no embryos in the distal tip of the catheter.
6. Advance the catheter approximately 2.5 cm into the oviduct via the previously identified (Subheading 3.6, **Step 4**) ostium.
7. Slowly dispense embryos, withdrawing catheter slightly as syringe plunger is depressed. The last portion dispenses air to "push" embryos the remaining distance before removing catheter from infundibulum.
8. Check to ensure that there was no backflow from oviduct. If so, attempt to collect fluid, examine for embryos, and retransfer if needed.
9. Check tom cat catheter under a microscope to confirm all embryos were dispensed, may see presence of ciliated cells on the outside of the catheter (brushed off from oviductal lumen).
10. Fold bursa back over the ovary.

3.7 Uterine Horn Embryo Transfer

1. Monitor estrus as in Subheading 3.1, **Step 1** to time synchronization of recipient bitch with the stage of embryos.
2. Exteriorize both uterine horns of recipient via laparotomy (*see* **Note 22**).
3. Using an #11 blade scalpel, make a small stab incision (avoiding blood vessels, as possible) into the cranial portion of the uterine horn.
4. Draw up embryos in a minimal volume of Blastocyst Culture Medium using a large-orifice 200 μl pipette tip (*see* **Note 23**).
5. Insert pipette tip, angled caudally, as far as possible into the uterine lumen from the stab incision. Slowly but firmly depress plunger to expel embryos, as you will likely encounter backpressure.
6. When plunger is fully depressed, leave pipette tip embedded in the lumen but remove pipette. Release plunger then reattach pipette to tip. Expel air while slowly backing tip out of the lumen and holding the sides of the incision closed to prevent backflow of embryos.
7. Check the pipette tip to make sure no embryos are adhered to the side of the tip.
8. Press sides of stab incision site closed for an additional minute following transfer to prevent backflow of embryos and blood entering lumen of the uterine horn.
9. If desired, close incision site with suture.

4 Notes

1. Add $CaCl_2$ last or dissolve first in 1 ml water and slowly add to media, to avoid forming calcium phosphate precipitate.
2. Depending on the size of the dog (and therefore oviduct), a different gauge needle may be required. Twenty-three gauge was utilized for work with beagles (~9 to 14 kg). Winged infusion sets, as in Fig. 3f, are easier to maneuver than straight needles for flushing.
3. Sodium lactate syrup sometimes "goes bad," even under recommended storage conditions, and will make the CCM toxic to the sperm.
4. Use of PVA coated tubes are preferred for the sperm incubation to reduce sperm adherence to the tube walls during capacitation. Uncoated sterile tubes can also be used if needed. To prepare PVA-coated tubes, dissolve 2 g polyvinylalcohol in 50 ml ultrapure water (4% w/v) by stirring while heating up to 98 °C. Slowly add 50 ml isopropanol and continue to stir and heat until solution becomes clear. Cool to room temperature, then fill ultra-clear sterile polystyrene tubes with the solution and let sit for 15 min. Aspirate the solution (will leave a clear film on the tube walls), then leave the tubes open to dry overnight at room temperature. Rinse tubes with distilled water then fill with distilled water and allow to stand for 1–2 h at room temperature. Rinse once more with distilled water, tap to remove excess liquid, then leave to dry.
5. Embryo transfer medium might be adjustable—this high protein version was utilized in [28] as embryos were postthaw. The critical portion is to avoid adding amino acids, as this might cause inflammation at ET site. This is also the rationale for minimal medium volume used during the transfer.
6. Some bursae are thicker or more "fatty" than others, making it difficult to follow where the oviduct is running. Rather than attempting to isolate the oviduct solely with tools, grasp the bursa through the opening created through the window with thumb and forefinger (with sterile glove) and gently roll the tissue between them to feel for the oviduct. With this position, once a shallow slice is made into the bursa, you can then also use your thumb to pull away the excess tissue to expose the oviduct (Fig. 3c).
7. It is helpful to reduce the amount of excess tissue or fat left on the oviduct while you are isolating it, as this will make it easier to cannulate for flushing later and reduce the amount of lipid droplets in the flushing dish later. However, this must be balanced with time to get the oocytes collected quickly postsurgery.

8. Flushing from the center toward the infundibulum or utero-tubule junction is preferred because cannulation of the lumen from a clean cut is straightforward. Oocytes are unlikely to be in this region anyway, as typically by 48 h postovulation all oocytes would have passed through the middle of the oviduct [63]. It is recommended to leave the most proximal portion of the uterine horn on the isthmic end of the oviduct as oocytes tend to gather near the UTJ. It will require more pressure to flush, as the UTJ is restrictive. Flush 2–3 times with gentle but steady pressure, as too much pressure runs the risk of exerting negative shear forces on any oocytes trapped in this region (It should also be noted that abnormally shaped or elongated oocytes, likely the result of too-intensive flushing, still cleaved to 8+ cell stages in our IVF system.).
9. As previously noted [18], dog ova and even early stage embryos maintain layer(s) of cumulus cells. While this obscures observations, it does not preclude successful fertilization or development (*see* Fig. 1 images).
10. Once the oviducts have been flushed, most work should be done in a dim room to reduce the potential of light damage to the oocytes or embryos.
11. If performing IVF with freshly collected semen and in vivo-matured oocytes, it is best to collect sperm and start the capacitation at least 1 h prior to surgery, or about 1.5 h prior to obtaining the reproductive tract. This provides time to collect oocytes within the 3 h capacitation period.
12. Males were collected 3 days prior to the sperm capacitation for IVF as a "flush," to improve the quality of sperm in the ejaculate on the IVF day.
13. Concentrations as low as 5×10^6 sperm/ml are sufficient for capacitation, but domestic dog sperm are very sensitive to both high and low concentrations. If working with low sperm count samples, reduce the volume of the capacitation medium rather than incubating at a concentration below 5×10^6 sperm/ml.
14. All of the published IVF-derived embryos producing puppies were fertilized and cultured in cNCSU supplemented with 10 mM progesterone in DMSO, for a final concentration of 6 μg/ml [28], but no significant difference in embryo production rates were observed in the presence or absence of progesterone. As no nonsupplemented embryos were transferred into recipient oviducts, it is not known if progesterone had an impact on final pregnancy or birth outcome, but based on the embryo results progesterone supplementation is considered "optional" to the method.
15. This concentration should be very close to the 7.5×10^6 sperm/ml, which results in 13 μl being added to 87 μl

cNCSU; however, sperm tend to clump together and may be either difficult to count, or the clumps fall quickly to the base of the incubation tube even after gentle mixing. As result, the density of sperm counted after 2 h of incubation was typically lower, so it was more common to add 15–20 μl capacitated sperm to IVF drops.

16. The lowest sperm concentration successfully producing embryos was 200,000 sperm/ml in the 100 μl fertilization droplet, wherein five of six oocytes developed into embryos.
17. It is possible to reduce the time of coincubation of sperm and eggs for fertilization. More recently, we have produced embryos from the same IVF protocol, but using a 12 h fertilization incubation.
18. Some loosely attached sperm and outer cumulus cells will detach during this gentle washing, but the majority will remain closely adhered. If the presumptive zygote is readily denuded at this time, it may have degenerated.
19. After 14 h of coincubation, embryos are typically spinning or rotating in the dish due to motility of attached sperm, as has also been noted in in vivo-derived zygotes [54].
20. Described culture conditions based on short-term (42 h) culture in which in vivo-derived morulae developed into blastocysts (Fig. 2).
21. Although shrinkage of the blastocoel cavity within the zona is typically attributed to changes in osmolarity, we have also noted fluctuations that appear to be due to mechanical pressure on blastocysts. Cavities will still reexpand after reequilibration and continued incubation.
22. Avoid use of nonsteroidal anti-inflammatory drugs for pain management of recipient bitch during and after embryo transfer, to reduce risk of miscarriage or pregnancy complications, if similar to humans [64].
23. It is useful to have embryos closely accessible to surgical suite for ET. Set up cart with microscope and warming stage for embryos in room adjacent to surgery, but wait until last moment to pipet up embryos. Hold pipette or catheter horizontally to prevent loss of embryos out the end while moving over to transfer site. We have typically utilized three individuals for a transfer—a surgeon preparing the recipient bitch (completely sterile), an assistant drawing up embryos into the pipette or catheter (clean, but not sterile—unless access to surgical-suite level sterile microscope or setup is available), and an embryo transfer technician (half sterile—hand which receives and pipets embryos from assistant becomes nonsterile but able to stand over surgical site to pipet or depress plunger to transfer embryos).

References

1. Tsai K, Clark L, Murphy K (2007) Understanding hereditary diseases using the dog and human as companion model systems. Mamm Genome 18(6):444–451. https://doi.org/10.1007/s00335-007-9037-1
2. Nicholas FW (n.d.) OMIA – online Mendelian inheritance in animals. http://omia.angis.org.au/home/. Accessed 1 May 2017
3. Hong SG, Kim MK, Jang G, Oh HJ, Park JE, Kang JT, Koo OJ, Kim T, Kwon MS, Koo BC, Ra JC, Kim DY, Ko C, Lee BC (2009) Generation of red fluorescent protein transgenic dogs. Genesis 47(5):314–322. https://doi.org/10.1002/dvg.20542
4. Kim MJ, Oh HJ, Park JE, Kim GA, Hong SG, Jang G, Kwon MS, Koo BC, Kim T, Kang SK, Ra JC, Ko C, Lee BC (2011) Generation of transgenic dogs that conditionally express green fluorescent protein. Genesis 49 (6):472–478
5. Jeong YW, Lee GS, Kim JJ, Park SW, Ko KH, Kang M, Kim YK, Jung EM, Hyun SH, Shin T, Jeung EB, Hwang WS (2012) Establishment of a canine model of human type 2 diabetes mellitus by overexpressing phosphoenolypyruvate carboxykinase. Int J Mol Med 30(2):321–329
6. Zou Q, Wang X, Liu Y, Ouyang Z, Long H, Wei S, Xin J, Zhao B, Lai S, Shen J, Ni Q, Yang H, Zhong H, Li L, Hu M, Zhang Q, Zhou Z, He J, Yan Q, Fan N, Zhao Y, Liu Z, Guo L, Huang J, Zhang G, Ying J, Lai L, Gao X (2015) Generation of gene-target dogs using CRISPR/Cas9 system. J Mol Cell Biol 7 (6):580–583. https://doi.org/10.1093/jmcb/mjv061
7. IUCN (2017) IUCN Red list of threatened species. www.iucnredlist.org. Accessed 3 Oct 2017. Vol 2017.2
8. Concannon PW (2011) Reproductive cycles of the domestic bitch. Anim Reprod Sci 124 (3/4):200–210. https://doi.org/10.1016/j.anireprosci.2010.08.028
9. Bouchard G, Youngquist RS, Vaillancourt D, Krause GF, Guay P, Paradis M (1991) Seasonality and variability of the interestrous interval in the bitch. Theriogenology 36(1):41–50. https://doi.org/10.1016/0093-691X(91)90432-D
10. Kutzler MA (2005) Induction and synchronization of estrus in dogs. Theriogenology 64 (3):766–775. https://doi.org/10.1016/j.theriogenology.2005.05.025
11. Shille V, Thatcher M, Simmons K (1984) Efforts to induce estrus in the bitch, using pituitary gonadotropins. J Am Vet Med Assoc 184(12):1469–1473
12. Kooistra H, Okkens A, Bevers M, Popp-Snijders C, Van Haaften B, Dieleman S, Schoemaker J (1999) Bromocriptine-induced premature oestrus is associated with changes in the pulsatile secretion pattern of follicle-stimulating hormone in beagle bitches. J Reprod Fertil 117(2):387–393
13. Concannon PW, Temple M, Montanez A, Newton L (2006) Effects of dose and duration of continuous GnRH-agonist treatment on induction of estrus in beagle dogs: competing and concurrent up-regulation and down-regulation of LH release. Theriogenology 66 (6–7):1488–1496. https://doi.org/10.1016/j.theriogenology.2006.02.007
14. Verstegen J, Onclin K, Silva L, Concannon P (1999) Effect of stage of anestrus on the induction of estrus by the dopamine agonist cabergoline in dogs. Theriogenology 51 (3):597–611
15. Volkmann D, Kutzler M, Wheeler R, Krekeler N (2006) The use of deslorelin implants for the synchronization of estrous in diestrous bitches. Theriogenology 66(6):1497–1501
16. Trigg T, Wright P, Armour A, Williamson P, Junaidi A, Martin G, Doyle A, Walsh J (2001) Use of a GnRH analogue implant to produce reversible long-term suppression of reproductive function in male and female domestic dogs. J Reprod Fertil Suppl 57:255–261
17. Romagnoli S, Stelletta C, Milani C, Gelli D, Falomo ME, Mollo A (2009) Clinical use of deslorelin for the control of reproduction in the bitch. Reprod Domest Anim 44:36–39. https://doi.org/10.1111/j.1439-0531.2009.01441.x
18. Reynaud K, Fontbonne A, Marseloo N, Thoumire S, Chebrout M, de Lesegno CV, Chastant-Maillard S (2005) *In vivo* meiotic resumption, fertilization and early embryonic development in the bitch. Reproduction 130 (2):193–201. https://doi.org/10.1530/rep.1.00500
19. Rodrigues BA, Rodrigues JL (2010) *In vitro* maturation of canine oocytes: a unique conundrum. Anim Reprod 7(1):3–15
20. Chastant-Maillard S, Viaris de Lesegno C, Chebrout M, Thoumire S, Meylheuc T, Fontbonne A, Chodkiewicz M, Saint-Dizier M, Reynaud K (2011) The canine oocyte: uncommon features of *in vivo* and *in vitro* maturation. Reprod Fertil Dev 23(3):391–402. https://doi.org/10.1071/RD10064

21. Pereira L, Bicudo S, Lopes M (2012) Oocyte maturation in bitches. Anim Reprod 9:205–209
22. Chastant-Maillard S, Chebrout M, Thoumire S, Saint-Dizier M, Chodkiewicz M, Reynaud K (2010) Embryo biotechnology in the dog: a review. Reprod Fertil Dev 22 (7):1049–1056
23. Reynaud K, Saint-Dizier M, Chastant-Maillard S (2004) *In vitro* maturation and fertilization of canine oocytes. Methods Mol Biol 253:255–272
24. Kim Y, Travis AJ, Meyers-Wallen VN (2007) Parturition prediction and timing of canine pregnancy. Theriogenology 68(8):1177–1182
25. Kutzler MA, Mohammed HO, Lamb SV, Meyers-Wallen VN (2003) Accuracy of canine parturition date prediction from the initial rise in preovulatory progesterone concentration. Theriogenology 60(6):1187–1196. https://doi.org/10.1016/S0093-691X(03)00109-2
26. Wildt DE, Panko WB, Chakraborty PK, Seager SWJ (1979) Relationship of serum estrone, estradiol-17β and progesterone to LH, sexual behavior and time of ovulation in the bitch. Biol Reprod 20(3):648–658. https://doi.org/10.1095/biolreprod20.3.648
27. Hase M, Hori T, Kawakami E, Tsutsui T (2000) Plasma LH and progesterone levels before and after ovulation and observation of ovarian follicles by ultrasonographic diagnosis system in dogs. J Vet Med Sci 62(3):243–248
28. Nagashima JB, Sylvester SR, Nelson JL, Cheong SH, Mukai C, Lambo C, Flanders JA, Meyers-Wallen VN, Songsasen N, Travis AJ (2015) Live births from domestic dog (*Canis familiaris*) embryos produced by *in vitro* fertilization. PLoS One 10(12):e0143930. https://doi.org/10.1371/journal.pone.0143930
29. Bouchard GF, Solorzano N, Concannon PW, Youngquist RS, Bierschwal CJ (1991) Determination of ovulation time in bitches based on teasing, vaginal cytology, and elisa for progesterone. Theriogenology 35(3):603–611. https://doi.org/10.1016/0093-691X(91)90456-N
30. Kim M, Oh H, Park J, Hong S, Kang J, Koo O, Kang S, Jang G, Lee B (2010) Influence of oocyte donor and embryo recipient conditions on cloning efficiency in dogs. Theriogenology 74(3):473–478
31. Hossein MS, Kim MK, Jang G, Fibrianto HY, Oh HJ, Kim HJ, Kang SK, Lee BC (2007) Influence of season and parity on the recovery of *in vivo* canine oocytes by flushing fallopian tubes. Anim Reprod Sci 99(3):330–341
32. Jang G, Kim M, Oh H, Hossein M, Fibrianto Y, Hong S, Park J, Kim J, Kim H, Kang S (2007) Birth of viable female dogs produced by somatic cell nuclear transfer. Theriogenology 67(5):941–947
33. Zou Q, Wang X, Liu Y, Ouyang Z, Long H, Wei S, Xin J, Zhao B, Lai S, Shen J (2015) Generation of gene-target dogs using CRISPR/Cas9 system. J Mol Cell Biol 7 (6):580–583
34. Hori T, Tsutsui T (2003) *In vitro* fertilisation of mature canine ova. Vet Rec 152(22):688
35. Schini SA, Bavister BD (1988) Two-cell block to development of cultured hamster embryos is caused by phosphate and glucose. Biol Reprod 39(5):1183–1192. https://doi.org/10.1095/biolreprod39.5.1183
36. Takahashi Y, First NL (1992) *In vitro* development of bovine one-cell embryos: Influence of glucose, lactate, pyruvate, amino acids and vitamins. Theriogenology 37(5):963–978. https://doi.org/10.1016/0093-691X(92)90096-A
37. Rodrigues BÁ, Santos LC, Rodrigues JL (2004) Embryonic development of *in vitro* matured and *in vitro* fertilized dog oocytes. Mol Reprod Dev 67(2):215–223. https://doi.org/10.1002/mrd.10394
38. Saikhun J, Sriussadaporn S, Thongtip N, Pinyopummin A, Kitiyanant Y (2008) Nuclear maturation and development of IVM/IVF canine embryos in synthetic oviductal fluid or in co-culture with buffalo rat liver cells. Theriogenology 69(9):1104–1110. https://doi.org/10.1016/j.theriogenology.2008.01.024
39. Yamada S, Shimazu Y, Kawano Y, Nakazawa M (1993) *In vitro* maturation and fertilization of preovulatory dog oocytes. J Reprod Fertil 47:227–229
40. Cruz M, Garrido N, Herrero J, Pérez-Cano I, Muñoz M, Meseguer M (2012) Timing of cell division in human cleavage-stage embryos is linked with blastocyst formation and quality. Reprod Biomed Online 25(4):371–381
41. Dal Canto M, Coticchio G, Renzini MM, De Ponti E, Novara PV, Brambillasca F, Comi R, Fadini R (2012) Cleavage kinetics analysis of human embryos predicts development to blastocyst and implantation. Reprod Biomed Online 25(5):474–480
42. Wong CC, Loewke KE, Bossert NL, Behr B, De Jonge CJ, Baer TM, Pera RAR (2010) Non-invasive imaging of human embryos before embryonic genome activation predicts development to the blastocyst stage. Nat Biotechnol 28(10):1115–1121
43. Muñoz M, Cruz M, Humaidan P, Garrido N, Pérez-Cano I, Meseguer M (2013) The type of

GnRH analogue used during controlled ovarian stimulation influences early embryo developmental kinetics: a time-lapse study. Eur J Obstet Gynecol Reprod Biol 168(2):167–172
44. Basile N, Morbeck D, García-Velasco J, Bronet F, Meseguer M (2013) Type of culture media does not affect embryo kinetics: a time-lapse analysis of sibling oocytes. Hum Reprod 28(3):634–641
45. Chavez SL, Loewke KE, Han J, Moussavi F, Colls P, Munne S, Behr B, Pera RAR (2012) Dynamic blastomere behaviour reflects human embryo ploidy by the four-cell stage. Nat Commun 3:1251
46. Reynaud K, Fontbonne A, Marseloo N, de Lesegno CV, Saint-Dizier M, Chastant-Maillard S (2006) *In vivo* canine oocyte maturation, fertilization and early embryogenesis: a review. Theriogenology 66(6):1685–1693
47. Tsutsui T (1989) Gamete physiology and timing of ovulation and fertilization in dogs. J Reprod Fertil Suppl 39:269
48. Tsutsui T, Takahashi F, Hori T, Kawakami E, Concannon PW (2009) Prolonged duration of fertility of dog ova. Reprod Domest Anim 44:230–233. https://doi.org/10.1111/j.1439-0531.2009.01450.x
49. Bysted B, Greve T (2000) Activation of the embryonic genome in the dog. Theriogenology 53:269
50. Otoi T, Shin T, Kraemer DC, Westhusin ME (2004) Influence of maturation culture period on the development of canine oocytes after *in vitro* maturation and fertilization. Reprod Nutr Dev 44(6):631–637
51. Hatoya S, Sugiyama Y, Torii R, Wijewardana V, Kumagai D, Sugiura K, Kida K, Kawate N, Tamada H, Sawada T, Inaba T (2006) Effect of co-culturing with embryonic fibroblasts on IVM, IVF and IVC of canine oocytes. Theriogenology 66(5):1083–1090. https://doi.org/10.1016/j.theriogenology.2005.12.015
52. Otoi T, Murakami M, Fujii M, Tanaka M, Ooka A, Suzuki T (2000) Development of canine oocytes matured and fertilised *in vitro*. Vet Rec 146:52–53
53. Lindeberg H, Jalkanen L, Savolainen R (1993) *In vitro* culture of silver fox embryos. Theriogenology 40(4):779–788. https://doi.org/10.1016/0093-691X(93)90213-O
54. Renton JP, Boyd JS, Eckersall PD, Ferguson JM, Harvey MJA, Mullaney J, Perry B (1991) Ovulation, fertilization and early embryonic development in the bitch (*Canis familiaris*). J Reprod Fertil 93(1):221–231. https://doi.org/10.1530/jrf.0.0930221
55. Abe Y, Suwa Y, Yanagimoto-Ueta Y, Suzuki H (2008) Preimplantation development of embryos in labrador retrievers. J Reprod Dev 54(2):135–137
56. Tsutsui T, Hori T, Okazaki H, Tanaka A, Shiono M, Yokosuka M, Kawakami E (2001) Transfer of canine embryos at various developmental stages recovered by hysterectomy or surgical uterine flushing. J Vet Med Sci 63 (4):401–405
57. Hiraoka K, Hiraoka K, Kinutani M, Kinutani K (2004) Blastocoele collapse by micropipetting prior to vitrification gives excellent survival and pregnancy outcomes for human day 5 and 6 expanded blastocysts. Hum Reprod 19 (12):2884–2888
58. Kamath MS, Mangalaraj AM, Muthukumar K, Cullinan R, Aleyamma TK, George K (2011) Blastocyst cryopreservation using solid surface vitrification: a preliminary study. J Hum Reprod Sci 4(3):114–120. https://doi.org/10.4103/0974-1208.92284
59. Abe Y, Suwa Y, Asano T, Ueta YY, Kobayashi N, Ohshima N, Shirasuna S, Abdel-Ghani MA, Oi M, Kobayashi Y, Miyoshi M, Miyahara K, Suzuki H (2010) Cryopreservation of canine embryos. Biol Reprod 84(2):363–368. https://doi.org/10.1095/biolreprod.110.087312
60. Whittingham D (1974) Embryo banks in the future of developmental genetics. Genetics 78 (1):395
61. Mahi CA, Yanagimachi R (1978) Capacitation, acrosome reaction, and egg penetration by canine spermatozoa in a simple defined medium. Gamete Res 1(2):101–109. https://doi.org/10.1002/mrd.1120010203
62. Holmquist L (1982) Surface modification of Beckman ultra-clear centrifuge tubes for density gradient centrifugation of lipoproteins. J Lipid Res 23(8):1249–1250
63. Tsutsui T (1975) Studies on the reproduction in the dog. V. On cleavage and transport of fertilized ova in the oviduct. Jpn J Anim Reprod 21(2):70–75. https://doi.org/10.1262/jrd1955.21.70
64. Li D-K, Liu L, Odouli R (2003) Exposure to non-steroidal anti-inflammatory drugs during pregnancy and risk of miscarriage: population based cohort study. BMJ 327(7411):368–368

Chapter 19

Culture of Mink Preimplantation Embryos

Jane C. Fenelon and Bruce D. Murphy

Abstract

The reproduction of the mink, *Neovison vison*, has been extensively studied over the past 70 years. The endocrine control of pregnancy is reasonably well understood, but our understanding of early embryo development is limited. The mink is one of the best characterized mammals for the study of embryonic diapause, but in order to unravel the complex interactions that occur between the blastocyst and the uterus during diapause and reactivation, a defined culture media system that supports growth is essential. Until recently, culture of the mink blastocyst has been relatively unsuccessful. This chapter will describe a method for successfully obtaining and culturing mink blastocysts and will highlight some of the unique challenges in working with this species. Methods to age match prediapause embryos in a mammal that exhibits superfetation, and to synchronize collection of reactivation from diapause stages using prolactin will be discussed. Finally, a quantitative method to determine the extent of cell proliferation in the blastocyst, a hallmark of reactivation from diapause, will be detailed.

Key words Mink, Embryonic diapause, Culture, Carnivore, Blastocyst

1 Introduction

Embryonic diapause is a period of temporary developmental arrest of the embryo that in most mammals occurs at the blastocyst stage. The study of how an embryo is maintained in diapause and subsequently reactivated provides a unique opportunity to advance our understanding of the endometrial–embryo cross talk required for successful implantation and pregnancy and many of the molecular signals appear to have been conserved across mammals [1]. However, fundamental questions remain about the molecular control of embryonic diapause in this species, as well as in all mammals.

The mink breeds in early to mid-March and exhibits an obligatory diapause that is seasonally induced, lasting from a few days to a few weeks [2]. Ovulation is induced and females can reovulate at 6 or 7 day intervals, and breed to different males. Embryos from different ovulatory events can be found in a litter, a phenomenon known as superfetation [3]. Following fertilization, the mink zygote takes approximately 6 days to form a blastocyst, which

Jason R. Herrick (ed.), *Comparative Embryo Culture: Methods and Protocols*, Methods in Molecular Biology, vol. 2006, https://doi.org/10.1007/978-1-4939-9566-0_19,

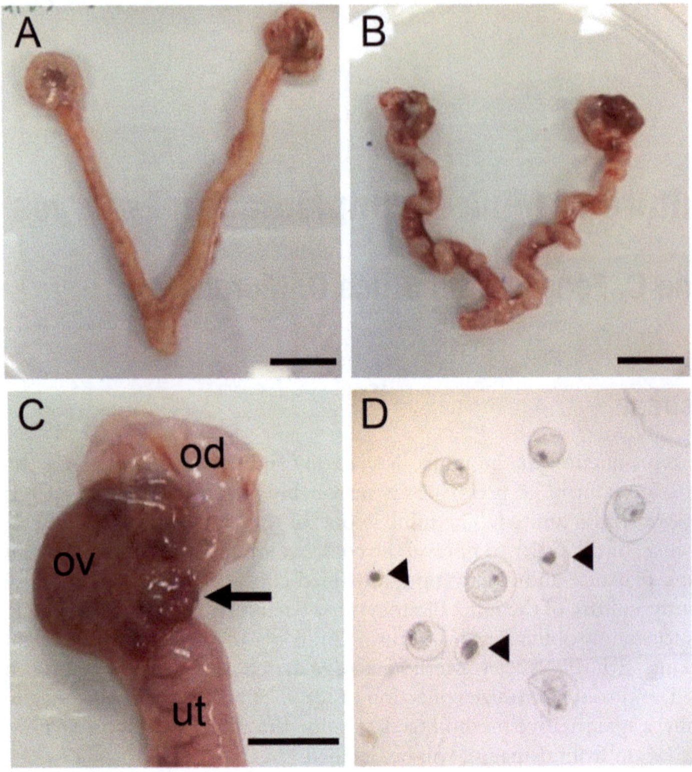

Fig. 1 (**a**) Nonimplanted uteri with intact ovaries and oviducts; (**b**) d13 PRL implanted uteri with intact ovaries and oviducts; (**c**) ovary (ov) with oviduct (od) lifted to show corpora lutea (arrow); (**d**) Flushed blastocysts showing examples of degenerate embryos (arrowheads) and "good" blastocysts. Note the shrinkage of the "good" blastocysts that occurs upon flushing, blastocysts will re-expand after 24 h in culture. Scale bar **a–b** = 1.0 mm, **c** = 0.5 mm

then enters into diapause ([4], Fig. 1a). The mink uterus is bicornuate and the blastocysts cluster in the cranial portion of the uterus until after reactivation from diapause [2]. During diapause, the mink blastocyst maintains a basal metabolism, RNA and protein synthesis, and an almost complete arrest of the cell cycle. The mink blastocyst in diapause has a diameter of around 230 μm, consisting of up to 500 cells with an inner cell mass of between 28 and 164 cells [2, 5]. Reactivation of the blastocyst from diapause occurs after the spring equinox (March 21, Northern Hemisphere) in response to the increasing photoperiod that reduces the nocturnal secretion of melatonin from the pineal gland (*reviewed in* [6]). Unlike in other mammals where the control of embryonic diapause is under the influence of progesterone and/or estrogen, in the mink, pituitary prolactin is the proximal signal responsible for reactivation of the blastocyst from diapause [7–10]. Exogenous prolactin can induce reactivation from diapause and is commonly used to synchronize reactivation stages for embryo collection

[5, 9]. The first signs of reactivation of the mink blastocyst from diapause are increased DNA and protein synthesis on the third day after prolactin injection (day 3 PRL), with the first expansion of the blastocyst occurring at d4 PRL (350–400 μm) and the largest embryos collected just before implantation (2000 μm) at d13 PRL [5]. Approximately half of the animals at d13 PRL have evidence of implantation chambers ([5] Fig. 1b).

Previous attempts to culture mink embryos in vitro have been relatively unsuccessful. The first attempt by Enders and Pearson [11] in either Tyrode's or Ringers solutions, showed that despite the trophoblast shrinking away from the zona pellucida and no obvious growth, the embryo could survive for 5 months [11]. Since then, various other culture media have been tested but with little success [12–14]. Investigations with F10 medium supplemented with different proportions of mink and rabbit serum and other various components were mostly unsuccessful [12, 13]. The only slight improvements in growth were observed after the addition of either estrone or ergosterol, but this could not be sustained [12]. Similarly, when first flushed from the uterus, most blastocyst and later stage mink embryos collapse with the zona pellucida shrunk in away from the shell coat but recover after a period in culture ([12], *personal observations*). This appears to occur regardless of the medium used, but subsequent growth is not sustained (*personal observations*). The first successful reactivation of mink diapause blastocysts in vitro occurred in a co-culture system with immortalized endometrial cells (stromal or epithelial) cultured in INRA Menezo B2 medium [15]. Culture of mink diapause blastocysts in DMEM on a mouse embryonic fibroblast feeder layer was also able to sustain embryonic survival [16]. However, until recently no monoculture media had been identified that was able to sustain embryonic growth in the mink.

Diapause blastocysts cultured in groups of 4–5 in 200 μl M16 medium can expand and reactivate, but only after the addition of the polyamine putrescine [14]. Since M16 was one of the original mouse embryo culture media that, although successful, was unable to support development from one cell to blastocyst, it was predicted that its replacement KSOM (potassium simplex optimized media) could also improve mink embryo development. However, there was no additional benefit to culturing mink embryos in KSOM compared to M16 (unpublished data).

However, there are multiple variations of embryo culture media now available that have been specifically optimized to maximize embryo development in other mammals. One of these based its design on the physiological requirements for the domestic cat preimplantation embryo, FOCM (feline optimized culture medium) [17]. Two variations were developed; FOCM1 for initial cleavage stages and FOCM2 for 32 cell to blastocysts. Culture of mink diapause blastocysts in FOCM2 significantly improved

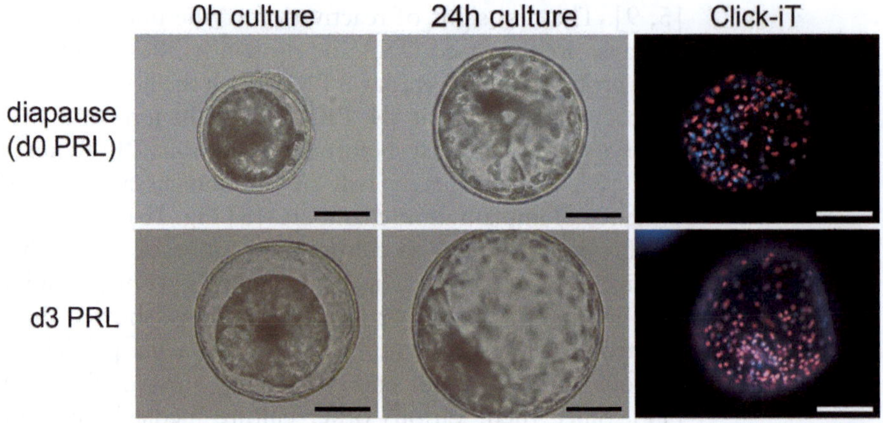

Fig. 2 Embryo stages that were successfully cultured. Diapause (d0 PRL) and d3 PRL (reactivated) blastocysts at collection (0 h), after culture (24 h) and after assay with the Click-iT® proliferation kit. Scale bar = 100 μm

embryo development in comparison to M16. Diapause blastocysts were able to increase their diameter an average of 45% from their original size, and this was confirmed by increases in both cell number and cell proliferation (Fig. 2). Similarly, the majority of the d3 PRL cultured embryos expanded by an average of 23%. Embryos were observed and measured for a maximum of 7 days or until their death as determined by embryo collapse and the onset of cell opacity. The maximum diameter obtained by any cultured embryo prior to collapse was 657.2 μm after 5 days in culture. However, no cleavage stage embryos developed in FOCM1 media and d6 PRL blastocysts did not develop in FOCM2 (older embryo stages have not been tested) (Fig. 3). Furthermore, no d6 PRL or older stage embryos were able to recover from the initial shrinkage (Fig. 3). The procedures below are based on these recent findings.

2 Materials

2.1 Prolactin Injections

1. Ovine prolactin (luteotropic hormone, 30 i.u./mg): 2 mg/ml in sterile mineral oil. Store 0.5 ml aliquots (1 mg) in 3 ml syringes at 4 °C for up to 1 month.
2. 25 G 5/8 needles.

2.2 Embryo Collection

1. DEPC-treated 1× phosphate buffered saline (PBS).
2. Embryo flushing Tissue Culture Medium 199 (TCM199): HEPES-buffered TCM199, 10% fetal bovine serum (FBS), and 50 U/ml penicillin and streptomycin. Store in 15 ml aliquots at 4 °C for 1 month.
3. Embryo culture medium, FOCM2: 100 mM NaCl, 8 mM KCl, 1 mM KH_2PO_4, 1 mM $MgSO_4{\cdot}7H_2O$, 1.5 mM glucose, 2 mM $CaCl_2{\cdot}2H_2O$, 0.1 mM Na pyruvate, 6 mM L-lactate, 0.5×

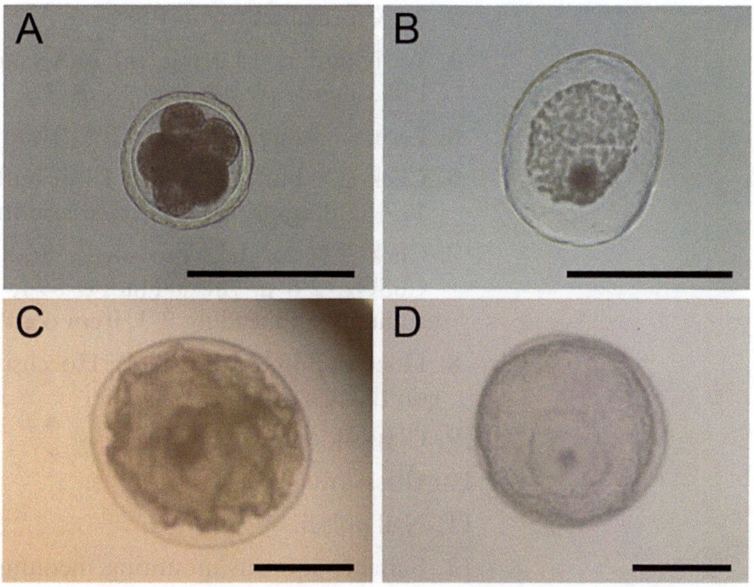

Fig. 3 Embryo stages that did not develop in culture. (**a**) d4 postmating (prediapause) embryo; (**b**) d6 PRL embryo; (**c**) d9 PRL embryo; and (**d**) d13 PRL embryo (just prior to implantation)

minimum essential medium (MEM) essential amino acids, 1× MEM nonessential amino acids, 0.075 mM myoinositol, 1× MEM vitamins, 5 μg/ml insulin, 2.75 μg/ml transferrin, 3.4 ng/ml selenium, 20 μg/ml gentamicin, 2 mg/ml recombinant human albumin or bovine serum albumin (BSA). Filter (0.22 μm) and store at 4 °C for 1 month.

4. Sterilized mineral oil.
5. Blunt 25 G 5/8 hypodermic needle.
6. 1 ml syringe.
7. Dissection scissors and forceps.
8. 35 mm petri dishes.
9. Pulled glass micropipettes.
10. Dissecting stereomicroscope.
11. Heated stage.
12. Standard tissue culture incubator set to 5% CO_2 and 95% air at 37 °C.

2.3 Embryo Cell Proliferation Assay (Click-iT® EdU Imaging Kit, Invitrogen)

1. Click-iT® EdU Medium: The provided 10 mM EdU solution is first diluted to 200 μM (2×) with FOCM2 (preequilibrated for at least 6 h—5% CO_2 and 95% air at 37 °C) and then diluted 1:1 with additional preequilibrated FOCM2. Prepare on the day of use.
2. DEPC-treated PBS.

3. 3.7% formaldehyde in PBS.
4. Wash buffer: 30.0 mg/ml BSA (suitable for cell culture) in PBS, filtered (0.22 μm).
5. Permeabilization buffer: 0.5% Triton X-100 in PBS.
6. Click-iT® buffer additive: 13 μl buffer additive (provided) in 117 μl ultrapure water. Prepare on the day of use.
7. Click-iT® cocktail (provided): 860 μl 1× RX buffer, 40 μl $CuSO_4$, 2.5 μl Alexa Fluor® 594, 100 μl Click-iT® buffer additive (Subheading 2.3, **item 6**). Prepare on the day of use.
8. Hoechst 33342 (provided): Hoechst 5 μg/ml in PBS. Prepare on the day of use.
9. Glass slides.
10. Glass coverslips.
11. Nail polish.
12. Suitable aqueous mounting medium for fluorescence microscopy (e.g., PermaFluor™).
13. Fluorescence microscope.

3 Methods

3.1 Synchronization of Embryo Collection

1. Breed each female to two fertile males within 1 week with the last mating occurring at least 1 week prior to the 21st of March to ensure the blastocysts have entered diapause as per usual husbandry practices (*see* **Note 1**).
2. Calculate embryo age before and during embryonic diapause from the last mating to the day of collection.
3. To synchronize embryo reactivation from embryonic diapause, inject (intramuscular) mated females (average body weight = ~1 kg) with 1 mg/kg/day ovine prolactin daily beginning on the 21st of March, as previously described ([8], *see* **Note 2**).
4. The first day of prolactin injection is designated d0 PRL of embryo reactivation (equivalent to embryonic diapause).

3.2 Preparation of Culture Dishes

1. Prepare enough FOCM2 dishes for the number of embryos required to be cultured (one embryo/dish) and 1× Click-iT® Edu dish for each female.
2. For each dish, pipet four drops of FOCM2 or Click-iT® EdU Medium (5 μl/drop) arrayed on a 35 mm sterile petri dish.
3. Immediately cover with 2 ml mineral oil to prevent evaporation.
4. Equilibrate dishes in a tissue culture incubator for at least 6 h at 5% CO_2 and 95% air at 37 °C (*see* **Note 3**).

3.3 Collection of Blastocysts

1. If uteri are not flushed immediately after collection, store in PBS at 37 °C for up to 1 h.
2. Clean uterus of excess fat (*see* **Note 4**).
3. Remove ovaries and oviduct (Fig. 1).
4. The bicornuate uterus can either be flushed from the cervix through both uterine horns or through each uterine horn individually.
5. Insert a blunt syringe needle through the cervix and into the lumen of the uterus. Hold the needle in place with forceps as 1 ml prewarmed TCM 199 is released from the syringe with gentle but firm pressure to flush the embryos into an empty 35 mm sterile petri dish.
6. Scan for blastocysts under a stereomicroscope (*see* **Note 5**, Fig. 1d).
7. Collect blastocysts by micropipette and store in a fresh 50 μl drop of prewarmed TCM 199 on a heated stage (*see* **Note 6**).
8. Transfer through at least two more 50-μl drops of TCM 199 to remove debris and epithelial cells.
9. Transfer embryos individually to the preequilibrated plate of FOCM2 media drops. Wash through three drops of FOCM2 to remove any remaining traces of TCM 199 and HEPES, and culture in the fourth drop.
10. Culture in a tissue culture incubator (5% CO_2 and 95% air at 37 °C).
11. The next day, confirm reexpansion has occurred (*see* **Note 7**, Fig. 2).
12. Change media every 2 days (*see* **Note 8**). Prepare a new dish of four 5-μl drops of FOCM2 and equilibrate (as in Subheading 3.2). Wash embryos through three drops and culture in the fourth.

3.4 Evaluation of Embryo Cell Proliferation

As per manufacturer's instructions with modifications as listed below.

1. Incubate embryos overnight in a tissue culture incubator (5% CO_2 and 95% air at 37 °C) with preprepared 100 μM of Click-iT®EdU Medium (Subheading 2.3, **item 1**).
2. The next day, fix embryos in 3.7% formaldehyde for 15 min.
3. Wash twice (1 × 5 min, 1 × 3 min) in wash buffer (*see* **Note 9**).
4. Permeabilize embryos in 0.5% Triton X-100 for 20 min.
5. Wash twice in wash buffer for 3 min each.
6. Remove Click-iT® reagents from −20 °C and allow to thaw on ice.

7. Prepare Click-iT® buffer additive and Click-iT® cocktail.
8. Incubate embryos in the Click-iT® reaction cocktail for 30 min in the dark (*see* **Note 10**).
9. Wash once in wash buffer for 3 min.
10. Wash once in PBS for 3 min.
11. Counterstain nuclei with Hoechst 33342 for 30 min.
12. Wash twice in PBS for 3 min each.
13. Mount embryos on glass slides in a 10 μl drop of mounting medium.
14. Add a drop of Vaseline to each corner of a coverslip and press down slowly until embryo is just flattened then seal the coverslip with nail polish.
15. Keep slides at 4 °C until visualization.
16. The signals are visualized by fluorescence microscopy (*see* **Note 11**, Fig. 2).

4 Notes

1. If the last mating occurs within 1 week of the 21st of March, blastocysts will not enter into embryonic diapause. This timeframe is suitable for ranch mink of the Pastel variety in Québec, Canada, although the experimental procedures have been successfully employed on other color phases.
2. It is essential to remove as much fat as possible to aid in collection of blastocysts, as fat will float to the top of the media and prevent observation of blastocysts which will tend to rest at the bottom of the dish.
3. A minimum of 4–6 h is required to equilibrate culture media dishes, but standard laboratory practice is to equilibrate overnight.
4. Standard practice requires one person to hold the mink down on top of the cage and one person to inject into the leg muscle. The injection is very viscous and must be injected slowly.
5. An average of 7.3 embryos are flushed per mink but are of varying quality with an average of 5.4 "good" embryos. A "good" embryo is classified as having no signs of opacity, with the expected diameter for the pregnancy stage (Fig. 1d).
6. It is important to minimize changes in temperature and to maintain embryos at 37 °C as much as possible to minimize stress to the embryos.
7. When first flushed from the uterus, most blastocysts and later stage embryos collapse with the zona pellucida shrunk in away

from the shell coat (Fig. 1–3). However, after 24 h in culture, the majority of diapause and d3 PRL blastocysts are able to reexpand ([12], *personal observations*) (Fig. 2).

8. Once embryos increase to greater than 450 μm in diameter, it is necessary to change media every day and increase the culture media volume to 7 μl.
9. Ensure formaldehyde is completely removed.
10. Keep protected from light from now on.
11. Proliferating cells are labeled red, nonproliferating cells are labeled blue.

References

1. Renfree MB, Fenelon JC (2017) The enigma of embryonic diapause. Development 144:3199–3210
2. Enders RK (1952) Reproduction in the mink (*Mustela vison*). Proc Am Philos Soc 96 (6):691–755
3. Hansson A (1947) The physiology of reproduction in mink (*Mustela vison*, Schreb.) with special reference to delayed implantation. Acta Zool 28:1–136
4. Llerena-Vargas E (2011) Embryonic and uterine characteristics of diapause. Mémoire de maîtrise Université de Montréal
5. Desmarais JA et al (2004) The escape of the mink embryo from obligate diapause. Biol Reprod 70:662–670
6. Fenelon JC, Lefevre PL, Banerjee A, Murphy BD (2017) Regulation of diapause in carnivores. Reprod Domest Anim 52(Suppl 2):12–17
7. Fenelon JC, Banerjee A, Murphy BD (2014) Embryonic diapause: development on hold. Int J Dev Biol 58:163–174
8. Murphy BD, Concannon PW, Travis HF, Hansel W (1981) Prolactin: the hypophyseal factor that terminates embryonic diapause in mink. Biol Reprod 25:487–491
9. Papke RL, Concannon PW, Travis HF, Hansel W (1980) Control of luteal function and implantation in the mink by prolactin. J Anim Sci 50:1102–1107
10. Stoufflet IM, Mondain-Monval M, Simon P, Martinet L (1989) Patterns of plasma progesterone, androgen and oestrogen concentrations and in-vitro ovarian steroidogenesis during embryonic diapause and implantation in the mink *(Mustela vison)*. J Reprod Fertil 87(1):209–221
11. Enders RK, Pearson AL (1946) Mink blastocysts in vitro. Anat Rec 96(4):570
12. Daniel JC (1967) Studies on growth of the mink blastocyst. J Embryol Exp Morphol 17 (2):293–302
13. Daniel JC, Krishnan RS (1969) Studies on the relationship between uterine fluid components and the diapausing state of blastocysts from mammals having delayed implantation. J Exp Zool 172:267–282
14. Fenelon JC, Banerjee A, Lefevre PL, Gratian F, Murphy BD (2016) Polyamime-mediated effects of prolactin dictate emergence from mink obligate embryonic diapause. Biol Reprod 95(1):6
15. Moreau GM et al (1995) Development of immortalized endometrial epithelial and stromal cell lines from the mink (*Mustela vison*) uterus and their effects on the survival in vitro of mink blastocysts in obligate diapause. Biol Reprod 53:511–518
16. Polejaeva IA, Reed WA, Bunch TD, Ellis LC, White KL (1997) Prolactin induced termination of obligate diapause in the mink (*Mustela vison*) blastocysts *in vitro* and subsequent establishment of embryonic stem-like cells. J Reprod Fertil 109:229–236
17. Herrick JR, Bond JB, Magarey GM, Bateman HL, Krisher RL, Dunford SA, Swanson WF (2007) Toward a feline-optimized culture medium: impact of ions, carbohydrates, essential amino acids, vitamins, and serum on development and metabolism of in vitro fertilization-derived feline embryos relative to embryos grown in vivo. Biol Reprod 76(5):858–870

Chapter 20

Culture of Marsupial Oocytes and Conceptuses

Lynne Selwood

Abstract

Marsupial oocytes and conceptuses provide special challenges to scientists wanting to develop reliable in vitro techniques. Yet these techniques are essential to the study of development. Such techniques also provide tools to help prevent further decline in marsupial biodiversity using assisted reproductive technology. Specific marsupial characters have made development of in vitro techniques difficult. Some of these are the high-energy requirements of cleavage and blastocyst formation and maintenance because of cell–zona adhesion; the essential nature of the shell coat for most of development; the prevalence of embryonic arrests in vivo and in vitro during cleavage and unilaminar blastocyst stages; and the fragility of blastocysts leading to precocious shell loss and developmental failure in vitro. The advantages of marsupial culture during development are that the gestation period is very short, that the implantation is superficial in many, and that the neonates are altricial. This chapter outlines solutions to some of these problems in a representative, polytocous, dasyurid marsupial, *Sminthopsis macroura*. It is a natural superovulator with the shortest gestation period of any mammal, which has been cultured to within 5 h of birth. The other representative marsupial, the monovular phalangerid *Trichosurus vulpecula*, has a very superficial implantation that allows vesicles to readily detach from the endometrium.

Key words Marsupial, *S. macroura*, *T. vulpecula*, Culture, Follicle, Oocyte, Cleavage, Blastocyst, Embryo, Fetus

1 Introduction

Several marsupial groups have some seriously endangered species chiefly because of loss of habitat and feral animals, such as the cat. Dasyurid marsupials, a widespread group, also have significant loss of diversity, with 35% of species vulnerable or endangered [1]. This situation is expected to worsen with the effects of climate change. Unfortunately, research in reproductive technology for marsupials is relatively sparse, partly because basic information on in vitro techniques to handle marsupial gametes and conceptuses has not been gleaned for most species. The methods described here, to allow for investigation of the fascinating developmental biology of

Jason R. Herrick (ed.), *Comparative Embryo Culture: Methods and Protocols*, Methods in Molecular Biology, vol. 2006, https://doi.org/10.1007/978-1-4939-9566-0_20, © Springer Science+Business Media, LLC, part of Springer Nature 2019

marsupials, also will allow for some reproductive technology to be applied to the management of endangered populations of dasyurid and phalangerid marsupials.

1.1 Terminology

In this chapter, I will use the terminology proposed by Johnson and Selwood [2] to standardize the terminology for the early cell lineages in all mammals. Relevant definitions for this study follow. The "conceptus" is the product of conception and covers all stages from the zygote to the fetus immediately before birth. The terminology correctly allows the term "embryo" to be restricted to the stage where the body form and developing germ layers will give rise to the fetus. The "trophoblast" (TB) describes the first stem cell population to emerge to nourish the developing conceptus, later differentiating into the markedly polarized epithelial phenotype, the "trophectoderm" (TE). The "pluriblast" (PB) is the stem cell population that does not contribute further to the TB and contributes to the cell lineages of the conceptus proper and the hypoblast (HB). The "hypoblast" is equivalent to the primitive or primary endoderm in the mouse and some other eutherian mammals and is the second nutritional lineage to contribute to the yolk sac. After the PB has given rise to the HB, the residual tissue is the "epiblast" (EB), which subsequently gives rise to the germ layers.

1.2 Unusual Features of Marsupials with Respect to Culture

1.2.1 Media Requirements

Marsupial conceptus culture constitutes a particular challenge to marsupial developmental biologists for a variety of reasons. It is not possible to hold the conceptuses for extended periods of time in phosphate buffered saline (PBS) without damage or to hold in culture media suitable for the mouse. This applies particularly to cleavage and blastocyst stage conceptuses. Suitable culture media for the stages are identified in each section.

1.2.2 Embryonic Coats

In addition to the zona pellucida (ZP), the conceptus is covered by a mucoid and shell coat deposited by the oviduct and utero tubal junction then the uterus respectively [3]. Cleavage begins in the uterus, not the oviduct as in eutherian mammals, and the oviductal transit is very brief, around 4 h or less [4], where it has been accurately timed. Cleavage begins after the mucoid coat and some shell coat have been deposited [5–8]. The mucoid coat provides an osmotic buffer zone to the conceptus until the bilaminar blastocyst stage, and ultrastructural evidence suggests that it provides sustenance to the conceptus [9, 10]. The permeable shell coat [11], composed of an unusual protein related to a-enolase and t-crystallin [5] is shed during embryonic stages at between 75% and 88% of gestation [6].

1.2.3 Body Temperature

The body temperature of some marsupials is several degrees lower than eutherian mammals [12] and although very short-term culture can be maintained at a eutherian body temperature of 37 °C in these species, its long-term use is deleterious [13].

Additional features that are relevant to culture are noted below in Subheading 1.3.

1.3 Marsupials In Vitro

All marsupial stages developed in vitro were reviewed in 1997 [14] and the following is a brief review of relevant findings and culture since then in *Sminthopsis macroura* (Dasyuridae) and *Trichosurus vulpecula* (Phalangeridae), the two species used as models for this chapter, and other marsupials where relevant. Emphasis is given to features in each stage that have impact on culture.

1.3.1 Cleavage Stages and Unilaminar Blastocyst (ULB) Formation

From early cleavage, as early as the zygote in some species, the conceptus attaches first to the ZP by microvillous processes that pass through the perivitelline space. When the ZP and mucoid coat are lost they hook into the shell coat [8–10, 17]. The shell coat is required for normal development [15, 16]. Because cell–zona adhesion precedes cell–cell adhesion, no morula or inner cell mass forms, and the result of cleavage is a ULB by the 64-cell stage, consisting of small PB and larger TB regions [7, 9, 17]. During subsequent stages the epithelial surface cells, including those of the PB, have microvillous projections that hook into the zona, then the mucoid coat and finally the shell [10, 17]. This has implications for culture that are outlined for each stage. Finally, and most importantly, normal cleavage and blastocyst formation is dependent upon vigorous blastomeric activity associated with cell–zona adhesion and then TB migration to complete the ULB; thus, energy requirements are high, and a glucose-rich medium is required during cleavage and blastocyst formation, and blastocyst maintenance [18–20].

Another feature that has relevance to culturing cleavage and ULB stages in marsupials is deutoplasmolysis, in which up to a third of the content of the zygote cytoplasm is shed into the cleavage cavity as small vesicular bodies, and larger membrane-bound yolk masses [YM)] [7]. Deutoplasmolysis of vesicular bodies begins in the late zygote and continues into early ULB stages and is more frequent in the abembryonic pole of the conceptus especially during cleavage cytokinesis [21]. They contribute to the hyaluronate-rich extracellular matrix, which fills the cleavage cavity.

Shedding of the membrane-bound YM or YMs begins at the 2-cell stage and is usually completed by the 4-cell stage and is orthogonal to the first cleavage furrow. Where one YM is formed, as in *S. macroura* [8, 9], it remains connected to a blastomere by a mid-body. This may occur in the *T. vulpecula* in which 8–12 YMs are shed [17], but it is not recorded. Deutoplasmolysis must also require energy. The membrane-bound YM are used by the beginning of the bilaminar blastocyst (BLB) stage when the residual YM appear to be attached to the PB and may contribute ECM for hypoblast (HB) migration. Both the vesicular bodies and the membrane-bound yolk masses contribute to the hydroscopic extracellular matrix (ECM), some of which is hyaluronate [8, 9, 17]. It is the hydroscopic nature of the ECM that allows the blastocoel to undergo preliminary expansion [7] before the epithelium is completed during cleavage. Once the TB lineage is separated at the 8-cell stage in *T. vulpecula* [17] and 16-cell stage in *S. macroura* [9], it is the major contributor of hyaluronate to the blastocoel.

It is relatively easy to culture cleavage stage embryos into early ULB stages once their energy requirements are recognized and two major systems have been developed, one based on DMEM with 4.5% glucose (DMEMG) and 10% FCS [18] and a serum-free medium, in a study that demonstrated the high energy requirements of cleavage [20]. It was possible to culture over the 4-cell stage arrest in *S. macroura* but not in *A. agilis* [21] or *S. crassicaudata* [14]. Culturing cleavages stages of *A. agilis* and *S. crassicaudata* to the early ULB stages, requires additional supplements.

1.3.2 Unilaminar Blastocysts (ULB)

ULB in marsupials have certain features that have implications for successful culture. One is the nature of blastocyst formation and expansion and the other is that some blastocysts undergo developmental arrest or diapause. A knowledge of the expansion process, which seems to be common to all marsupials [7], allows the experimenter to easily age the conceptuses without resorting to cell counts or measurements. Toward the end of cleavage, preliminary expansion occurs by uptake of water by hyaluronate of the ECM in the blastocoel [9, 10, 17]. Preliminary expansion first causes compression of the zona and then the mucoid coat but the outer diameter stays the same [7]. TB cell exocytosis further contributes to the blastocoel ECM as expansion proceeds, which can be excessive in shell-free conceptuses [16]. Initially TB and PB cells are equal in number, but TB cells exceed PB cells as definitive expansion mainly involves TB cell proliferation and expansion [22]. Usually at the end of the ULB stage, the PB occupies about one third of the blastocyst surface. Culturing ULB further into the preliminary expansion stages has been achieved in a medium containing serum [18, 19] and one without [20].

Definitive expansion' i.e. increasing the outer diameter of the conceptus [7] is dependent on active transport by TB cells and begins between the 64- and 100-cell ULB [10]. Further definitive expansion is accompanied by progressive disappearance of the zona and mucoid coats, initially by compression and later by uptake by TB cells and to a lesser extent PB cells. As these coats disappear, extensive TB apical processes hook into the shell coat [10, 16,

17]. More limited studies in other marsupials suggest that the processes are essentially the same, with some variation in the stage when expansion is initiated [7]. Culturing successfully into definitive ULB expansion with continued cell proliferation is easier in *M. domestica* and *T. vulpecula*, in which the mucoid coat is very wide (about 50–80 μm respectively) and cell numbers reach 100–300 in the blastocyst. It seems likely that the mucoid coat acts as an osmotic buffer and also could provide additional energy supplements. There is electron microscopy evidence in *S. macroura* [9] and other marsupials that TB cells take up mucoid material by intense endocytotic activity during ULB expansion. So far, it has not been possible to culture far into the period of definitive expansion ([14] for review, [18–20]), suggesting a change in the uterine environment at this stage. At this stage, blastocysts can be particularly fragile. Occasionally, one encounters blastocysts that look darker or slightly crumpled and do not do well in culture, suggesting that they have begun to fail in utero.

ULB from some species may enter diapause and will not survive in vitro for more than 6 h. This applies to blastocysts that are in a fully arrested diapause, such as the tammar, or undergoing diapause with arrest of the PB and slow growth of the TB, as in some dasyurid marsupials (for review [14]). Analysis of uterine fluids and endometrial signaling molecules may help with this problem. Some marsupials, which retain discernible differences between the PB and TB, do not appear to have the ULB enter diapause, for example *T. vulpecula* [16] or *M. domestica* [23]. In the former, ULB grow in diameter and cell number in vitro, but it is worth noting that they are still undergoing preliminary expansion and have a broad mucoid coat. It is not possible to identify whether it is the absence of diapause or the presence of the mucoid coat that allowed normal development to occur in *T. vulpecula* in vitro.

1.3.3 Bilaminar Blastocysts (BLB)

Late ULB and early BLB, formed when the hypoblast migrates to line the ULB [7, 10] require some specific features of the uterine environment for survival, as these stages usually collapse after short term exposure to culture media used on *S. macroura, A. stuartii* (=*agilus*) that is suitable for earlier stages [14, 19, 20]. Transition to the BLB stage, also involves increased expansion with further cell division of the blastocyst in the polytocous *S. macroura* [24, 25] and the monovular *T. vulpecula* [16]. ULB have not been cultured to the BLB and early stages of BLB have been difficult to culture [14, 16, 19]. The transition from ULB to BLB and the maintenance of these stages in vivo also appears to be difficult because failing blastocysts are found on collection at these stages. These blastocysts are also sensitive to handling damage.

The suggestion that the late ULB and early BLB in *S. macroura* and *T. vulpecula* either require specific uterine signals to proceed and/or the uterine milieu has changed is supported by studies on early pregnancy factor (EPF). Pregnant *S. macroura* [26], *M. domestica*, and *T. vulpecula* [27] have high levels of EPF at this time, but nonpregnant marsupials at an equivalent time in the cycle have none. In addition, at this time of increasing progesterone in the plasma [5], the uterus shows signs of increasing secretory activity in *S. macroura* [28], *T. vulpecula* [29].

Yousef and Selwood [19] did try a number of other media and different levels of FCS or substitutes on a small number of unilaminar and bilaminar blastocysts and found that DMEG +10% FCS and human amniotic fluid (HAF) gave the best results. The other media Whittingham's T6 medium with 10% or 15% FCS, Monomed +5% FCS, and sheep amniotic fluid were less successful. Unfortunately, HAF which gave good results in these tests is now very difficult or impossible to obtain, because amniocentesis is no longer used to test fetal chromosomes. It is also variable in quality and cannot be frozen [30, 31].

1.3.4 Trilaminar Blastocysts (TLB)

It has not proved possible to culture early BLB into TLB, but late BLB can be cultured well into TLB until the formation of the node, notochord, neural plate, and a long primitive streak in *S. macroura* [19] or to earlier stages in mesoderm formation in *T. vulpecula* [16]. This suggests that if a signal is required for initiation of mesoderm formation, it is given in the late bilaminar blastocyst stages. At these stages, one sees an accumulation of cells, presumably mesodermal, to one side and under the round EB or at one end if the EB is oval shaped. The signal for the transition to the TLB from the late BLB seems more likely to be a uterine signal, rather than a change in the uterine milieu, since late BLB can develop in vitro to late TLB in the same medium, namely DMEMG +10% FCS or HAF [19]. Where it has been timed, the TLB is short, about 12 h from the appearance of the streak to the appearance of the first somite in *S. macroura* [25, 32] and seems slightly longer in *T. vulpecula*.

1.3.5 Embryonic and Fetal Stages

The early embryonic stages in marsupials show precocious development of the myelencephalon, branchial arches, lungs, anterior somites, ear primordia, and forelimb buds compared to the chick [32]. Because the embryo curves around a relatively small vesicle compared to the size of the chick egg, the effect of the cranial and cervical flexures lifts the body back to the anterior margin of the forelimb bud ridge off the yolk sac without major torsion and the embryo remains relatively flat until this time. Thus, embryonic stages in marsupials can be subdivided into two quite distinct types:

1. Flat embryos which include formation of early somites, neural tube, the heart, and forelimb buds as long elevated ridges. Cranial flexure lifts the head off the subcephalic pocket and

the early cervical flexure lifts the embryo off the yolk sac. This is facilitated by the curvature of the vesicle and the beginnings of torsion.

2. C-shaped embryos that are caused by further progress of the cervical and dorsal flexures and eventually the caudal flexure and torsion, which converts the C-embryo to a tubular form held to the yolk sac by an umbilical cord.

The transition between these two stages is when shell loss and attachment to the uterus occurs. It is easy to handle the vesicles in such a way that precocious shell loss occurs with resultant abnormal epithelium in TE, HB and embryonic endoderm during culture [16]. From a review of those species where it has been examined [6], normal shell loss in polytocous species begins above the anterior midline line of the conceptus and the adhesive section of the yolk sac is gradually revealed, while in monovular species where adhesion to adjoining blastocysts is not a problem, shell loss begins in the bilaminar or trilaminar yolk sac. In addition to damage of embryonic and extraembryonic epithelia, precocious loss of the shell means that the shell does not apply closely to the developing amnion which blows out to form large vesicles and does not close.

Fetal stages are when the body form is attained and hence the major changes achieved in fetal stages generally are those of growth and further differentiation of the organ systems. Marsupial neonates are altricial, but there is relative advancement of anterior structures and related sense organs associated with breathing, sucking, climbing to the pouch in some marsupials (e.g., as in *T. vulpecula*) [34] or climbing to the teat in the pouch in those that are squirted into the pouch (e.g., as in *S. macroura* (Selwood, unpublished) [35] or *Dasyurus hallucutus* [36]). The marsupial neonate also develops a variously complex oral shield and epitrichium that masks the face and sense organs during birth [37]. The limbs also have deciduous claws that fall off after birth.

The pioneering work on culturing embryonic and fetal stages was performed by New and Mizell [38] and New et al. [39] using Medium 199 in which they identified the importance of homologous serum and oxygen for successful culture, high oxygen (95% O_2/5% CO_2) being necessary for early embryonic stages. This was followed by a number of smaller studies and observations (review [14]), but more recently Cruz et al. [24] in *S. macroura* and Hickford et al. [40] in *Macropus eugenii* have made some progress. In *M. eugenii*, culture from primitive streak stage to early somite stages and regressing node and in some cases development of the forelimb ridges was improved by the use of 95% O_2/5% CO_2 in either Medium 199 or DMEMG with added sodium pyruvate and homologous serum, although interference with the mesodermal lineages of the yolk sac and some darkening of the embryos occurred. To move forward to get development of C-embryos and fetal stages in vitro in *M. eugenii* will require some modification of the methods, perhaps the use of roller culture and lower oxygen levels similar to the techniques used in the opossum [38, 39]. Cruz et al [24] in *S. macroura* achieved development in vitro from a C-embryo to early fetal stages on the day of birth and early fetal stages to within 5 h of birth using roller culture and air in the gas phase. Some modification of the gas phase might take these fetuses to birth. The great advantage of using *S. macroura* rather than any other marsupial or *Sminthopsis* species is that it has the shortest gestation of any mammal and C-embryonic and fetal stages last for 1.5–2 days [25] and implantation is relatively superficial.

1.3.6 Follicles and Oocytes

Culture of various stages of development in addition to use of different species are summarized in [14]. Since then progress has also been made on culture of gametes. Culture of preantral follicles showing growth but not progression to antral stages using follicles from prepubertal females was achieved in *M. domestica* [41]. Nation and Selwood [42] demonstrated that it was possible to obtain mature oocytes following primary follicle culture from adult ovaries in *S. macroura* to provide an in vitro tool for conservation of dasyurid marsupials. Adult ovaries were deliberately chosen for this study because many dasyurid marsupials are endangered, thus obtaining immature ovaries, which have often been used in the mouse would be very unlikely if not impossible in endangered marsupials.

1.3.7 In Vitro Fertilization

Since Moore and Taggart achieved in vitro fertilization in *M. domestica* [43], ICSI techniques have also progressed to obtain fertilization and early development in the tammar wallaby [44]. For both IVF and ICSI conceptuses, it would be interesting to transfer these conceptuses to a surrogate mother to see whether shell secretion in utero [5] is sufficient to rescue conceptuses derived from tubal oocytes that lack a shell coat or those from uterine oocytes whose shell was damaged by ICSI. Removal of the shell [15, 16], or even precocious detachment from the shell in vitro causes disruption to the ECM, and discharge of extra ECM into the cleavage cavity, cleavage of TB/TE cells, and also, from the ULB onward, damage to epithelial integrity especially affecting the TB and HB.

This chapter will concentrate on culture in the polytocous dasyurid, *S. macroura*, the stripe-faced dunnart, because of the wealth of data in that species. In addition, *S. macroura* has a large mean number of conceptuses of 15.3 and can have up to 39 ([33], Selwood, unpublished), a timetable for development that is accurate to within 6 h [25], and reliable methods for detection [33] and induction of ovulation [45] are available. For comparison, where information is known, methods are given for culture of a monovular phalangerid marsupial, *T. vulpecula*, the brushtail possum, which also has a reliable method to detect ovulation [46] and conceptuses can be cultured to the late gastrulation stages [16]. The methods used here should be readily applicable to other species, once the culture methods are adapted for different sizes of oocytes and conceptuses.

2 Materials

2.1 Preparation

All solutions are prepared with ultrapure water at a sensitivity of 18 MΩ cm at 25 °C and analytical grade reagents. Prepare solutions at room temperature, store at 4 °C, and then warm to the basal body temperature of the marsupial concerned before use (*see* **Note 1**).

2.2 Collecting Instruments (See Note 2)

1. Oocytes and cleavage stage conceptus can be collected with small micropipette tips cut to size or by pulling a glass Pasteur pipette and polishing the tip after breaking.
2. To collect ULB, BLB and some TLB, glass Pasteur pipettes can be cut down and polished across the 5.0 mm tube which can then be attached to a mouth tube for blastocysts 4.0 mm or less. Alternatively, large pipette tips can be cut to the right size and used with the pipette.
3. Commercial non–toxic disposable spoons with three holes drilled in the bowl are convenient for quick transfer of embryos or vesicles greater than the diameter of the cut down Pasteur pipette base (5.0 mm). Any rough edges in the spoons can be polished off with sand-paper. Use the spoons as follows. Teaspoons, blastocysts up to 1.5 mm diameter or dissected embryo or fetal stages; Tablespoons, blastocysts 1.6–9.0 mm; Larger scoops (e.g., coffee scoops), vesicles, 9.5–36 mm. The holes prevent the blastocysts or vesicles escaping back into the medium, but transfer should be quick so they do not drain entirely. The size of the spoon should be such that the blastocyst or vesicle flattens but does not collapse during transfer.

2.3 Handling Instruments (See Note 3)

A suitable tool for rolling specimens around in the dish is the eyelash stick.

1. Pluck several eyelashes from your eyelid, wash in 70% alcohol, and allow to dry in a sterile petri dish. It is worth seeking donations from friends with longer eyelashes!
2. Trim the end of an orange, or cuticle, stick so that one side is flat.
3. Melt some dental wax and attach the follicle end of the eyelash to the flat end of the stick with wax.

4. Hold the eyelash sticks in a flat covered container held down in a strip of plasticine so that the eyelash does not touch the bottom of the dish.
5. A length of glass rod with the end polished to a small knob can be used for larger vesicles.

2.4 Culture Dishes with Sylgard Base for Late BLB, TLB, Early Embryos (See Note 4)

Sylgard 184 Silicone Elastomer (Dow Corning) is used to line the base of a 35 mm petri dish for culture to facilitate observation, photography, and manipulation of the epiblast. A funnel-shaped hollow is made in the Sylgard to a maximum depth about 0.5 mm greater than the largest blastocyst [47]. The dish is filled with medium and the blastocyst gently rolled into the well with eyelash sticks or fine glass polished rods and positioned so that the epiblast is visible.

2.5 Washing Solution for All Glassware

10 g sodium hexametaphosphate, 90 g sodium metasilicate, 945 ml deionized water. Dissolve by stirring and let stand overnight.

1. Dilute 1 in 100 with ultrapure H_2O before use.
2. Immerse glassware and soak for 4–5 h overnight.
3. Rinse in running tap water seven or eight times.
4. Final rinse in deionized water and dry overnight at 42 °C.

2.6 Testing Tissue Culture Plates for Follicle Culture

Plates should be tested with droplets each containing a follicle before use to confirm that the droplet is retained when inverted and after media changes. As follicle size varies in different species, the droplet size should also be tested, keeping the same proportion of vesicle volume to droplet volume.

2.7 Containers

Where possible we use commercially available tissue or embryo culture plastics. Occasionally we use glass, which requires careful washing by hand.

Preparation of new glassware

1. Soak new glassware in 1% aqueous solution of sodium hydroxide for 24 h at 42 °C. Keep lids on while soaking.
2. Rinse in running tap water seven or eight times.
3. Dry overnight at 42 °C.

2.8 Anesthesia

1. For *S. macroura*. Halothane in oxygen and nitrous oxide delivered at the rate of 200 cc/min for each gas by a Cyprane TEC vaporizer. Anesthesia is induced using 4.25–4.5% halothane for 4 min and maintained at 2.5% until blood collection and cervical dislocation. This method is suitable for other small 15–30 g dasyurid marsupials.
2. For *T. vulpecula* Zoletil 50: 125 mg Tiletamine, 125 mg Zobzepan, 5 ml sterile water. This makes a 25 mg/ml solution.

Store for 3 months at 4 °C then discard residue. Deliver at 3.0 mg/kg using insulin syringe into thigh. Give double dose after 2 months storage. Anesthesia complete after 10–15 min. Terminal dose is 125 mg/kg Lethobarb delivered by cardiac puncture.

2.9 Culture and Handling Media and Solutions

Any supplements to the media are dissolved in ultrapure water prior to preparation.

1. *Dulbecco's phosphate buffered saline, calcium and magnesium free (PBS)*: 8.0 g NaCl, 0.2 KCl, 0.2 g KH_2PO_4, 1.15 g Na_2HPO_4, $7H_2O$, 1 l pure water. Dissolve in this order and autoclave. For field work it is acceptable to dissolve a PBS pellet in water.
2. *Dulbecco's modified Eagle's medium (DMEM) with 4.5 g/l glucose (DMEMG)*: DMEM, 4.5 g/l glucose, 100 IU/ml penicillin, 50 μg /ml streptomycin, 100 μg/ml kanamycin, 100 μg/ml glutamine (added immediately prior to culture from frozen concentrated stock). For commercially prepared media, other additives are dissolved in the water during preparation. For field work without access to a CO_2 incubator it is possible to use an oven, with moist atmosphere and use HEPES-buffered DMEMG made up as shown below.
3. *HEPES-buffered DMEMG*: 2.5 ml 0.1 M HEPES solution, 7.5 ml DMEMG previously held in incubator for several hours to get appropriate pH.
4. *Follicle culture medium*: DMEMG, insulin–transferrin–selenium (ITS-X: 10 μg/ml insulin, 5.5 μg /ml transferrin, 6.7 ng/ ml selenium), 100 μg /ml ascorbic acid, 1.0 IU /ml recombinant human follicle-stimulating hormone (FSH), 5% homologous serum obtained from the experimental female and heat-inactivated before use.
5. *Follicle and oocyte collection medium for S. macroura and for T. vulpecula conceptuses*: HEPES-buffered DMEMG, 3 mg/ml bovine serum albumin (BSA).
6. *Type 2 Oocyte maturation medium*: Follicle culture medium, 10 μg/ml porcine pituitary luteinizing hormone, 10 mM lithium chloride.
7. *Cleavage to early ULB medium*: DMEMG, 10% fetal calf serum (FCS).
8. *Mid-cleavage to ULB with completed preliminary expansion medium (DMEMGL)*: DMEMG, 2.76 g/l sodium lactate, 10% FCS.
9. *HEPES-buffered DMEMGL*: 2.5 ml 0.1 M HEPES solution, 7.5 ml DMEMGL previously held in an incubator for several hours to get appropriate pH.

10. *Induced ovulation solution*: 1000 IU equine serum gonadotropin (eSG), 1 ml sterile PBS. This provides 1000 IU/ ml solution.

For *S. macroura*: give a total of 0.1 IU eSG/g delivered in 2 doses the first at time 0, a dose of 0.06 IU eSG/g (Folligon) followed on Day 4 by a second injection of 0.04 IU eSG/g. The volume required per injection was determined by the weight of individual females at Time 0 and hand-warmed prior to i.p. injection.

For *T. vulpecula*: Prepare the 1000 IU eSG/g as outlined above. Give at a rate of 7.5 IU/kg or 0.08 ml/kg, hand warmed and delivered i.p.

3 Methods

3.1 Collection of Tract of S. macroura

1. The basal body temperature of *S. macroura* is 35 °C, so warming plates on the microscope and bench and the incubator are set to this temperature well before collection begins (*see* **Note 1**).
2. To obtain samples in a sterile fashion, all procedures are done under a dissecting microscope, using a heated stage in a laminar flow hood. The surgeon should wear gloves, mask, and a clean laboratory coat during the collection.
3. Apply anesthetic (Halothane, Subheading 2.8, **item 1**) followed, after blood collection, by cervical dislocation to kill females.
4. Before cervical dislocation, swab the ventral surface of the abdomen with 70% alcohol.
5. Make a mid-ventral incision and hold the sides back with artery forceps.
6. Take a 3 ml syringe with 21 G needle attached. Hold so that the needle aperture is uppermost and bend the last 0.5–1.0 mm of the needle toward you by 30°.
7. Collect blood from the posterior vena cava by holding the bent part, after insertion, parallel with the vessel and slowly pull on the plunger without touching the side of the vessel. To provide the homologous serum required for experiments, allow the blood to clot for 30 min at room temperature, before spinning for 6 min at 1500 × *g*, and then storing in small aliquots in glass vials at −20 °C. Serum is heat inactivated at 56 °C for 30 min immediately before use.
8. Remove the two sides of the tract by cutting across each cervix. Blot with sterile gauze. Place in warm PBS in a 35 mm Petri dish. Remove excess blood by stroking each vessel toward the

ovary. Blot-dry with sterile gauze. Hold in fresh PBS in a Petri dish in the incubator (*see* **Note 5**).

3.2 Collection of Tract of T. vulpecula

1. Set warming plates in incubator or oven at 36.2 °C (*see* **Note 1**).
2. Choose a draft free area, swab bench with 70% alcohol, and dress for sterile work.
3. Weigh animal and examine the pouch to determine reproductive status. Hold animal in the bag, closed with a cord, until anesthetized.
4. Using a sterile insulin syringe deliver Zoletil (hand-warmed before delivery) into the thigh (Subheading 2.8, **item 2**).
5. Position the animal on its back and using a Vacutainer with a 27 G needle take 2–10 ml of blood from the heart and prepare as serum as in Subheading 3.1, **step 7**.
6. Replace blood collection device with a 2 ml syringe and deliver Lethobarb (Subheading 2.8, **item 2**).
7. Swab the ventral surface anterior to the pouch with 70% alcohol.
8. Make a mid-ventral incision and hold the sides back with artery forceps.
9. Remove the two sides of the tract by cutting across each cervix (*see* **Note 5**). Blot with sterile gauze. Place in warm PBS in a 35 mm Petri dish. Remove excess blood by stroking each vessel toward the ovary. Blot-dry with sterile gauze. Hold in fresh PBS in a Petri dish in the incubator or in an oven for field samples.

3.3 Follicle Isolation and Inverted Microdrop Culture in S. macroura

1. Following on from Subheading 3.1, **step 8**, separate each side of the tract and wash to remove blood.
2. Separate the ovaries under a laminar flow hood and transfer immediately into HEPES-buffered DMEG +3 mg/ml bovine serum albumin.
3. Microdissect individual follicles from the ovary at 40× magnification using 29 G insulin syringes with the plungers removed.
4. Select isolated preantral Type 3b primary follicles with a single layer of cuboidal granulosa cells for culture under a dissecting microscope with reflected and transmitted light on the basis of size (early, mean 49.7 × 57.5 μm to late, mean 89.5 × 106.1 μm diameters) (*see* **Note 6**). Confirm identification in a few samples by confocal microscopy, then base selection on translucency and size.
5. Place 100 μl microdrops of Follicle Culture Medium containing one follicle in 96-well Sarstedt round-bottomed suspension

tissue culture plates, and invert the plate prior to incubation (*see* **Note 7**). The medium is held in place by surface tension and the follicle drops to lie on the media–gas interface, thereby maximizing oxygen access.

6. Culture follicles for 6 days and replace half the volume of medium every 48 h.
7. Incubate in a humidified gas environment of 5% CO_2 in air.
8. Examine follicles prior to culture (time 0) and then every 24 h using an inverted microscope with a heated stage.
9. Evaluate follicular development and viability by measuring the diameter to determine growth and assess general morphology and structure, including ZP deposition, antral fluid accumulation, and theca development (*see* **Note 8**).

3.4 Preparation for Culture in Microdrops Under Paraffin Oil: Use for Oocytes, Cleavage Stages, and Isolated Cells

1. The day before you expect to collect follicles or cleavage and blastocyst stages, pour equal parts of paraffin oil for culture and gassed medium into plastic, media bottles, close, shake, and leave for 24 h.
2. The equilibrated paraffin floats to the top and can be removed, if it is free of bubbles of media and the medium is the appropriate color, and held in smaller containers, if necessary.
3. Transfer the equilibrated paraffin (~2 ml) to half fill a 35 mm Petri dish.
4. Transfer droplets of media to the base of the Petri dish. The volume of the droplet and number of specimens in the droplet varies according to what is to be cultured.

3.5 Type 2 Oocyte Maturation In Vitro

1. In advance, prepare 30 μl drops of oocyte maturation medium under equilibrated oil and hold in incubator.
2. Remove oocytes of cumulus–oocyte complexes (COCs) from viable follicles >300 μm using an insulin syringe with the plunger removed.
3. Incubate COCs at 35 °C in 6% CO_2 in air for 24 h in microdrops under oil.
4. After 24 h, assess oocyte maturity (*see* **Note 9**) as Type 2 (GV stage) or Type 3 (Telophase/MII).

3.6 Collection of Known Age Stages of Estrus and Pregnancy (See Notes 10)

For brevity in the following text the species are not identified unless procedures are different. To provide a relatively sterile collection of conceptuses from *S. macroura*, follow instructions under Collection of Tract (Subheading 3.1). For larger animals such as *T. vulpecula*, 1.5–4.0 kg, collected under field conditions, sterile collection can be achieved by following the guidelines outlined in Subheading 3.2. During the breeding season, analysis of daily weight and microscopic analysis of the contents of urine sample in *S. macroura* allows collections to be specifically timed (*see* **Note 11**):

1. During the estrus period (cornified epithelial cells in the urine, weight increase).
2. On the day of mating (sperm in the urine or a plug).
3. At ovulation (appearance of polymorphonucleocytes in the urine, weight fall).
4. At particular stages of pregnancy (time since ovulation and profile of weight changes).

For *T. vulpecula*, a modification of this technique by analysis of the daily urine samples alone allows detection of estrus, mating, ovulation, and stages of pregnancy.

3.7 Induced Ovulation Protocol

This is a protocol developed for adult females, in their first or second year of breeding, and can be used to rescue about one-third of nonreproductive females in the colony (*see* **Note 12**).

1. Females are monitored as described previously to separate cycling and noncycling females. The latter can be induced to cycle following the induction protocol and are used on their first or second cycle (*see* **Note 13**).
2. Females are best induced during the follicular, late luteal (Day 6–10 of gestation) phases and the intermediate phase of the estrous cycle using the induced ovulation solution of eSG (Subheading 2.9, **item 10**).
3. Two injections are given, both at 1600 h, and the dose is calculated from the weight of the female on the day of the first injection. The total injected is 0.1 IU eSG/g delivered i.p. in two doses.
4. The time of the first injection of 0.6 eSG/g is Time 0 and the second injection of 0.4 IU eSG/g is given on Day 4. After calculating the dose, the sample is drawn up into an insulin syringe and hand warmed before injection.
5. Examine females at the days after Time 0 as indicated to collect the desired stages.

3.8 Collection and Culture of Ovulated Oocytes, Tubal and Uterine Zygotes, and Cleavage Stages (See Note 14)

1. Collection is done using warmed solutions, with warming plates on both the bench and the microscope. Material not being used is held in the incubator.
2. Ovulating eggs are collected at Time 0 (ovulation) in *S. macroura* in a natural cycle and at the end of Day 6 in an induced cycle and at time 0 (sperm plug) in *T. vulpecula*. Zygotes to 32-cell stages are collected on Day 0.5–3.0 in both *S. macroura* [25] and *T. vulpecula* [29] in natural cycles and on Day 7.0–9.5 in *S. macroura* induced cycles [45].

3. The paired ovaries, oviducts, and uteri are separated and mesenteries are removed before placing each side in PBS in a labeled 35 mm petri dish (*see* **Note 15**).
4. In *S. macroura*, briefly examine the ovary from each side to determine whether it contains follicles, ovulation sites, or CL and their degree of development (*see* **Note 16**). Make the same assessment in *T. vulpecula* to determine which is the active side of the tract.
5. If collection is made at Time 0, the time of ovulation is detected in *S. macroura* or the time a plug or sperm in the urine is detected in *T. vulpecula*, the ovary could contain antral follicles, ovulated follicles with ovulation sites, and ovulated oocytes or zygotes in the oviduct. Count the ovulation sites to determine how many oocytes will be found in the oviduct (*see* **Note 17**).
6. If mature oocytes are wanted, antral follicles can be collected and cultured as in Subheadings 3.3–3.5.
7. If an ovulation site is visible, it is possible that eggs are still in the oviduct. Examine the oviduct by transmitted light to see whether eggs are visible and note location. Whether they are visible or not, carefully sever the oviduct from the uterus and ovary, and place each in PBS in separate labeled 35 mm Petri dishes. Hold uterus and ovary in the incubator. Eggs in the oviduct are usually visible in *S. macroura*, but not in *T. vulpecula* (*see* **Notes 15** and **16**).
8. Any eggs remaining in the oviduct are flushed using warmed medium. In advance, the sharp tip of sterile 25 G hypodermic needles are first rubbed with a file to smooth the tip and sterilized by autoclaving. The blunt ended needle is attached to a 1 ml hypodermic syringe filled with warmed PBS. Take up a small amount of air so that the path of the PBS can be followed in the oviduct from the bubbles. The needle is inserted into the infundibular end of the oviduct and held firmly with a pair of curved, soft-tipped forceps. Collect any eggs and place in warmed culture medium, noting the location within the oviduct.
9. Repeat flushing until the number of eggs matches the number of ovulation sites. Should the numbers be less, some eggs may have already reached the uterus. They will be sticky at this stage because of recent shell secretion (*see* **Note 18**).
10. Place the uterus in warmed PBS and ensure it is free of red blood cells. Insert small iridectomy scissors into the cervical end of the uterus and, while holding the cervical side up, cut along the mid-line to the oviduct end. Evert the uterus and flush several times with warm PBS.

11. Count the number of eggs, which are recognizable by the round shape and the shiny shell coat. Confirm that the number of uterine eggs plus the oocytes and zygotes from the oviduct (if any) match the number of ovulation sites. Separate into unfertilized and fertilized eggs (*see* **Note 19**).
12. Transfer the eggs in a very small amount of PBS into fresh DMEMG +10% FCS in a Petri dish and then transfer in batches of 3–4 eggs to 30–50 μl medium in either microdrops or 96-well plates, which have been previously prepared and held in the incubator. A sample of the eggs is held briefly in PBS for photography, and afterward transferred to fresh medium, fixed, or frozen.
13. The plate is covered and transferred to the incubator. The embryos are examined and photographed at daily intervals, or less, when half of the medium is changed, or they can be examined by time-lapse cinematography (*see* **Note 20**). Some cleavage stages may be undergoing parthenogenetic development (*see* **Note 21**).
14. Cell numbers are counted in whole embryos up to about the 128-cell stage by gently rolling the embryos around in culture with eyelash or glass sticks (Subheading 2.2). This is extremely easy in the more transparent *S. macroura* but less easy in other dasyurids or *T. vulpecula*.
15. Repeat ovarian, oviduct, and uterine recoveries for the other side.
16. Cleavage stages to 32-cells should be cultured in DMEMG +10% FCS to obtain completion of cleavage and early preliminary expansion of the unilaminar blastocyst. If completion of preliminary expansion (obliteration of the zona and mucoid coat) and early definitive expansion is desired in *S. macroura* then the appropriate media is DMEMGL (*see* **Note 22**).

3.9 Collection and Culture of Unilaminar Blastocysts (UBL)

1. Collection is done using warmed solutions, with warming plates on both the bench and the microscope. Material not being used is held in the incubator.
2. UBL of *S. macroura* are found between days 3–6 of the natural cycle and days 10–12 in the induced cycle. UBL of *T. vulpecula* are found 6–8 days pc. Because handling can induce collapse, collection of conceptuses about 12–24 h before the beginning of preliminary expansion improves the chances of development in vitro (*see* **Note 23**).
3. Place the uterus in warmed DMEMGL plus 10% FCS and confirm it is free of blood cells.
4. Cut the uterus, evert, and carefully flush as described in Subheading 3.9, **step 3** (*see* **Note 24**).

5. Count the blastocysts and confirm that the number of CL match that number. If not flush again. Examine the unfertilized eggs to check for parthenogenetic development.
6. Set up blastocysts in groups of three in **100** μl medium in either microdrops or 96 well plates, which have been previously prepared and held in the incubator. Culture for 24 h or longer, if successful, in the incubator, checking every 8 h (*see* **Note 25**). Change half the media after 24 h.
7. Blastocysts should be assessed for cell number, both of pluriblasts and trophoblasts, where visible, and size. The widths of the coats should be taken to estimate progress.

3.10 Collection and Culture of Bilaminar Blastocysts (BLB)

1. Prepare a series of collection pipettes of different sizes and sterilize prior to surgery (*see* **Note 26**).
2. Collection of early BLB is done on Day 6.0–6.5 in *S. macroura* and Day 9 in *T. vulpecula*. Late BLB are collected in *S. macroura* on Day 7.0–7.5 and in *T. vulpecula* on Day 10. After induced ovulation in *S. macroura*, early BLB are collected on Day 12.5–13.5 and late BLB on Day 13.5–14.5 after the first injection.
3. Follow instructions in Subheading 3.9, **steps 1** and **3** (*see* **Note 27**).
4. Separate each side of the tract and hold in warmed DMEMG +10% FCS for *S. macroura* and HEPES-buffered DMEMG +10% FCS for *T. vulpecula* in the field. Handle each uterus separately and hold the other in the incubator or oven (*see* **Note 28**).
5. Cut the uterus along the midline to avoid the thickened endometrium, evert, and allow the blastocysts to roll out into the medium. If their number does not match the number of CL, lightly flush the uterus with medium (*see* **Note 29**). Unfertilized eggs can still be found at this stage and become increasing brown and opaque.
6. Transfer each blastocyst to a well of a 24-well plate containing 1 ml of medium or the 35 mm Sylgard-lined Petri dishes, as necessary, and hold in an incubator (*see* **Note 30**).
7. In early BLB, examine every 12 h, but examine late bilaminar blastocysts at 6 h intervals including the beginning and the end of each day (*see* **Note 31**).
8. Assess the stage of development of the BLB by examining the degree of migration of the hypoblast cells (*see* **Note 31**). Note the dimensions of the blastocyst and the pluriblast/epiblast, the general appearance, and the width of the shell. In early BLB, examine every 12 h, but examine late bilaminar

blastocysts at 6 hourly intervals including the beginning and the end of each day.

3.11 Collection and Culture of Trilaminar Blastocysts (TLB)

1. In advance, prepare equipment as indicated in Subheadings 2.1–2.3 (*see* **Note 32**).
2. TLB are collected at Day 7.75 and 8.25 in naturally cycling *S. macroura* and Day 14.75 and 15.25 in an induced cycle. In *T. vulpecula* they are collected on Day 9.
3. Follow instructions Subheading 3.10, **steps 1** and **3–5**.
4. When the uterus is first opened, check whether the anterior-- posterior (A/P) axis of the blastocyst (primitive streak) is aligned with the A/P axis of the uterus.
5. When removing TLB from the uterus, they can lie in hollows in the endometrium and are removed by gentle flushing or by pulling on the adjacent endometrium, not the TLB. Handle very carefully to avoid precocious shell loss (*see* **Note 33**).
6. Transfer each TLB to a Sylgard-lined petri dish or organ culture dish containing medium and move it with the eyelash stick so that the epiblast lies on top. Hold in the incubator.
7. TLB assessment (*see* **Note 34**) should be made every 3 h, so maintenance at the body temperature is very important and individual dishes are more successful than multiwell plates.
8. Assess each blastocyst for transparency and normal development (*see* **Note 32**). Measure the blastocyst diameter, the size of the epiblast, the extent and width of the extraembryonic mesoderm, the length of the primitive streak, and the presence or absence of the node, neural plate, and notochord. Note the shape of the epiblast. All of these features will help determine the relative age of the blastocyst.
9. Stages that are likely to be found are as follows: T1. Oval to slightly pear-shaped epiblast with increased density at the end; T2. Appearance and extension of the primitive streak, but no node; T3. Primitive streak and node; T4. Extension of the notochord seen through the epiblast and appearance of the neural plate; T5. Appearance and anterior extension of the extraembryonic mesoderm to form a rim around the epiblast. Shortening of the primitive streak as the node moves centrally and the lengthening of the notochord and neural plate lengthen; T6. Late trilaminar blastocyst with the transparent proamnion lying anterior to the edge of the epiblast and developing arch of the cephalic fold. Some specimens show early detachment of the shell in the anterior midline (*see* **Note 35**).

3.12 Collection and Culture of Embryonic Stages Before Implantation in *S. macroura*

These and later stages have not been cultured in *T. vulpecula*.

1. Prepare collection and handling devices and filled culture dishes as indicated in Subheading 2.1–2.3 (*see* **Note 35**).
2. Embryonic stages in natural cycles are collected between Days 8.5–9.5 and Days 15.5–16.5 of the induced cycle (*see* **Note 36**). Fetal stages are collected on Days 10–11.0 days in natural cycles or on Day 16–17 after induced ovulation.
3. Follow instructions provided in Subheading 3.11, **steps 3** and **4**.
4. Remove the vesicles from the uterus by cutting along the midline, everting the uterus, and allowing the vesicles with the flat embryo to roll out into the media. Confirm that the number of embryos matches the number of CLs.
5. Flat embryos are best placed on Sylgard-lined Petri dishes or organ culture dishes. The medium, DMEMG +10% FCS, should cover the vesicles. Look for early signs of normal shell loss (*see* **Note 33**).
6. Half of the medium is replaced every 12 h or every 6 h after appearance of the somites. This method facilitates observation and manipulation.
7. Culture can be continued in roller culture (*see* **Note 37**) if the shell detaches (*see* Subheading 3.13).
8. You can expect to see three stages as follows: The early flat embryo has no cranial flexure and 1–15 somites, the mid flat embryo has cranial and early cervical flexures and up to 25 somites without torsion, and the late flat embryo converts to the C-embryo by completing the cervical flexure and beginning the dorsal flexure.
9. Measure vesicle diameter, diameter/s of epiblast, length of primitive streak, notochord, and neural plate and note the number of somites, presence and width of shell, and other developing features (*see* Note **38**).

3.13 Collection and Culture of Postimplantation C-Embryonic and Fetal Stages in *S. macroura*

1. Prepare prebored teaspoons and tablespoons for collection (Subheading 2.1, **item 3**) and pulled glass rods topped with a ball for rolling conceptuses (Subheading 2.2).
2. Maintenance at body temperature is essential during collection and dissection, so work on a warm plate at all times (*see* **Note 1**).
3. Follow instruction provided in Subheading 3.11, **steps 3** and **4**.
4. As speed is essential, counting CL, unfertilized eggs, and abnormal fetuses should be done after all the normal conceptuses are placed in culture.

5. Hold the medial surface of the uterus upward and cut along the midline from the cervical to the oviductal end making sure the scissors point upward and penetration of the uterine lumen is as superficial as possible.
6. C-Embryo stages, when collected early in implantation should be allowed to roll out of the uterus, assisted by gentle irrigation, before being placed in roller culture.
7. Using transmitted light, visualize the location of late C-embryos or fetal stages to assist in their removal. This is very important when you are looking at monovular species, such as *T. vulpecula*, and variably helpful in polyovular species, such as *S. macroura*.
8. There will be differences in size, especially if the uterus is crowded, but if the appearance is normal the fetus can be used for culture.
9. Remove the embryonic/fetal stages so that the amnion/chorion is intact, the allantois is untouched and as much of the yolk sac as possible is attached.
10. The roller culture chamber (BTC Engineering) using roller culture tubes (Nunc) is best for later C-embryo and fetal stages, or earlier for more continuous culture (*see* **Note 37**).
11. Transfer embryos individually to tubes half filled with warmed, gassed medium and culture for 24 h at 35 °C in air, making observations every 6 h with minimal disturbance. It is best to leave embryos in the tube in a temperature controlled environment during observation.

4 Notes

1. Marsupials may have a different body temperature to that of comparable eutherian mammals [12, 13], so check the basal body temperature of the chosen marsupial before setting incubators and warming plates. The basal body temperature of *S. macroura* is 35 °C and of *T. vulpecula* is 36.2 °C.
2. Marsupial oocytes and cleavage stages are often larger than in other mammals [7], so collection may involve special instruments. Secondary oocytes are 120–250 μm, cleavage stages with the coats 200–400 μm, ULB, 0.5–1.5 mm, BLB, 0.7–8.5 mm, TLB 1.4–9.8 mm, vesicles with embryos and fetuses, 3.5–29.0 mm. Before collection of samples, check the literature to get an estimate of diameters inside the zona pellucida to estimate the size of unfertilized oocytes from the oviduct or uterine zygotes during deposition of the mucoid coat and shell. An estimate of total diameter of cleavage stages is also desirable for the same reason. This allows for advance

preparation of the pulled pipette or pipette tips trimmed to a particular size.

3. ULB, BLB, and TLB are fragile and readily damaged. If handled carelessly, the shell can detach precociously and development will fail. They should be gently handled with eyelash sticks or polished glass rods.
4. The extra weight of the developing epiblast causes the blastocyst to roll and the epiblast to lie on the bottom of the conceptus, facing the culture surface, so it needs to be examined through the other side of the blastocyst. Sometimes this is convenient, when counting HB cells or determining the state of the HB epithelium. To hold them with the epiblast on the top of the embryo (away from the culture surface), they can be cultured in Sylgard-lined 35 mm Petri dishes. Smaller organ culture dishes or 24-well plates can also be used. In the media, usually the blastocysts lie with the epiblast on the bottom of the conceptus/closest to the culture dish surface in *S. macroura* from the late BLB and all TLB stages and in early BLB to early TLB in *T. vulpecula*.
5. Marsupials have two uteri and only one side contains the conceptus in monovular species. An examination of follicles and CL on the ovary will let you know in polytocous species, whether oocytes have ovulated and you need to flush the tube (recent ovulation site in the ovary and fimbria still wrapped round the ovary), how many specimens will be found in each uterus, and the approximate stage of the conceptus/es. In polytocous species, prior assessment of ovarian characteristics and labeling dishes containing parts of the tract as left or right is important in effective collection and adequate preparation for culture of the particular stage. It is easy to miss specimens if you do not do this. In monovular species, ovarian condition matched with the state of the fimbria and appearance of the uterus allows prediction of what will be found. In *T. vulpecula* on rare occasions, the conceptus will be found in the inactive side uterus. Occasionally in both species, an extra specimen will be found because of twin oocytes in one follicle.
6. Preantral, primary follicle definition follows the criteria established by Pedersen and Peters [48] for Type 3b follicles. The size range from early to late primary follicles for *S. macroura* was determined by Kress et al. [49].
7. Of the several methods tested, inverted culture [50] with some modification was the most successful. Twelve different tissue culture plates were tested to identify which was most successful for inverted follicle culture in *S. macroura* [42], which has one the largest marsupial secondary oocytes [7]. Inverted culture using the Sarstedt plates was the most successful. As other

marsupial oocytes can be less than half the size of *S. macroura* oocytes, which could influence follicle sizes, plate performance with respect to retention of the droplet and follicle, droplet size, and frequency of media change should be tested.

8. For assessment of follicles in vitro note that viable follicles have clear granulosa cells and an intact follicular structure, regardless of growth rate. Nonviable (atretic) follicles have one or more of the following characteristics: an absence of growth; an irregularly shaped, retracted or darkening oocyte; dark, granular (apoptotic) granulosa cells; a damaged (thinning/budding) or ruptured thecal layer.
9. *S. macroura* oocytes are assessed on the criteria of Merry et al. [51] as Type 2 or Type 3 oocytes. Type 2 oocytes, recovered from early antral follicles (>300, but <350 μm), are in meiotic arrest at the germinal vesicle (GV) stage, and retain centrally concentrated cytoplasmic vesicles, and adherent cumulus cells. Culture of the intact cumulus–oocyte complex (COC) is necessary to get mature oocytes (see below). Type 3 oocytes, obtained from late antral follicles, are mature and show marked cytoplasmic polarity and extrusion of the first polar body (PBI) following germinal vesicle breakdown (GVBD) and meiosis I. They can be obtained from Type 2 oocytes by addition of 10 μg/ml porcine pituitary luteinizing hormone to the follicle culture medium [52].
10. Collection of cleavage stages can be done in PBS because procedures are relatively fast. Stages requiring long work at the microscope are collected in HEPES-buffered DMEMG collection medium and stages from the ULB onward that will be cultured immediately after collection are collected in the culture medium used (variation of DMEMG as noted).
11. Collection of known age estrus and conceptus stages is desirable. Ovulation is spontaneous in many marsupials. Collecting at 12 h intervals from ovulation is possible in naturally cycling *S. macroura* [25], *Antechinus agilis* [53], and *S. crassicaudata* [54]. Daily monitoring of the female weight and the presence of urinogenital sinus cells and sperm in the urine sample during the breeding period allows detection of whether females are in estrus, have mated, ovulated, and their stage of pregnancy by time from ovulation. Sperm storage for up to 5 days occurs in *S. macroura*, and can be as long as 21 days in some dasyurids, so detection of ovulation is essential. The same method, without using weights, has been adapted for use with *T. vulpecula* [46] to detect the onset of development and allow collection of conceptuses at chosen intervals post coitus (PC). Sperm storage can occur for up to 2 days in *T. vulpecula*. In addition, a reliable method of induced ovulation [45] has allowed

collection of known-age stages in *S. macroura* [25]. In *M. domestica*, where ovulation is induced and a distinctive male behavior allows for determination of the time of coitus and hence accurate aging of conceptuses, Baggott et al. [55] developed a video detection system that has been used to collect accurately aged conceptuses [14, 23]. Without the video detection system in *M. domestica* or monitoring of females during estrus to detect mating and ovulation in *S. macroura*, conceptuses are not collected in about 40% of females, the experimenter has less idea of the expected age of the conceptuses [13], animals are killed unnecessarily and experiments take longer to complete.

12. Induced ovulation in *S. macroura* was originally developed as an aid for conservation in dasyurids to get nonbreeding animals in a colony to breed, but has since been applied to breeding animals [45]. It requires monitoring of the reproductive status with urine samples or weight profiles, which allow the researcher to determine when an animal is in estrus, has mated, ovulated, is pregnant, and when to collect conceptuses at specific developmental stages.
13. Unsuccessful induction can occur. Ovulation can be induced in adult and nonbreeding animals, but it should be used within the breeding season so that males producing sperm are available for mating, and the environment is suitable for rearing a litter. Induction is not successful during the middle of the luteal phase between days 1 and 5 after ovulation when body weight and progesterone levels are rising [56]. For this reason, animals should be monitored so that injection does not occur in this period [4]. Noncycling animals can be induced to ovulate [45].
14. For culture of all stages of *S. macroura*, the incubator is maintained with a moist environment with 5% CO_2 in air at 35 °C , unless stated otherwise. When culturing *T. vulpecula* in the field, the medium is HEPES-buffered and culture is done in a small oven at 36.2 °C humidified with large Petri dishes full of distilled water in the bottom for at least 24 h before the initiation of culture.
15. For early stages, that is, oocytes to TLB, it is best to collect uterine and oviduct contents in a small 35 mm dish to avoid tedious searches through a larger volume of PBS if larger dishes are used. Dishes should be labeled so the left side (LHS) and right side (RHS) remain separate. Once the oviducts are separated from the uterus and ovary, which is best done by transmitted light so oocytes can be visualized in the oviduct in *S. macroura,* some oocytes may roll out of the oviduct or ovulate

from the ovary. All dishes should be kept in case additional searching is necessary.

16. Ovulation in *S. macroura* is not always synchronized, both between and within the two ovaries. Ovulation occurs in a series of bursts of about six oocytes. Ovulation in both ovaries takes about 7.5 h [4]. In addition, fertilized eggs roll down the oviduct one at a time [57]. As a result, normal stages about 12 h apart can be found in the same female.
17. Ovulation in *T. vulpecula* occurs late in the afternoon and into the evening (Selwood, unpublished) and in *S. macroura* it occurs around midnight in both naturally cycling and induced animals [4]. For other marsupials, if stages at ovulation are desired, it is worth checking to narrow the time of ovulation by doing a few trial examinations in late afternoon and evening.
18. Zygotes in the early stages of shell deposition from the isthmus or uterus are sticky, so always treat collecting pipettes with silicon. Shell secretion in both *S. macroura* and *T. vulpecula* [3, 5] also occurs during cleavage ULB, BLB, and TLB stages, which can make the coats temporarily sticky. These specimens can be recognized by having a bumpy surface. It is always worth having siliconized collecting pipettes available at each stage for these specimens.
19. Separation of fertilized or unfertilized oocytes requires microscopic examination. Fertilization occurs in the ampullary region as the first layers of the mucoid coat are being deposited [6, 57]. Oocytes and zygotes collected in this region will have a visible second polar body in the cleavage cavity but it becomes less visible as deutoplasmolysis progresses. Further down the oviduct it may be possible to see the developing pronuclei. A marked characteristic of fertilized eggs is that they have many sperm trapped in the mucoid coat near the zona pellucida.
20. Cleavage length varies between 6 and 12 h in marsupials, so more frequent assessment may be necessary to time the stages. *S. macroura*, like other dasyurid marsupials can have a cleavage arrest that occurs in vivo and in vitro at the 4-cell stage in some [14].
21. Recognizing parthenotes is important. Parthenogenetic activation has been reported in a number of marsupials both in polytocous and monovular species. Parthenogenetic development can replicate normal cleavage at least until the 8-cell stage. It is also a problem in wild-caught animals because ovulation is spontaneous in most marsupials. A routine procedure to follow is to examine whether sperm can be detected in the mucoid coat especially near the zona pellucida, the blastomeres are of abnormal size and whether deutoplasmolysis of the yolk mass/es is normal. If sperm are absent or rare in the

mucoid coat then the other abnormalities are usually present. Later in development the unfertilized eggs become brown and eventually opaque and in some species, can have uterine cells stuck to the surface.

22. The serum-free medium of Gardner et al. [20] is also a suitable medium for this stage.
23. Less collapse of UBL and BLB occurs if they are collected from the uterus in the culture medium.
24. Blastocysts are very fragile at ULB stages and are liable to collapse if handled carelessly so have the proper sized collection instruments available. Frequently when the ULB is forming or when first collected the cells can temporarily round up and appear to pull away from the coats, but some do recover, so it is worth leaving them for 12 h to check.
25. Blastocysts can occasionally be individually identified by distinctive blebs on the shell, but if they are collected at the time of shell secretion [5] these blebs can disappear. In *S. macroura*, a good marker for the pluriblast area is the yolk mass which remains attached to a cell at the side of the pluriblast until hypoblast formation begins.
26. Larger collection pipettes are needed for later stages. For the models used here the diameter range for BLB is 1.0–3.0 mm in *S. macroura* and 0.54–1.7 mm and in *T. vulpecula*. Moving blastocysts of this size with a cut down Pasteur pipette and bulb can be difficult so a mouth pipette or a cut down micropipette can be easier.
27. Pay particular attention to the state of the CL. In marsupials, the growth of the conceptus significantly accelerates during BLB stages [58], and this in turn relates to the vascular invasion and expansion of the CL [33] and the thickness and glandular development of the endometrium and increasing progesterone levels [28]. If the CL is retarded for some reason, the uterine changes are inhibited and the BLB development falters. These retarded blastocysts usually do not culture well.
28. BLB are sensitive to temperature changes, so warmed medium, warming plates, and working near the incubator are essential. They are less likely to collapse if collected in full medium.
29. Failing BLB tend to have rounding cells and much of the epithelium pulling away from the shell. On occasion a collapsed shell coat, recognized from its shiny appearance, may also be found. The shell is not normally detached from the conceptus at this stage. BLB size can vary so smaller blastocysts may be normal.
30. Increased weight in epiblast occurs during hypoblast migration from the pluriblast and causes the blastocyst to float with the

pluriblast/epiblast side down. They can be gently manipulated to lie with that surface uppermost using eyelash sticks to roll them.

31. To stage the BLB, and locate and count hypoblast cells under the pluriblast, examine the BLB with pluriblast lying underneath and look through the wall to examine the lower side of the pluriblast. As the hypoblast migrates under the surface epithelium of the BLB it progressively transforms from a reticulum to a complete epithelium. As the reticulum consolidates, the blastocyst yolk sac becomes more opaque in that region. A good rule of thumb is that a BLB with hypoblast cells only under the pluriblast is early, a BLB opaque to the equator is incomplete, and a BLB opaque to the opposite pole is complete. When hypoblast migration is ended, the pluriblast is transformed to an epiblast, which can have a shiny glassy appearance.
32. The size of TLB is variable. Increased size does not necessarily mean a more advanced blastocyst, so the descriptions and measurements are most important. Darkening of part of or all of the blastocyst is a bad sign. *S. macroura* are 2.5–4.6 mm in diameter and are usually slightly oval. Those of *T. vulpecula* are 3.0–7.5 mm in diameter. Minimum handling works best in these large blastocysts. For culture and any manipulations, they are best held on Sylgard plates, with the epiblast facing upward and covered with the culture medium.
33. Fragility of the blastocysts due to precocious loss of the shell in TLB is a problem. The shell is usually lost between 12 and 24 h before implantation [6] and shell-free TLB develop abnormally [16]. The shell is held closely attached to the TE by apical processes that hook into the shell meshwork. These can be seen under the microscope when the shell is detaching precociously. Less frequent processes occur on the apical surface of epiblast cells. Normal shell loss occurs in embryonic stages over the anterior mid line of the epiblast in polyovular species and over the yolk sac in monovular species. In situ at BLB, TLB, and early embryonic stages, close apposition of the shell to the apical surface of blastocyst cells is maintained by the apical processes and the swollen endometrium closely applied to the shell surface.
34. Range of stages in one animal is normal because of the extended ovulation period and eggs rolling separately down the oviduct so normal conceptus stages can be 8–12 h apart. Note if the shell is detaching from the conceptus in early stages before T6 because of rough handling, abnormal development is likely.

35. The size of vesicles at the flat and C-embryo stages varies. Flat embryo vesicle diameters vary from 3.0 to 5.0 mm in *S. macroura* and 9.0–19.0 mm in *T. vulpecula*. At the C-embryo stage in *S. macroura* vesicles have implanted and it is no longer possible to measure them. *T. vulpecula* C-embryo vesicles are 12.0–27 mm and also are in the early stages of implantation but if held in medium will slowly roll off the endometrium.
36. Because dissection is complicated and slow the chance of temperature falls is high.
37. Using the roller culture system, Cruz et al. [23] were able to culture late C-embryos to within 5 h of birth. They recommended that it would be worthwhile to try homologous serum or rat serum and optimizing the gas phase by increasing the O_2 levels higher than 20% to push culture success to birth stage. Mortola et al. [59] have shown that the related dasyurid *S. douglasi* neonate uses the skin for gas exchange. *S. macroura* may well do the same because the neonate is very red.
38. Remember that some features in embryos will appear relatively early because of precocious development of the highly altrical marsupial pouch young.

Acknowledgments

I am indebted to the Australian Research Council and the New Zealand Department of Agriculture and Foundation for Research Science and Technology for support over many years and to Unimutual, Australia in more recent years. Many postgraduate students, colleagues, and support staff contributed to development of these culture and induced ovulation systems including Angela Nation, Danielle Hickford, Aida Yousef, Ellen Menkhorst, Nicole Ezard, Yolanda Cruz, Isabel Douglas Lisa Mansini, and Kamani Nanayakkara, and I thank them all for their dedication to improving culture systems for marsupial conceptuses.

References

1. Menkhorst P (2001) A field guide to the mammals of Australia. Oxford University Press, Oxford
2. Johnson MH, Selwood L (1996) The nomenclature of early development in mammals. Rep Fertil Dev 8:759–764
3. Menkhorst EM, Selwood L (2008) Vertebrate extracellular postovulatory egg coats. Biol Reprod 79:790–797
4. Menkhorst EM, Ezard N, Selwood L (2007) Induction of ovulation and natural oestrous cycling in the stripe-faced dunnart, *Sminthopsis macroura*. Reproduction 133:495–502
5. Menkhorst EM, Selwood L, Cui S (2009) Uterine expression of *cp4* gene homolog in the stripe-faced Dunnart, *Sminthopsis macroura*: relationship with conceptus development and progesterone profile. Mol Reprod Dev 76:863–872
6. Selwood L (2000) Marsupial egg and embryo coats. Cells Tissues Organs 166:208–219

7. Selwood L (1992) Mechanisms underlying the development of pattern in marsupial embryos. In: Pedersen RA (ed) Current topics in developmental biology, vol 27. Academic, New York, NY
8. Kress A, Selwood L (2003) Conceptus polarity and cell-zona adhesion during early cleavage (fertilized tubal egg to 8-cell stage) in the marsupial *Sminthopsis macroura*. Cells Tissues Organs 175:9–24
9. Kress A, Selwood L (2004) Precedence of cell-zona adhesion over cell-cell adhesion during marsupial blastocyst formation prohibits morula formation and ensures that both the pluriblast and trophoblast are superficial. Cells Tissue Organs 177:87–103
10. Kress A, Selwood L (2006) Marsupial hypoblast: formation and differentiation of the bilaminar blastocyst in Sminthopsis macroura. Cells Tissues Organs 182:155–170
11. Hughes RL, Shorey CD (1973) Observations on the permeability properties of the egg membranes of the marsupial *Trichosurus vulpecula*. J Reprod Fertil 32:25–32
12. Dawson TJ, Hulbert AJ (1969) Standard energy metabolism of marsupials. Nature 22:383
13. Selwood L, VandeBerg JW (1992) The influence of incubation temperature on oocyte maturation, parthenogenetic and embryonic development in vitro of the marsupial *Monodelphis domestica*. Animal Reprod Sci 29:99–117
14. Selwood L, Robinson ES, Pedersen RA, VandeBerg JL (1997) Development *in* of marsupials: a comparative review of species and a timetable of cleavage and early blastocyst stages of development in *Monodelphis domestica*. Int J Dev Biol 41:397–410
15. Selwood L (1989) Development *in vitro* of investment-free marsupial embryos during cleavage and early blastocyst formation. Gamete Res 23:399–413
16. Casey N, Selwood L (2003) Removal of the shell coat affects maintenance of epithelia in blastocysts of the brushtail possum *in vitro*. Cells Tissues Organs 173:21–45
17. Frankenberg S, Selwood L (1998) An ultrastructural study of the role of an extracellular matrix during normal cleavage in a marsupial the brushtail possum. Mol Reprod Dev 50:420–433
18. Selwood L, Young GJ (1983) Cleavage in vivo and in culture in the dasyurid marsupial *Antechinus stuartii*. J Morphol 176:443–460
19. Yousef A, Selwood L (1993) Embryonic development during gastrulation in *Antechinus stuartii* (MacLeay) and *Sminthopsis macroura* (Spencer) in culture. Reprod Fertil Dev 5:445–458
20. Gardner DK, Selwood L, Lane M (1996) Nutrient uptake and culture of *S. macroura* (stripe-faced dunnart) embryos. Reprod Fertil Dev 8:685–690
21. Selwood L, Smith D (1990) Time-lapse analysis and normal stages of development of cleavage and blastocyst formation *in vitro* in the marsupials the Brown Antechinus and the Stripe-faced Dunnart. Mol Reprod Dev 26:53–62
22. Selwood L (1996) The blastocyst epithelium is not a protoderm in dasyurid marsupials. A review of the evidence. Reprod Fertil Dev 8:711–724
23. Mate KE, Robinson JL, VandeBerg JL, Pedersen RA (1994) Timetable of *in vivo* embryonic development in the grey short-tailed opossum (*Monodelphis domestica*). Mol Reprod Dev 39:365–374
24. Cruz YP, Hickford D, Selwood L (2000) A staging scheme for assessing development *in vitro* of organogenesis stage embryos. J Reprod Fertil 120:99–108
25. Selwood L, Hickford D (1999) Early cell lineages in marsupial embryo. In: Moody SA (ed) Cell fate and cell determination. Academic, New York, NY
26. Cruz YP, Selwood L, Morton H, Cavanagh AC (2001) Significance of serum early pregnancy factor concentrations during pregnancy and embryonic development in *Sminthopsis macroura* (Spencer) Marsupialia: Dasyuridae. Reproduction 121:933–939
27. Cruz YP, Morton H, Cavanagh H, Selwood L, Wilson SD, Sasaki M (2006) Early pregnancy factor in marsupials. Austr J Zool 54:211–215
28. Cruz YP, Selwood L (1997) Histological differences between gravid and non-gravid uteri in the dasyurid marsupial, *Sminthopsis macroura* (Spencer). J Reprod Fertil 111:319–325
29. Fletcher TP, Selwood L (2000) Possum reproduction and development. In: Montague T (ed) The Brushtail Possum in New Zealand. Landcare Press, NZ, Auckland
30. Gianaroli L, R S, P FA, Flamigni C, Bovicelli L (1986) The successful use of human amniotic fluid for mouse embryo culture and human *in vitro* fertilization, embryo culture and transfer. Fertil Steril 46:907–913
31. Dorfmann AD, Bender SN, Robinson P, Fugger EF, Bustillo M, Reed G, Schulman JD (1989) Cell-free human amniotic fluid as culture medium for mouse and human embryos. Fertil Steril 51:671–674

32. McCrady E Jr (1938) The embryology of the opossum. Am Anat Mem 16:1–233
33. Selwood L, Woolley PA (1991) A time-table of embryonic development and ovarian and uterine changes during pregnancy in the stripe-faced dunnart *Sminthopsis macroura* (Marsupialia: Dasyuridae). J Reprod Fert 91:213–227
34. Hughes RL, Hall LS (1984) Embryonic development in the common brushtail possum, *Trichosurus vulpecula*. In: Smith AP, Hume ID (eds) Possums and gliders. Australian Mammal Society, Sydney, NSW
35. Gemmell RT, Selwood L (1994) Structural development in the marsupial, the stripe-faced dunnart, *Sminthopsis macroura*. Acta Anat 149:1–12
36. Nelson JE (1992) Developmental staging in a marsupial *Dasyurus hallucatus*. Anat Embryol 185:335–354
37. Schneider NY, Gurovich Y (2017) Morphology and evolution of the oral shield in marsupial neonates including the newborn monito del monte (*Dromiciops gliroides*, Marsupialia Microbiotheria) pouch young. J Anat 231:59–83
38. New DAT, Mizell M (1972) Opossum foetuses grown in culture. Science 175:533–536
39. New DAT, Mizell M, Cockroft DL (1977) Growth of opossum embryos *in vitro* during organogenesis. J Embrol Exp Morphol 41:111–123
40. Hickford D, Shaw G, Renfree MB (2008) *In vitro* culture of peri-gastrulation embryos of a macropodid marsupial. J Anat 212:180–191
41. Butcher L, Ullmann SL (1996) Culture of pre-antral ovarian follicles in the grey-short tailed opossum *Monadelphis domestica*. Reprod Fertil Dev 8:535–549
42. Nation A, Selwood L (2009) The production of mature oocytes from adult ovaries following primary follicle culture in a marsupial. Reproduction 138:247–255
43. Moore HDM, Taggart DA (1993) In vitro fertilization and embryo culture in the grey short-tailed opossum, *M. domestica*. J Reprod Fertil 98:267–274
44. Richings NM, Shaw G, Temple-Smith PD, Renfree MB (2004) Intra-cytoplasmic sperm injection in a marsupial. Reproduction 128:595–560
45. Au PCK, Nation A, Parrott M, Selwood L (2010) Induced ovulation mimics the timetable of natural development in the stripe-faced dunnart, *Sminthopsis macroura*, and results in the birth of fertile young. Reproduction 139:419–425
46. Duckworth JA, Scobie S, Jones DE, Selwood L (1998) Determination of oestrus and mating in captive female Brushtail Possums, *Trichosurus vulpecula* (Marsupialia: Phalangeridae) from urine samples. Austr J Zool 46:547–555
47. Cruz Y, Yousef A, Selwood L (1996) Fate map analysis of the epiblast of the dasyurid marsupial, *Sminthopsis macroura*. Reprod Fertil Dev 8:779–788
48. Pedersen T, Peters H (1968) Proposal for a classification of oocytes and follicles in the mouse ovary. J Reprod Fertil 17:555–557
49. Kress A, Merry NE, Selwood L (2001) Oogenesis in the marsupial the stripe-faced dunnart, *Sminthopsis macroura*. Cells Tissues Organs 168:188–202
50. Wycherley G, Downey D, MT K, Hynes A (2004) A novel follicle culture system markedly increases follicle volume, cell number and oestradiol secretion. Reproduction 127:669–677
51. Merry N, Johnson M, Gehring C, Selwood L (1995) Cytoskeletal organization in the oocyte, zygote, and early cleaving embryo of the stripe-faced dunnart (*Sminthopsis macroura*). Mol Reprod Dev 41:212–224
52. Maleszewski M, Selwood L (2004) Induced parthenogenetic activation of oocytes of the marsupial *Sminthopsis macroura*. Reprod Fertil Dev 15:559–604
53. Selwood L (1980) A timetable of embryonic development of the dasyurid marsupial *Antechinus stuartii* (Macleay). Austr J Zool 28:649–668
54. Selwood L (1987) Embryonic development in culture of two dasyurid marsupials, *Sminthopsis crassicaudata* (Gould) and *Sminthopsis macoura* (Spencer) during cleavage and blastocyst formation. Gamete Res 16:355–370
55. Baggott LM, Davis-Butler S, Moore HD (1987) Characterisation of oestrus and timed collection of oocytes in the grey short-tailed opossum, *Monodelphis domestica*. J Reprod Fertil 79:105–114
56. Menkhorst EM, Hinds LA, Selwood L (2009) Progesterone concentration in the marsupial *Sminthopsis macroura*: relationship with the conceptus, uterine glandular regeneration and body weight. Reproduction 137:107–117
57. Selwood L (1982) A review of maturation and fertilization in marsupials with special reference to the dasyurid marsupial *Antechinus stuartii*. In: Archer M (ed) Carnivorous marsupials. Royal Zoological Society, Sydney, NSW
58. Selwood L (1989) Marsupial pre-implantation embryos in vivo and in vitro. In: Yoshinaga K, Mori T (eds) Development of pre-implantation embryos and their environment. A Liss, New York, NY
59. Mortola JP, Frappell P, Woolley PA (1999) Breathing through skin in a newborn mammal. Nature 397(6721): 660

Chapter 21

In Vitro Culture of Embryos from the Common Marmoset (*Callithrix jacchus*)

Jenna Kropp Schmidt and Thaddeus G. Golos

Abstract

The biology of fertility, early development, and pregnancy is variable across mammalian species. In addition, while the physiology and pathophysiology of human diseases can be investigated in other animal models (principally, rodents), differences between human and lower mammals often present limitations in the applicability of physiological processes from rodent models to human biology. Since 1984, when the first live birth from rhesus monkey in vitro fertilization and embryo transfer was reported (Bavister et al., Proc Natl Acad Sci 81:2218–2222, 1984), there has been progress in the implementation of assisted reproductive technologies with several nonhuman primate (NHP) species that play important roles in biomedical research. In recent years, the significance of this progress has been amplified by the development of genomic editing approaches for facile genetic manipulation of the embryo, including methods now applied to NHPs (Liu et al., Cell Stem Cell 14:323–328, 2014; Niu et al., Cell 156:836–843, 2014). In this review, we summarize current protocols and practices for the common marmoset. It is our intention to provide current state-of-the-art protocols for gamete procurement and in vitro fertilization techniques, so that laboratories wishing to implement experimental embryology in marmoset models will have a basic set of tools with which to initiate such studies.

Key words Marmoset, Controlled ovarian stimulation, Fertilization, Embryo culture, Nonhuman primate

1 Introduction

The common marmoset is a unique nonhuman primate in its robust reproductive capacity. Marmosets reach sexual maturity at about 18 months of age, ovulate two or more oocytes per cycle, and have a relatively short gestation length of 144 days. Additionally, the ovarian cycle can be readily managed by administration of pharmacological agents to induce luteal regression and also ovulation. These attributes provide a basis for which assisted reproductive technologies can be implemented to generate marmoset embryos in a timely manner. Considerable effort to procure marmoset embryos in vitro has established methods for ovarian hyperstimulation [1–3], in vitro maturation [1, 4], semen collection

Jason R. Herrick (ed.), *Comparative Embryo Culture: Methods and Protocols*, Methods in Molecular Biology, vol. 2006, https://doi.org/10.1007/978-1-4939-9566-0_21,

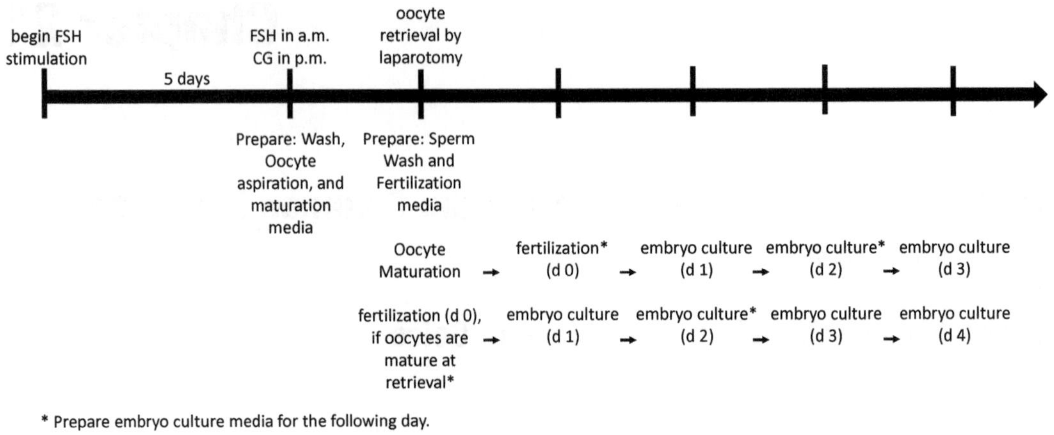

Fig. 1 Timeline of events for in vitro production of common marmoset embryos

[5, 6], fertilization by in vitro fertilization [1, 2, 4, 7] and intracytoplasmic sperm injection [2, 3], embryo culture [1, 4], and embryo transfer [8–11]; *see* review by Kropp et al. [12]. The facile management of the ovarian cycle allows for synchronization of oocyte donors and embryo recipients, in which, transferred in vitro-produced embryos have resulted in birth of live offspring. In vitro production of marmoset embryos provides a platform for genetic manipulation of an early embryo to develop nonhuman primate models of human disease with recent advances in genomic editing [13] and transgenesis [14] in the marmoset. Here, methods for establishing and managing the marmoset ovarian cycle, oocyte maturation, in vitro fertilization, and embryo culture are presented in detail (Fig. 1).

2 Materials

2.1 Hormones

1. Prostaglandin F2α (PGF2α) analog.
2. Recombinant human follicle-stimulating hormone (FSH) (*see* **Note 1**).
3. Recombinant human chorionic gonadotropin (CG).

2.2 Culture Media

1. Sodium pyruvate stock (100 mM): 22 mg sodium pyruvate in 2 mL PBS.
2. Glutamine solution (200 mM): L-alanyl-L-glutamine, 200 mM solution purchased commercially.
3. Wash Medium: 200 mL TL-Hepes (commercially available), supplemented with 1 mL Sodium Pyruvate Stock (100 mM), 25 mg/mL gentamicin, and 3 mg/mL fatty acid-free bovine serum albumin (BSA). Filter (0.22 μm) into a sterile bottle. Store at 4 °C for up to 2 weeks.

4. Oocyte Aspiration Medium: 100 mL Wash Medium with 1 IU/mL heparin. Store at 4 °C. Prepare fresh medium for each aspiration.
5. Maturation Medium: 675 μL CMRL-1066 (commercially available) supplemented with 5 μL sodium pyruvate stock (100 mM), 10 μL glutamine solution (200 mM), 20% FBS (v/v), 80 IU/mL FSH, and 10 IU/mL hCG. Filter (0.22 μm) and pipet four 30-μL drops per 35 mm petri dish and overlay with 3.5 mL paraffin oil. Place the plate in the incubator (*see* **Note 2**) and equilibrate overnight prior to day of oocyte retrieval. Prepare fresh medium for each retrieval.
6. Sperm Wash Medium: Origio Sequential Fert medium. Supplement with 5% FBS (v/v). Allow medium to equilibrate overnight in incubator.
7. Fertilization Medium: Origio Sequential Fert medium. Supplement with 5% FBS (v/v). Pipet four 25-μL drops per 35 mm petri dish and overlay with 3.5 mL paraffin oil. Place the plate in the incubator (*see* **Note 2**) and allow media to equilibrate overnight.
8. Embryo Culture Medium: Sequential medium is recommended and several are commercially available (e.g., Origio Sequential Medium). Pipet four 30-μL drops per 35 mm petri dish and overlay with 3.5 mL paraffin oil. Place the plate in the incubator (*see* **Note 2**) and allow media to equilibrate overnight prior to the day of moving embryos.

3 Methods

3.1 Ovarian Cycle Management and Stimulation

The follicular and luteal phase of the ovarian cycle can be established through ultrasound imaging and analysis of serum progesterone levels. Figure 2 illustrates the serum progesterone levels throughout the marmoset ovarian cycle. To begin, ultrasound evaluation of the oocyte donor is done to assess pregnancy status:

1. If the animal is early in pregnancy, administer PGF2α at a dose of 0.75 μg to induce luteal regression. Pregnancy diagnosis by ultrasound in the marmoset has previously been described [15]. Representative ultrasound images of a nonpregnant and early pregnant uterus are shown in Fig. 3. To ensure that the animal has undergone luteal regression, a 0.2 mL blood sample can be drawn on the day of PGF2α administration and days 2 and 5 post administration for analysis of serum progesterone levels.
2. If the animal is not pregnant, draw 0.2 mL of blood twice weekly to determine timing of ovulation through analysis of serum progesterone levels [16]. Continue to monitor the

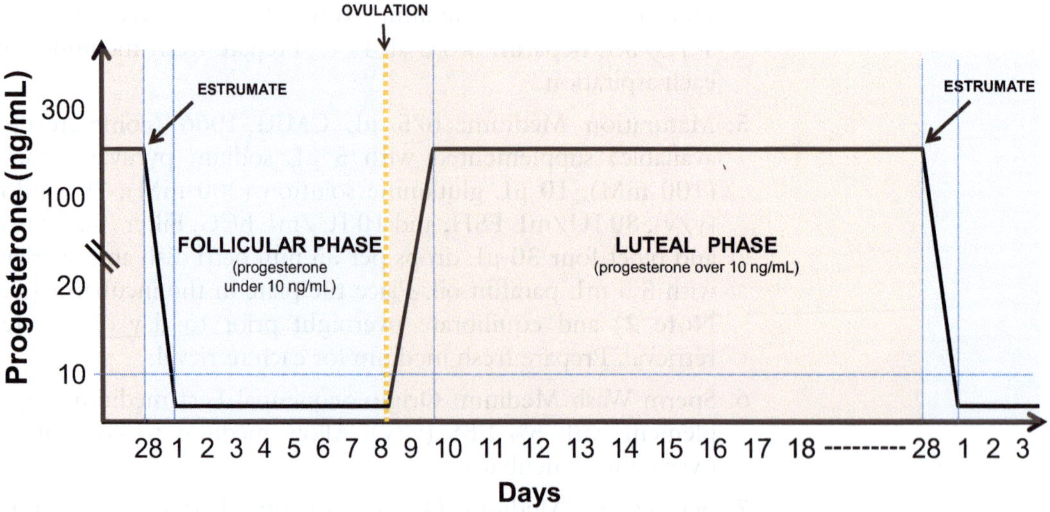

Fig. 2 The ovarian cycle of the marmoset. Total cycle length is about 28 days with a follicular phase of about 8 days and the luteal phase of about 19 days [16]. The follicular phase is defined as serum progesterone concentrations <10 ng/mL, while the luteal phase is characterized by concentrations >10 ng/mL [16]. A prostaglandin $F_{2\alpha}$ analog, cloprostenol (Estrumate), can be administered on days 10–17 after ovulation for luteal regression to synchronize oocyte donors and embryo recipients [17, 18]

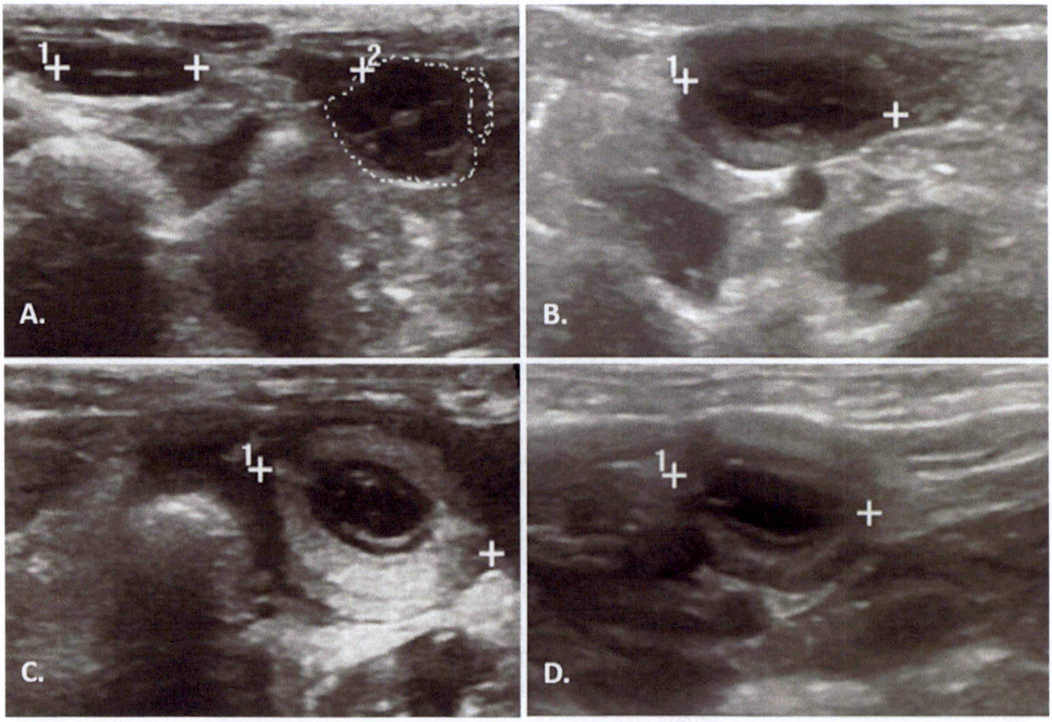

Fig. 3 Marmoset pregnancy detection by ultrasound. A nonpregnant uterus has a distinct echogenic line present in the center as observed in **a** and **b**. Follicles are present on the ovary denoted structure 2 in image **a**. In comparison, the pregnant uterus is fluid filled as shown in **c** and **d**. A uterine diameter can be measured as shown between the (+) and marked with a 1

	$PGF_{2\alpha}$ analogue	FSH		FSH		FSH		FSH		FSH		FSH AM	CG PM	Oocyte Retrieval
		AM	PM	AM	PM	AM	PM	AM	PM	AM	PM			
Day	0	1		2		3		4		5		6		7

Fig. 4 Marmoset ovarian hyperstimulation regimen. Following injection of $PGF_{2\alpha}$ analogue (deemed day 0), administration of FSH should begin the following day with twice daily 50 IU doses. On day 6, administer 50 IU FSH in the morning and CG in the afternoon. The oocyte retrieval laparotomy should occur at 20–24 h post CG injection

animal's serum progesterone levels until consistent cycling is observed.

3. PGF2α can be administered every 28 days to induce luteal regression, allowing for minimal variation in cycle length, to optimally time ovulation or begin ovarian stimulation. Monitoring of serum progesterone levels should be performed to ensure luteal regression has occurred. Maintaining animals on a 28-day schedule with PGF2α administration reduces the need for biweekly blood draws and allows for ovarian synchronization of oocyte donors and embryo recipients.
4. Timed ovulation can be achieved by administration of 50–75 IU CG on day 7 following PGF2α administration. Animals will ovulate within 48–72 h post-injection [17]. This injection may be given to embryo recipients to achieve synchronization with the oocyte donor.
5. The regimen for ovarian hyperstimulation is depicted in Fig. 4. Establish the ovarian cycle of the animal prior to beginning ovarian hyperstimulation. Administer a single injection of PGF2α on day 10–17 of the luteal phase to induce luteal regression [17, 18], deemed day 0. Beginning on day 1, start FSH stimulation by administering 50 IU FSH twice daily for 5.5 days. In the afternoon of day 6, administer 75 IU CG for induction of ovulation (*see* **Note 3**).

3.2 Oocyte Aspiration, Grading, and Maturation

1. Oocyte retrieval is to be performed at 20–24 h post-CG administration by laparotomy to expose the ovaries. To perform the aspirations, an example of a marmoset follicle aspiration apparatus is described by Marshall et al. [1].
2. Collect the cumulus–oocyte complexes in Wash Medium warmed to 37 °C.
3. Pipet the aspirate into a 60 mm petri dish.
4. Transfer the cumulus–oocyte complexes into fresh Wash Medium.

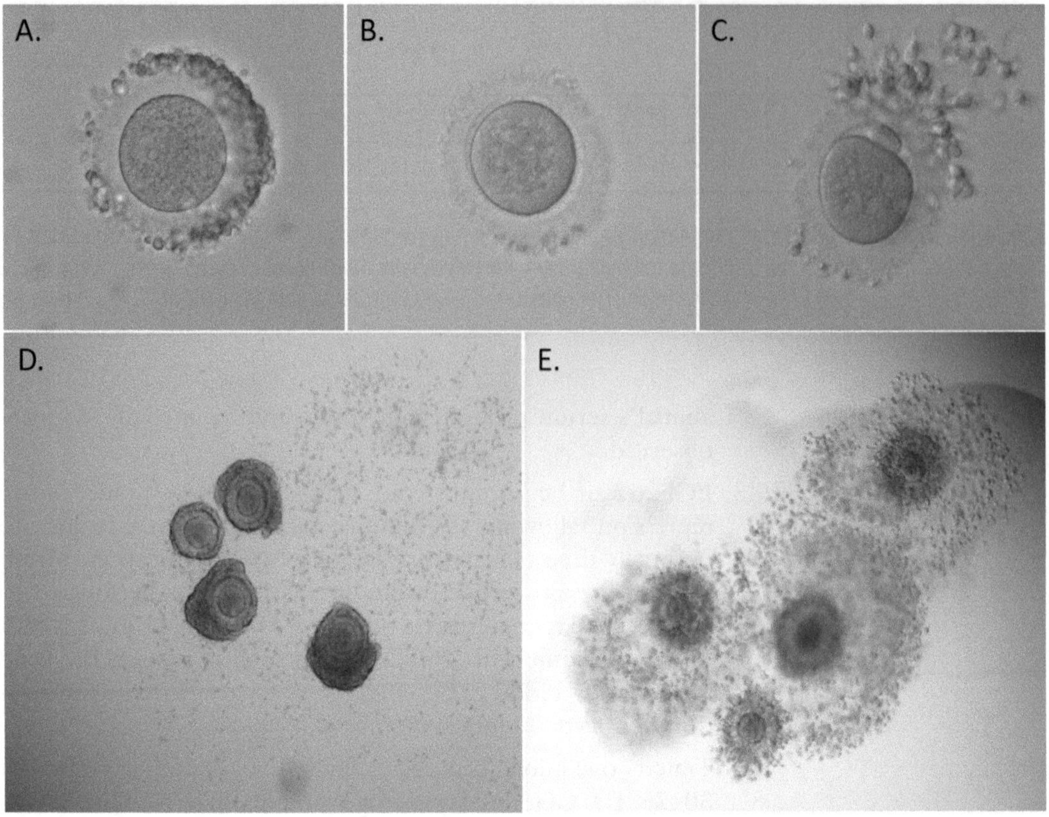

Fig. 5 Oocyte maturity at laparotomy. Oocyte maturity at the time of oocyte retrieval varies by oocyte donor where immature oocytes GV (**a**) and MI (**b**) or mature MII (**c**) oocytes may be recovered. Immature oocytes (**d**) at the time of retrieval will mature (**e**) within 24 h of in vitro maturation

5. Wash the cumulus–oocyte complexes twice by moving through 50 μL drops of Wash Medium.
6. Evaluate and grade oocyte maturity as described by Wilton et al. [19], grouping those of similar grade.
7. Transfer up to five cumulus–oocyte complexes of similar grade into 30 μL drop of Maturation Medium.
8. Incubate oocytes in Maturation Medium until fertilization (*see* **Note 4**). Fig. 5 displays representative morphology of immature and mature oocytes.

3.3 Semen Collection

1. Place the fully alert male into a restraint tube for collection by penile vibratory stimulation as previously described [1, 6], *see* Fig. 6.
2. Attach a 5 mm diameter, 0.4-mL microcentrifuge tube to a FertiCare Personal vibrator (ILTS, Inc., Evans, GA, USA) instrument.

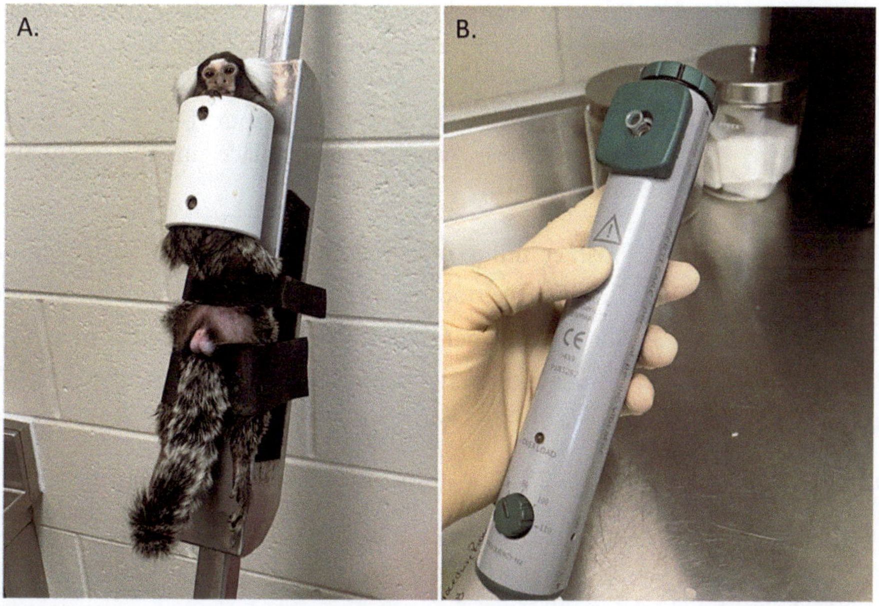

Fig. 6 Marmoset semen collection. A male marmoset is placed into the restraint tube (**a**) prior to collection by penile vibratory stimulation. A FertiCare Personal vibrator (**b**) affixed with a 0.4-mL microcentrifuge tube is used for the collection procedure

3. Position the collection tube over the penis. The tube will serve as an artificial vagina.
4. Apply FertiCare instrument to the penile area for vibratory stimulation.
5. Allow the FertiCare instrument to cycle through one stimulation cycle (*see* **Note 5**), which lasts approximately 2 min.

3.4 Sperm Preparation by the Swim-Up Method

1. Record the volume of ejaculate.
2. If the ejaculate is less than 50 μL, overlay the ejaculate with 50 μL of Sperm Wash Medium. For larger ejaculate volumes, overlay with 100 μL of Sperm Wash Medium.
3. Incubate the ejaculate for 30 min at 37 °C and sperm to swim up.
4. Remove 10 μL from the upper two-thirds of medium, deemed the swim-up region, and assess motility.
5. Mix the coagulum and swim-up fraction and transfer to a 0.7 mL tube (*see* **Note 6**).
6. Overlay with 200 μL of Sperm Wash Medium.
7. Incubate for at least 1.5 h at 37 °C.
8. Remove 10 μL from the swim-up region and assess motility.

9. Remove an additional 10 μL from the swim-up region, mix with 20 μL of water, and load onto a hemocytometer to count sperm.
10. Dilute the sperm sample to about 1–4 × 10^6 cells/mL in Sperm Wash Medium and place in the incubator until fertilization.

3.5 In Vitro Fertilization

1. Wash cumulus–oocyte complexes once in Wash Medium.
2. Transfer five oocytes into a drop of Fertilization Medium.
3. Add 5 μL of sperm diluted to 1–4 × 10^6 cells/mL (*see* **Note 7**).
4. Coincubate gametes for 18–20 h in an incubator at 5% CO_2, 5% O_2 and 37 °C.

3.6 Embryo Culture

1. At 18–20 h postfertilization (day 1), remove the presumptive zygotes from Fertilization Medium and place them into 50 μL drops of Wash Medium.
2. Transfer presumptive zygotes into an 80 IU/mL drop of hyaluronidase.
3. Remove cumulus cells and excess sperm by pipetting the presumptive zygotes up and down in hyaluronidase (*see* **Note 8**).
4. Assess fertilization by determining the presence of two polar bodies as illustrated in Fig. 7.
5. Wash the presumptive zygotes twice through 50 μL drops of Wash Medium.

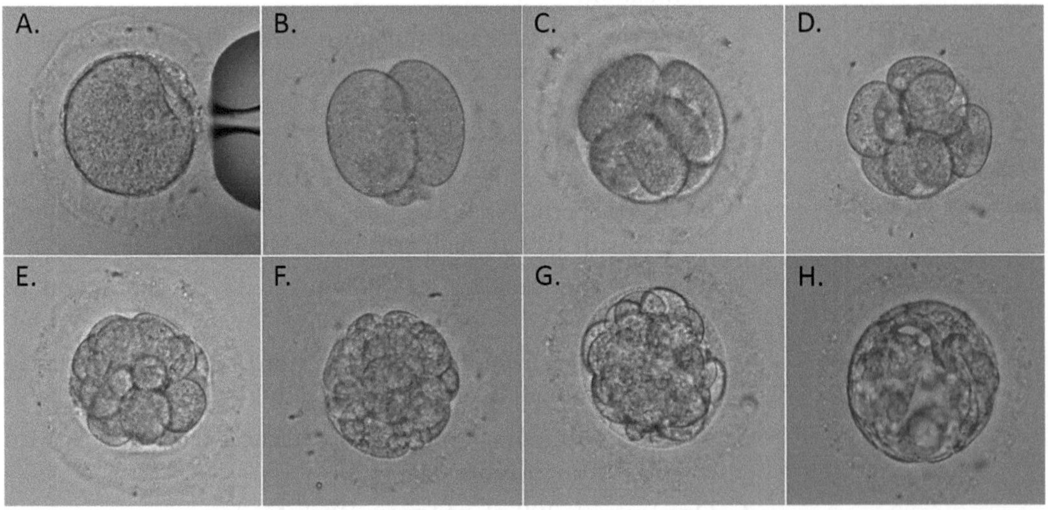

Fig. 7 Marmoset in vitro embryonic development. The stages of embryonic development are represented as follows: (**a**) Zygote, (**b**) 2-cell, (**c**) 4-cell, (**d**) 8-cell, (**e**) 16-cell, (**f**) 32-cell, (**g**) morula, and (**h**) an early blastocyst. A timeline of in vitro development and additional images are present in a review by Kropp et al. [12]

6. Transfer five presumptive zygotes into a 30 μL drop of the first step medium for cleavage-stage embryos.
7. Place embryos in incubator.
8. On day 3 of culture, wash oocytes once in 50 μL drop of Wash Medium.
9. Place cohorts of five embryos per drop into the second step medium for blastocyst formation.
10. Replace medium every other day, recording embryonic development as shown in Fig. 7.

4 Notes

1. The follitropin beta product Follistim/Puregon (manufactured for Merck Sharp & Dohme Corp, Whitehouse Station, NJ, USA) reliably yields a large cohort of medium- to large-sized follicles, whereas other hormone preparations gave variable results.
2. Incubator conditions should be maintained at 37 °C, 5% CO_2 and 5% O_2 in humidified air.
3. Various FSH/CG regimens have been implemented to achieve ovarian hyperstimulation as reviewed by Kropp et al. [12].
4. The timing of cumulus–oocyte incubation prior to fertilization is contingent on oocyte maturity at the time of retrieval, *see* Fig. 5. Oocytes with tight compact layers of cumulus cells are immature oocytes that should undergo 24 h of in vitro maturation to allow for germinal vesicle breakdown and progression of meiosis. Those with expanded cumulus cells can be fertilized after a few hours of incubation, allowing for sperm preparation. Oocyte donors with corpora lutea present at the time of the retrieval will have oocytes with expanded cumulus cells that are to be fertilized on the same day as the retrieval.
5. FertiCare Personal vibrator instrument settings for semen collection:
   (a) Start with a frequency 80 Hz and amplitude 1.0 mm.
   (b) At 30 s, increase amplitude to 1.5 mm.
   (c) At 1 min, increase frequency to 90 Hz.
   (d) At 1 min 30 s, increase amplitude to 2.0 mm.
   (e) At 2 min, discontinue stimulation.
   (f) If no sample is obtained, wait 2 min and repeat. Up to three attempts may be performed.

6. An alternative approach for sperm isolation is to utilize a density gradient as previously described by Hernández-López et al. [20].
7. Fertilization of marmoset oocytes can also be achieved by ICSI [2, 3].
8. A 125–135 μm pipette tip should be used for stripping away cumulus cells.

Acknowledgments

T.G.G. and J.K.S. would like to thank the veterinary and animal care staff at the Wisconsin National Primate Research Center, particularly Michele Schotzko and Dr. Nancy Schultz-Darken of the Scientific Protocol Implementation Unit, Megan Sosa of the Animal Services Unit, and Drs. Casey Fitz and Kevin Brunner of the Veterinary Services Unit, for assistance in synchronizing donors and collection of gametes to perform marmoset in vitro experiments to generate embryo images. Studies at the University of Wisconsin-Madison were supported by NIH grants R24OD019803 to T.G.G. and P51OD011106 to the WNPRC.

References

1. Marshall VS, Browne MA, Knowles L, Golos TG, Thomson JA (2003) Ovarian stimulation of marmoset monkeys (*Callithrix jacchus*) using recombinant human follicle stimulating hormone. J Med Primatol 32:57–66
2. Grupen CG, Gilchrist RB, Nayudu PL, Barry MF, Schulz SJ, Ritter LJ, Armstrong DT (2007) Effects of ovarian stimulation, with and without human chorionic gonadotrophin, on oocyte meiotic and developmental competence in the marmoset monkey (*Callithrix jacchus*). Theriogenology 68:861–872
3. Takahashi T, Hanazawa K, Inoue T, Sato K, Sedohara A, Okahara J, Suemizu H, Yagihashi C, Yamamoto M, Eto T, Konno Y, Okano H et al (2014) Birth of healthy offspring following ICSI in in vitro-matured common marmoset (*Callithrix jacchus*) oocytes. PLoS One 9:e95560
4. Gilchrist RB, Nayudu PL, Hodges JK (1997) Maturation, fertilization, and development of marmoset monkey oocytes in vitro. Biol Reprod 56:238–246
5. Kuederling I, Schneiders A, Sonksen J, Nayudu PL, Hodges JK (2000) Non-invasive collection of ejaculates from the common marmoset (*Callithrix jacchus*) using penile vibrostimulation. Am J Primatol 52:149–154
6. Schneiders A, Sonksen J, Hodges JK (2004) Penile vibratory stimulation in the marmoset monkey: a practical alternative to electro-ejaculation, yielding ejaculates of enhanced quality. J Med Primatol 33:98–104
7. Tomioka I, Takahashi T, Shimada A, Yoshioka K, Sasaki E (2012) Birth of common marmoset (*Callithrix jacchus*) offspring derived from in vitro-matured oocytes in chemically defined medium. Theriogenology 78:1487–1493
8. Summers PM, Shephard AM, Taylor CT, Hearn JP (1987) The effects of cryopreservation and transfer on embryonic-development in the common marmoset monkey, *Callithrix jacchus*. J Reprod Fertil 79:241–250
9. Marshall VS, Kalishman J, Thomson JA (1997) Nonsurgical embryo transfer in the common marmoset monkey. J Med Primatol 26:241–247
10. Ishibashi H, Motohashi HH, Kumon M, Yamamoto K, Okada H, Okada T, Seki K (2013) Efficient embryo transfer in the common marmoset monkey (*Callithrix jacchus*) with a reduced transfer volume: a non-surgical approach with cryopreserved late-stage embryos. Biol Reprod 88:115

11. Ishibashi H, Motohashi HH, Kumon M, Yamamoto K, Okada H, Okada T, Seki K (2013) Ultrasound-guided non-surgical embryo collection in the common marmoset. Reprod Biol 13:139–144
12. Kropp J, Di Marzo A, Golos T (2017) Assisted reproductive technologies in the common marmoset: an integral species for developing nonhuman primate models of human diseases. Biol Reprod 96:277–287
13. Sato K, Oiwa R, Kumita W, Henry R, Sakuma T, Ito R, Nozu R, Inoue T, Katano I, Okahara N, Okahara J, Shimizu Y et al (2016) Generation of a nonhuman primate model of severe combined immunodeficiency using highly efficient genome editing. Cell Stem Cell 19:127–138
14. Sasaki E, Suemizu H, Shimada A, Hanazawa K, Oiwa R, Kamioka M, Tomioka I, Sotomaru Y, Hirakawa R, Eto T, Shiozawa S, Maeda T et al (2009) Generation of transgenic non-human primates with germline transmission. Nature 459:523–U550
15. Oerke AK, Einspanier A, Hodges JK (1995) Detection of pregnancy and monitoring patterns of uterine and fetal growth in the marmoset monkey (*Callithrix jacchus*) by real-time ultrasonography. Am J Primatol 36:1–13
16. Harlow CR, Gems S, Hodges JK, Hearn JP (1983) The relationship between plasma progesterone and the timing of ovulation and early embryonic-development in the marmoset monkey (*Callithrix jacchus*). J Zool 201:273–282
17. Hodges JK, Cottingham PG, Summers PM, Liang YN (1987) Controlled ovulation in the marmoset monkey (*Callithrix jacchus*) with human chorionic-gonadotropin following prostaglandin-induced luteal regression. Fertil Steril 48:299–305
18. Summers PM, Wennink CJ, Hodges JK (1985) Cloprostenol-induced luteolysis in the marmoset monkey (*Callithrix jacchus*). J Reprod Fertil 73:133–138
19. Wilton LJ, Marshall VS, Piercy EC, Moore HD (1993) In vitro fertilization and embryo development in the marmoset monkey (*Callithrix jacchus*). J Reprod Fertil 97:481–486
20. Hernandez-Lopez L, Umland N, Mondragon-Ceballos R, Nayudu PL (2005) Comparison of the effects of Percoll and PureSperm (R) on the common marmoset (*Callithrix jacchus*) semen. J Med Primatol 34:86–90

Chapter 22

In Vitro Culture of Embryos from the Cynomolgus Macaque (*Macaca fascicularis*)

Eliza Curnow and Eric Hayes

Abstract

The cynomolgus macaque (*Macaca fascicularis*) has contributed significantly as an animal model in a variety of biological systems including reproductive biology. The first reported delivery of a cynomolgus infant born through IVF was in 1984, and since then the use of assisted reproductive technologies (ARTs) in the cynomolgus macaque for modeling primate early embryonic development has evolved to include studies on gene editing in primate embryos and embryonic stem cell chimerism. The significance of the cynomolgus as a model of primate reproduction not only lies in the similarity of their reproductive physiology to the human but also in their availability, moderate size, and year-round utility with no seasonal breeding constraints. While many of the ARTs developed and refined in the rhesus also apply to the cynomolgus there are a number of variations that have proven effective in this species. The following chapter provides an overview of the ART methodologies that support the efficient production of *M. fascicularis* embryos, their cryopreservation and their transfer to recipient females for the establishment of pregnancy.

Key words Cynomolgus, Fascicularis, Controlled ovarian stimulation, In vitro fertilization, Embryo culture, Embryo transfer, Cryopreservation, Nonhuman primate

1 Introduction

The cynomolgus, like other laboratory macaques, is an ideal model in which to explore the fundamentals of early embryonic biology with ovarian stimulation, preimplantation and periimplantation embryonic development, and ART protocols that parallel those employed in a human IVF clinic. Furthermore, with the advent of TALEN and CRISPR/Cas9 gene editing modalities the NHP will provide an important and appropriate translational bridge between lower mammals and the human as interventions addressing genetic disorders in the human advances.

Whether exploring early embryonic development or generating genetically modified NHP models it is necessary to have robust and repeatable protocols for efficient in vitro production of viable NHP embryos. In this chapter we will provide detailed methods specific

Jason R. Herrick (ed.), *Comparative Embryo Culture: Methods and Protocols*, Methods in Molecular Biology, vol. 2006, https://doi.org/10.1007/978-1-4939-9566-0_22,

to the cynomolgus for (1) the recovery of cynomolgus gametes, (2) the in vitro production, culture, and cryostorage of embryos, and (3) embryo transfer to recipient females for the establishment of pregnancy. As with all US Department of Agriculture (USDA) regulated laboratory species, appropriate Institutional Animal Care and Use Committee (IACUC) approval must be acquired prior to initiating any NHP studies. Furthermore, appropriate precautions and training need to be in place for the handling and disposal of biological materials derived from the NHP.

2 Materials

2.1 Supplies

1. Sterile culture dishes: 35, 60, and 100 mm petri dishes, organ culture dishes, and 4-well plates.
2. 5 and 14 ml round bottom tubes.
3. 15 and 50 ml conical tubes.
4. 1, 5, and 10 ml sterile syringes.
5. 21 G needles.
6. Scalpel (#10).
7. Scissors.
8. Blunt nose scissors.
9. Hemostat.
10. Forceps.

2.2 Culture Media

All culture media listed below are available commercially as IVF/embryo tested medium, unless otherwise indicated.

1. Bovine serum albumin (BSA), Fraction V (*see* **Note 1**).
2. Light mineral oil.
3. BENCH Medium: HEPES-buffered, modified human tubal fluid medium (mHTF) supplemented with 3 mg/ml BSA. Filter sterilized. Warm to 37 °C before use.
4. Aspiration medium (ASP): BENCH Medium supplemented with 1 IU/ml heparin. Prepare 50 ml for each oocyte collection.
5. Fertilization Medium (FERT): HTF (No HEPES) supplemented with 4 mg/ml BSA. Filter sterilized. Twenty-microliter drops of FERT are overlaid with mineral oil (3.5 ml in 35 mm dishes) and equilibrated at 6% CO_2 in humidified air at 37 °C overnight before use.
6. Maturation Medium (MAT): CMRL-1066 supplemented with 0.4 mM glutamine, 0.25 mM pyruvate, 10 mM sodium lactate, 50 μg/ml gentamycin sulfate, 3.0 mM glutathione ethyl ester,

Table 1
Stock solution for maturation medium

Component	Stock soln.	Notes
Pyruvic acid	25 mM	100× stock prepared in water. Make fresh before use
Sodium lactate	60% syrup	Add 14 μl per 10mls MAT media
Glutamine	200 mM	500× stock prepared in water. Freeze aliquots at −20 °C for up to 3 months.
Gentamycin sulfate	10 mg/ml	20× stock prepared in water. Freeze aliquots at −20 °C for up to 3 months
Glutathione ethyl-ester	150 mM	50× stock prepared in water. Freeze aliquots at −20 °C for up to 3 months

and 20% (v/v) FCS or BSA (4 mg/ml). Prepare supplemented media from stock solutions (Table 1) and filter before use. Store at 4 °C for 1 week. Equilibrate dish at 6% CO_2 in humidified air at 37 °C for at least 4 h before use.

7. CULTURE Medium (*see* **Note 2**): Prepare a dish with 4–6 50 μl drops overlaid with oil. Equilibrate at 6% CO_2, 5% O_2, 89% N_2 at 37 °C overnight before use.
8. Hyaluronidase Medium (HYAL): 40 IU/ml hyaluronidase in protein-free BENCH medium. Store 1 ml aliquots in 5 ml tubes at −20 °C for 6 months. Warm to 37 °C before use.
9. Sperm Activation Medium (ACT): FERT medium containing 0.5 M caffeine and 0.2 M dbcAMP. Caffeine Stock (75mM) is made just prior to use in sterile water and filtered. dbcAMP Stock (20mM) is made in FERT and frozen in aliquots at −20 °C for up to 6 months. Add 13.3 μl Caffeine Stock and 100 μl dbcAMP Stock to 1ml FERT. Return ACT media to 6% CO_2 incubator at 37 °C until needed.
10. Sperm Cryopreservation Medium: We use Irvine Scientific Sperm Freezing Medium containing 20% (v/v) egg yolk and 12% glycerol. The medium is aliquoted (5 ml) and stored at −20 °C until manufacturer's expiration date.
11. Sperm Separation Medium: 80% solution of density gradient.
12. Embryo Cryopreservation Solutions (I, II, and III, Table 2). Prepare all Cryopreservation solutions the day before in 14 ml round bottom tubes. Prepare BENCH media and Sucrose Stock (0.1 M) solution; 0.384 g sucrose in 10 ml BENCH. Filter-sterilize using a 0.2 μm filter. Store media in the dark at RT overnight.
13. Embryo Thaw Solutions; (A, B, and C, Table 3). All solutions are prepared the day before in 14 ml round bottom tubes and

Table 2
Preparation of embryo cryopreservation solutions

Solution #	BENCH, ml	1,2-propanediol (1.5 M), ml	Sucrose stock (0.1 M), ml
I	10	–	–
II	8.9	1.1	–
III	–	1.1	8.9

Table 3
Preparation of embryo thaw solutions

Solution #	BENCH, ml	1,2-Propanediol stock (1.0 M), ml	Sucrose (0.2 M), g
A	–	5	0.342
B	2.5	2.5	0.342
C	5	–	0.342
D	5	–	–

held at RT in the dark. All solutions require the addition of Propanediol Stock solution at 1.0 M: Add 0.8 ml 1,2-propanediol into 10 ml BENCH medium.

14. 10% polyvinylpyrrolidone (PVP) for immobilization of sperm during the intracytoplasmic sperm injection procedure. We use a commercial supply prepared in BENCH medium.

2.3 Hormones for Ovarian Stimulation

1. Recombinant human follicle stimulation hormone (r-hFSH).
2. Human chorionic gonadotropin (hCG): Lyophilized and resuspended in 10 ml of sterile PBS for final conc. of 1000 IU/ml. Store at 4 °C for 1 month. Can also be frozen at −20 °C in aliquots. Load sterile 1 ml insulin syringe with 1 ml (1000 IU) of hCG solution on the day of hCG injection. Warm to RT prior to injection.

2.4 Supplies for Oocyte Aspiration

1. Suction pump capable of at least 125 mmHg.
2. Suction tubing attached to a sterile 22 G needle (Fig. 1).
3. Tube warmer capable of holding both 50 and 15 ml tubes.
4. Warming block for 14 and 5 ml tubes.

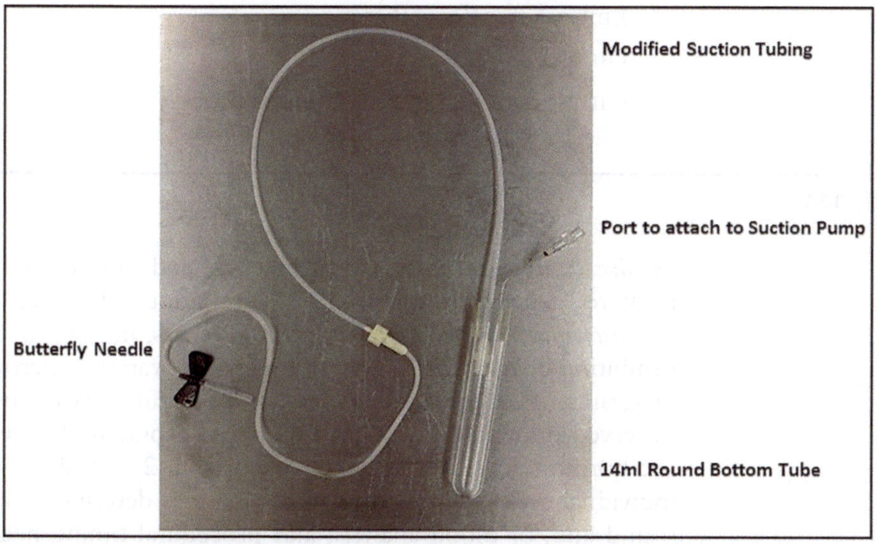

Fig. 1 Setup of the aspiration tubing

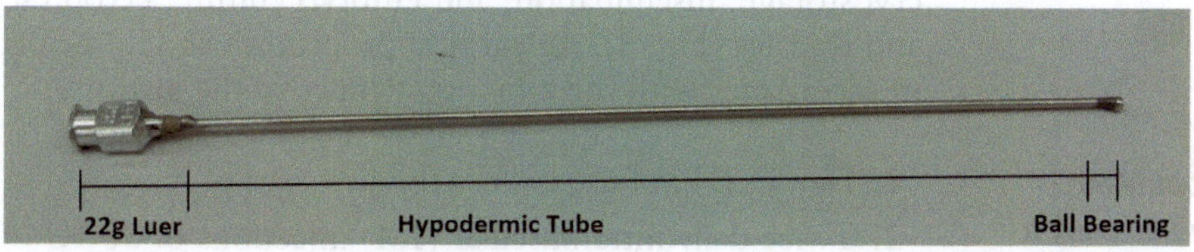

Fig. 2 In-house generated transfer stylet designed to traverse the convoluted *M. fascicularis* cervical canal for embryo transfer

2.5 Cryopreservation Supplies

1. 0.25 and 0.5 ml plastic straws for cryopreservation.
2. Heat sealer.
3. Styrofoam box for liquid nitrogen.
4. Styrofoam float: Small block of styrofoam approx. 5 cm height on which straws are placed to in order to freeze in LN_2 vapor prior to plunging.
5. Long handled forceps.
6. 30 and 33 °C water baths.
7. Storage canes and goblets.

2.6 Nonsurgical Embryo Transfer

1. Stylet (Fig. 2, *see* **Note 3**).
2. Nasal speculum.
3. Head lamp.
4. Sterile swab sticks.
5. BENCH (no protein).

6. Embryo transfer catheter.
7. Ultrasound gel.
8. Ultrasound with transabdominal probe.

3 Methods

M. fascicularis are not seasonal breeders and can be used for the in vitro generation of embryos throughout the year. Female *M. fascicularis* aged between 6 and 12 years, in good health and exhibiting regular cycles are optimal for ovarian superovulation programs. As a guideline, at least three regular cycles should be observed prior to an ovarian stimulation program. The length of the hormone regime can be modified by 1–2 days depending on individual cycle lengths, ovarian response as determined by ultrasound and/or blood analysis, and procedural timing needs. The following methods describe the steps necessary to produce and culture *M. fascicularis* embryos, including ovarian stimulation to recover multiple mature ova; sperm retrieval; preparation and cryostorage; insemination; and embryo culture, cryopreservation, and transfer (Fig. 3). All surgical procedures should be performed in accordance with the ethics guidelines of the institute.

3.1 Controlled Ovarian Stimulation

1. Day 1 (D1) of the cycle is the first sign of menses.
2. Twice daily intramuscular (IM) injections of 30 IU of rhFSH are given at 0900 and 1700 for 8–10 days starting on D2 of the cycle.
3. On D9, D10, or D11 of the cycle, 1000 IU hCG is given IM.
4. Oocyte aspiration should be performed 28–32 h after hCG injection (D10–D12 of cycle).

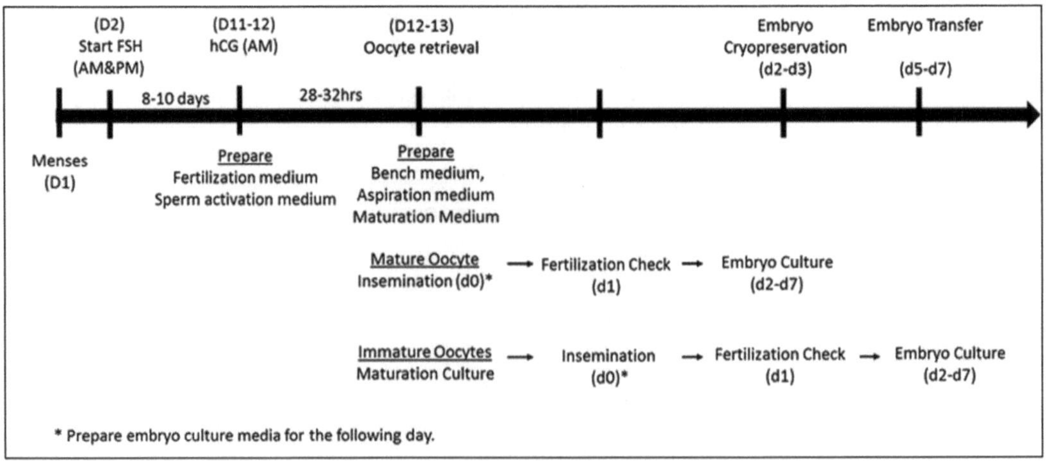

Fig. 3 Timeline of events for the in vitro production of *Macaca fascicularis* embryos

5. Ultrasound measurements of ovaries can also be performed. Visualization of follicles and enlargement of ovaries is a good indication of ovarian response to the gonadotropins.

3.2 Oocyte Aspiration

All dishes and media should be pre-warmed or equilibrated before use. Aseptic techniques should be practiced and personal protective equipment should be worn when handling nonhuman primate biologics (i.e., gloves, goggles, mask, lab coat).

1. The day before aspiration, prepare approximately 12–15× 5 ml Egg Pick Up (EPU) tubes each with 1 ml of FERT. Transfer loosely capped EPU tubes to a 6% CO_2 humidified incubator at 37 °C for overnight preequilibration.
2. On the day of oocyte aspiration (Day 0 of in vitro embryo development), set up the work bench with a stereomicroscope with warm plate, warming tray for dishes, glass pipettes, 6 × 35 mm dishes with 3 ml of BENCH, and approx. 12× 100 mm dishes.
3. Ovaries are visualized at laparotomy.
4. A sterile 22 G butterfly needle is attached to the suction tubing (Fig. 1). A small volume of ASP is flushed through the tubing into a 15 ml conical tube and discarded.
5. In a new 15 ml conical tube, aspirate 1–2 ml ASP and proceed with aspiration of individual follicles. Regularly flush the needle with ASP to ensure cumulus oocyte complexes (COC) do not remain in the suction tubing.
6. Empty the contents of the tube containing the follicular aspirate into the bottom of a 100 mm dish placed on the warm stage of the microscope.
7. Identify and retrieve COCs using a glass pipette. Expel the oocytes into a 35 mm dish with BENCH. Continue searching the aspirate.
8. Using a second pipette, transfer COCs into a second wash dish to remove blood, follicular fluid, and cellular debris.
9. Using a third glass pipette, transfer COCs into a 5 ml tube containing 1 ml of preequilibrated FERT and return tube to the incubator. When transferring the oocytes from the wash dishes to the 5 ml tubes it is important to do this as quickly as possible to avoid changes in pH and temperature of the medium.

3.3 Removal of Cumulus Cells

To identify the mature oocytes prior to insemination they need to have their cumulus cells removed. This should be started 1–2 h after oocyte aspiration. All dishes and media should be prewarmed. FERT should be prepared and equilibrated overnight before use.

1. The HYAL is removed from the freezer (−20 °C) at the start of oocyte aspiration and placed in a warm block.
2. Prepare a 4-well dish 15 min prior to use: Label well #1 with "H" to mark the HYAL. Place approximately 500–800 μl of HYAL into the well.
3. Place approximately 500–800 μl of BENCH into the other three wells.
4. Transfer up to ten oocytes from the EPU tubes into the well of HYAL.
5. Allow the oocytes to sit in the HYAL for ~30 s before starting to strip the cumulus.
6. Using a large bore pipette identify the oocytes and remove the surrounding cumulus with gentle pipetting. Once done, transfer the oocytes in a minimal amount of medium to the first wash well.
7. Continue this process until all oocytes have been placed into wash well #1. Avoid leaving the oocytes in the HYAL for more than 3 min.
8. Change to a pipette with a smaller inner diameter and continue to remove cumulus cells until the meiotic stage of the oocyte can be identified.
9. Transfer the oocytes using fresh pipettes through other wash wells so that all hyaluronidase has been removed.
10. Using a high-power objective, visualize the oocytes and group them according to the stage of meiosis, that is, MII—one polar body; MI—no polar body, no GV; GV—germinal vesicle (Fig. 4).
11. Oocytes identified as MII are either transferred to individual drops (20 μl) of FERT for microdrop insemination (one or two oocytes per drop) or held in FERT medium until ICSI.
12. Allow oocytes to rest for at least 1–2 h before insemination.

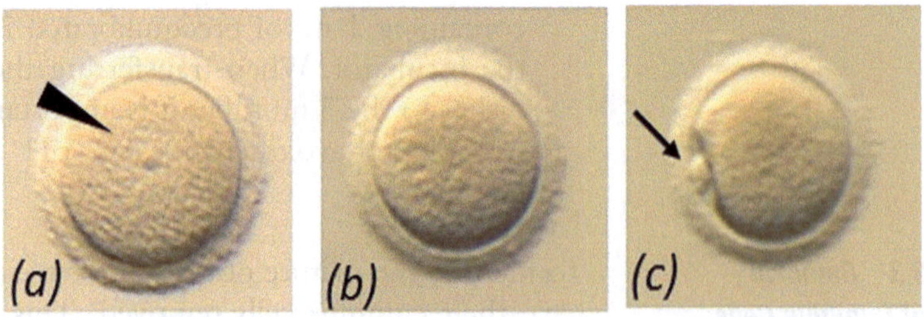

Fig. 4 *M. fascicularis* (**a**) germinal vesicle (GV), (**b**) Metaphase I (MI), and (**c**) Metaphase II (MII) oocytes retrieved from stimulated ovaries. Right arrow head indicates GV stage nucleus, right arrow indicates presence of polar body

13. MI oocytes should be transferred to MAT medium and reassessed at the time of insemination for progression to the MII stage.

3.4 Oocyte Maturation

Metaphase I (MI) oocytes collected from stimulated ovaries can be matured in vitro and contribute quality embryos to the cohort. All dishes and media should be warmed and equilibrated overnight [1, 2]. While GV oocytes will also progress to maturity during IVM, they do not tend to yield quality embryos and do not contribute significantly to the cohort.

1. Aliquot 2–3 ml of MAT into the outside well of the organ culture dish.
2. Aliquot 800 μl of MAT into the center well of the dish, but do not cover with oil, and return to the incubator for 2–4 h prior to use.
3. Transfer MI oocytes to the preequilibrated maturation dish.
4. Assess MI oocytes at 6 and 18 h postcollection for progression to MII.

3.5 PESA Sperm Retrieval (Fresh)

Retrieval of motile and viable sperm from the epididymis can be achieved by needle puncture. Sperm collected via this method still retain their cytoplasmic droplet but will mature with further in vitro incubation prior to insemination.

1. The male is sedated and weighed. The scrotum is shaved and scrubbed in preparation for PESA.
2. Place 1 ml of BENCH into 5 ml collection tubes and place in a warm block.
3. Using a 21 G needle attached to a 5 ml syringe, draw a small amount of medium into the needle and syringe.
4. Locate the cauda (tail) of the epididymis and puncture the skin drawing back on the syringe to retrieve sperm. Flush contents of needle back into tube and rinse several times. It is important to stretch the skin tightly over the puncture site to reduce blood contamination.
5. Take a 10–15 μl aliquot of the sperm suspension with a sterile tip and place onto a slide. Assess the sperm sample for concentration, motility, agglutination, debris and blood cells. Heavy red blood cell (RBC) contamination will cause a large proportion of sperm to become agglutinated prior to insemination, thereby reducing the availability of motile sperm in the final sample.
6. If adequate sperm has been retrieved, then continue with processing. If not, obtain a sample from the second testis or in some cases another male. No more than two or three

aspirations from each testis should be performed during the collection.

7. After sperm retrieval, the male should be treated with dexamethasone (2 mg/kg, IM) every day for 3 days, followed by injections every other day over 6 days. This reduces the potential of anti-sperm antibody formation as a result of the procedure.
8. Transfer the sperm suspension into a 15 ml conical tube with 2–3 ml of prewarmed BENCH. Centrifuge sample at 400 × *g* for 10 min.
9. Transfer supernatant to a fresh tube and resuspend remaining pellet by gently pipetting with a sterile tip. If the motility of the sample is less than 50%, perform a swim-up retrieval. Overlay the pellet with approximately 500 μl of BENCH medium and place in test tube rack at a 45° angle at room temperature for 15–30 min. If the sample is heavily contaminated with RBC, the swim-up procedure should be repeated several times at short intervals to retrieve a relatively clean sample.
10. The supernatant should be recentrifuged at 400 × *g* for 10 min for further recovery of sperm and kept as a backup. Often this sample contains less RBC, but just as many motile sperm as the first portion. However, this second sample tends to have a shortened motile life span.
11. At the end of the swim-up period, gently remove the top portion of the medium and transfer to a new 5 ml tube. Take a 15 μl sample with a clean pipette tip and place on a slide for motility and concentration assessment. At this point if the sample is adequate an accurate assessment of concentration (using a hemocytometer) should be made. Also note the degree of hyperactivation, agglutination, and debris in the sample. This should be left at room temperature for a further 1–3 h before undergoing activation and being used for insemination procedures.
12. For activation prepare a fresh solution of ACT. Gently mix equal aliquots of the final sperm suspension and ACT and return the tube to the incubator. This should be left for up to 30 min at 6% CO_2 in air at 37 °C. Check the sample at 10 min intervals to assess the sample for hyperactivation, overall motility, and agglutination. Male-to-male variation occurs. If the majority of sperm are showing hyperactivation, the sample is ready for use. If the degree of activation is still low after 30 min, a second sample should be activated with a higher concentration of caffeine (1 mM) and dbcAMP (0.2 mM).
13. Calculate the amount of sample to aliquot into each fertilization drop (50 μl final volume) to reach a final insemination concentration of 250,000 sperm/ml.

$$\frac{250{,}000\ \text{sperm/ml} \times \text{volume of FERT drop } (50\ \mu l)}{\text{Actual sperm concentration}} = \text{volume of sperm } (\mu l).$$

Actual sperm concentration = total sperm concentration × $X\%$ motile sperm.

3.6 Microdrop Insemination

Microdrop insemination is achieved by preparing 50 μl droplets overlaid with oil, in which a known concentration of sperm (250,000 ml^{-1}) and one or two oocytes are incubated for in vitro fertilization. This method is best used when fresh sperm samples are collected, that is, through PESA technique.

1. Calculate the required volume of medium to add to the drops to reach a final volume of 50 μl. For example, if 10 μl of sperm suspension is added to the original drop (20 μl), 20 μl of additional, preequilibrated FERT is added to reach a final volume of 50 μl. Avoid using low sperm concentrations in order to keep the volume of sperm suspension ≤10 μl, thereby avoiding exposure of oocytes to higher concentration of caffeine and dbcAMP.
2. Remove the hyperactivated sperm suspension and a small tube of preequilibrated FERT from the incubator. Tightly cap and place in the warm block.
3. Remove the dishes containing the FERT drops and the oocytes and place on the warm stage.
4. Add the additional FERT to each drop and then add the aliquot of sperm suspension.
5. Activate a second aliquot of sperm to ensure there is available, activated sperm if reinsemination of oocytes is required.
6. After 1 h, check the FERT drops to ensure there is good sperm binding and some motile sperm remaining. If the sperm binding is poor, reinseminate each drop with the freshly activated sperm. Return to the incubator and leave for an additional 12–15 h.

3.7 Retrieval and Cryopreservation of Epididymal Sperm

The testis from males that undergo either castration or euthanasia can be collected and sperm from the epididymis retrieved for either immediate use or frozen for long-term storage. Work in a biohazard hood. Treat all tissue as infectious. Place all required equipment in hood prior to working with the material.

1. Remove a 5 ml aliquot of the cryopreservation medium from the freezer and allow to thaw and reach RT before use.
2. Place testis in a 100 mm culture dish.
3. Ensure that the vas deferens is ligated before proceeding with the remainder of the protocol. If the vas deferens is not ligated, use a hemostat to occlude the vas deferens distal to the cauda epididymis.

4. Remove the scrotal sac with a pair of scissors making sure not to cut the epididymis or vas deferens.
5. Remove the epididymis from the testicular meatus using scissors. A cut across the corpus should be made distal to the cauda and sufficiently high so as not to risk cutting into the cauda or the vas deferens. Make cuts in a manner that avoids direct severing of blood vessels.
6. Once removed, the cauda epididymis and attached vas deferens (still occluded with a hemostat or suture) should be lightly rinsed in saline to remove any blood. Excess saline is removed and the tissue is placed in a 60 mm culture dish. Add 6–7 ml of BENCH medium to the dish and remove the hemostat or suture from the vas deferens.
7. Sperm are released by making radial cuts with a scalpel from the center of the cauda epididymis outwards to the epididymal border. Short sections of attached vas deferens should be squeezed lightly to liberate sperm from the open end. Larger sections of vas deferens can be cut into smaller sections and squeezed lightly with the scalpel blade.
8. Sperm should be allowed to swim out into the medium for 5–10 min.
9. Repeat procedure for the second testis.
10. Load the sperm suspension from each dish into a 15 ml tube. The sliced epididymis and the tissue culture dish should be rinsed with BENCH medium (3–4 ml) and the contents added to the 15 ml conical tubes.
11. Wipe the capped conical tubes with 70% ethanol before loading into the centrifuge.
12. Centrifuge sample at 500 × g for 10 min.
13. Remove supernatant from each tube and discard.
14. For cryopreservation of the sample, pool the recovered pellets into a 50 ml conical tube and resuspend pellet in a final volume of 5 ml BENCH.
15. Prepare a 1/100 dilution of this sample in a separate 5 ml tube (i.e., 10 μl of the sperm sample added to 990 μl of fresh media). Perform semen analysis and record motility and concentration.
16. Using a 1000 μl pipette, add the cryopreservation medium to the sperm sample dropwise with constant, gentle mixing until a 1:1 sample–medium ratio is achieved. Gently swirl the sperm sample for even mixing.
17. Load sperm suspension into straws leaving an air gap at the open end where the straw can be sealed.
18. Once all straws are loaded and sealed transfer them to the refrigerator (4 °C) and slow-cool over 90 min.

19. Set up a styrofoam box with LN_2 and insert the styrofoam float.
20. Place the straws horizontally on the float and cover loosely with the lid. Allow straws to freeze in LN_2 vapor for 10 min.
21. After 10 min, quickly tip the straws into the liquid nitrogen.
22. Label a cane and a goblet with the male's ID#, date, and any other pertinent information.
23. Submerge the cane and goblet into the liquid nitrogen and transfer the straws. Transfer the loaded cane to a liquid nitrogen tank for long-term storage.

3.8 Processing of Frozen-Thawed Epididymal Sperm for ICSI

1. Thaw straw in air for 10 s and then transfer to a 34 °C water bath for 1 min.
2. Remove straw and wipe with 70% ethanol.
3. Using sterile scissors, carefully cut the sealed end of the straw. Place this end inside a 5 ml tube and cut the plugged end of the straw to empty the contents.
4. Perform abbreviated semen analysis (motility and concentration) to check for recovery and continue with sperm density gradient separation.
5. Prepare a 15 ml conical tube with 1 ml of prewarmed Sperm Separation Medium. Overlay the gradient with the sperm sample.
6. Centrifuge at 500 × *g* for 10 min.
7. Carefully remove the pellet from the tube and transfer to a second 15 ml conical tube with 3 ml of prewarmed BENCH.
8. Centrifuge at 400 × *g* for 5 min.
9. The supernatant is gently removed, the pellet resuspended in 1 ml of BENCH, and motility and concentration of the sample assessed.

3.9 Intracytoplasmic Sperm Injection (ICSI)

1. Final sperm preparations do not need to be activated prior to ICSI.
2. To prepare an ICSI dish, place a 10 μl drop of 10% polyvinylpyrrolidone (PVP) and a 50 μl drop of BENCH medium in the lid of a 60 mm dish and cover with 9 ml mineral oil. Warm to 37 °C before use.
3. Add a small volume (1–3 μl) of the sperm suspension to the PVP drop.
4. Place mature oocytes (up to five at a time) for injection in the drop of BENCH.
5. Single sperm are selected for ICSI based on morphology and motility.

6. Sperm are immobilized by striking the tail, aspirated into an injection pipette, and injected into the cytoplasm of the oocyte.
7. The moment of sperm injection is taken as the time of insemination.

Injected oocytes are transferred to 20 μl drops of FERT medium under oil and cultured overnight in 6% CO_2 in humidified air at 37 °C for 12–15 h.

3.10 Evaluation of Fertilization

The first signs of successful fertilization are observed between 12 and 15 h after insemination. This is considered Day 1 of in vitro embryo development. The presence of two pronuclei (2PN) within the cytoplasm and two polar bodies (2PB) in the perivitelline space indicates fertilization. PBs are capable of fragmentation and are not a reliable confirmation of fertilization in the absence of PN. After approx. 20 h, the PN disappear as the fertilized oocyte commences syngamy. The presence of three or more PN represents polyploidy. The presence of a single PN indicates early syngamy or abnormal oocyte activation.

CULTURE drops (50 μl) in 60 mm dishes with 9 ml oil should be equilibrated overnight at 6% CO_2, 7% O_2, and 87% N_2 at 37 °C before use.

1. To assess fertilization following microdrop insemination, transfer presumptive zygotes to an organ well dish containing 1 ml BENCH. Gently remove any remaining cumulus and bound sperm from the zona to visualize PN and PB.
2. Assessment of fertilization following ICSI can be made directly from ICSI drops.
3. Wash zygotes three times in 50 μl drops of CULTURE and transfer groups of up to five embryos to CULTURE drops.
4. Return zygotes to the incubator and culture in 6% CO_2, 7% O_2, and 87% N_2 at 37 °C.

3.11 Embryo Culture

1. Embryos are assessed on Day 3 for cleavage. At this stage embryos are expected to have reached the 8-cell stage (Fig. 3).
2. Embryos can be transferred, cryopreserved [3], or remain in culture to the blastocyst stage.
3. Embryos are reassessed on Day 6–8 for development to the blastocyst stage.

3.12 Embryo Cryopreservation—- Early Cleavage Stage (Day 2–3)

This protocol supports the cryopreservation of embryos at the 2- to 8-cell stage at D2-D3 of in vitro culture. Embryos are dehydrated stepwise by exposure to sucrose solutions and cell membrane protection is facilitated by exposure to 1,2-propanediol. Embryos are loaded into straws and slowly frozen to −30 °C before plunging into LN_2 for long-term storage [4].

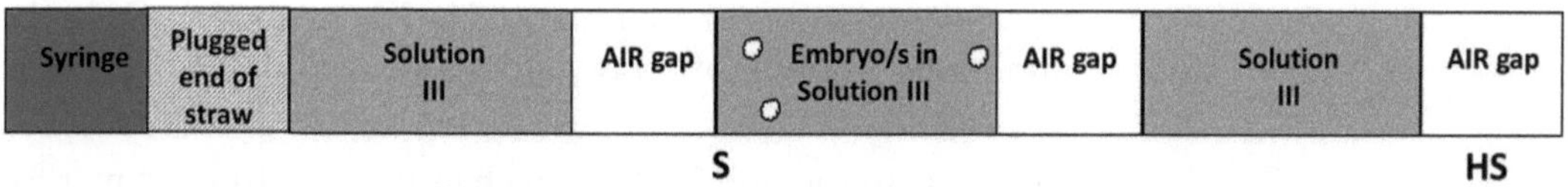

Fig. 5 Contents of straws for the cryopreservation early cleavage stage embryos (Day 2–3)

Controlled rate freezing program	
Starting temperature	20 °C
Ramp 1	−2 °C/min to −6 °C
Ramp 2	HOLD at −6 °C for 10 min (seed)
Ramp 3	−0.3 °C/min to −35 °C (hold for plunging)

1. This procedure is conducted at RT. Cryoprotectant solutions become highly embryo toxic when warmed.
2. Aliquot approximately 800 μl of BENCH (Soln. I) and Embryo Cryopreservation Solutions II, and III into wells 1, 2, and 3, respectively, of a 4-well dish. Aliquot Solution III into well 4 and use to wash and load straws prior to loading the embryos.
3. Transfer embryos from culture dishes into BENCH (Soln. I) for 5–10 min.
4. In a minimal amount of medium, transfer embryos to Solution II for 10 min.
5. Set up freezing machine (e.g., Freeze Control®, Cryologic). Machine should reach starting temperature prior to loading embryos.
6. Transfer embryos in groups of two or three into Solution III. Rinse the straw with Solution III. Immediately load embryos into straws.
7. Load each embryo into a straw as shown in Fig. 5.
8. Heat-seal at the air gap at the end of the loaded straw (*HS* in figure above) and maintain at RT while other straws are loaded.
9. Once all embryos are loaded, place straws into the freezing chamber.
10. Put freezing machine on to RUN and set timer for 10 min to coincide with the seeding temperature.
11. Seeding can be done once the program holds at −6 °C. To seed the straws, chill the tip of a pair of fine forceps or a cotton bud in LN_2. Touch the straw at the top of the cryoprotectant

column containing the embryo (*S* in figure above). Only the minimum amount of straw should be withdrawn from the freezing chamber to seed.

12. Continue to top up LN_2 level in chamber by setting a timer for 30 min.
13. Upon completion of freeze (−35 °C), remove straws from cryochamber and plunge directly into LN_2. Have the storage goblets prechilled in the same Dewar and load straws into the goblets and transfer to storage tanks.

3.13 Embryo Thawing

Embryo recovery is performed as a stepwise removal of cryoprotectant using methods previously described by Balmaceda et al. [5].

1. Set water bath at 30 °C. Just prior to thawing the straw turn water bath off to avoid excessive vibration during the thaw procedure.
2. Label the wells of the first 4-well dish A1, A2, A3 depending on the number of straws to be thawed, allowing one well per straw thawed. Place 700 μl of Embryo Thaw Solution A to each well and maintain this dish at RT.
3. Label the wells of the second dish B, C, and D1, fill with 700 μl Embryo Thaw Solutions B, C, and D, respectively, and keep at RT.
4. Label one well of the third dish D2, fill with 700 μl D solution, and place on a warming plate at 37 °C.
5. Remove one straw at a time, thaw in air for 10 s, and then transfer to the 30 °C water bath for another 10 s.
6. Remove straw and wipe-dry.
7. Cut plugged end of straw and attach syringe. Cut sealed end of straw and gently expel the first column and air bubble into an empty well. Expel column of media containing the embryo/s into well A1. Continue thawing straws until the required number of embryos are thawed.
8. Leave embryo in Solution A for 5–10 min.
9. Transfer embryos into Solution B for 10 min.
10. Transfer embryos through wells C and D1 respectively for 10 min in each solution.
11. Transfer embryo/s from D1 into D2 and place on 37 °C warm stage for 10 min.
12. Examine embryos for cellular lysis intermittently during the thawing procedure. If 50% or more of the cells survive the thaw, the embryo is considered suitable for further culture and/or transfer.

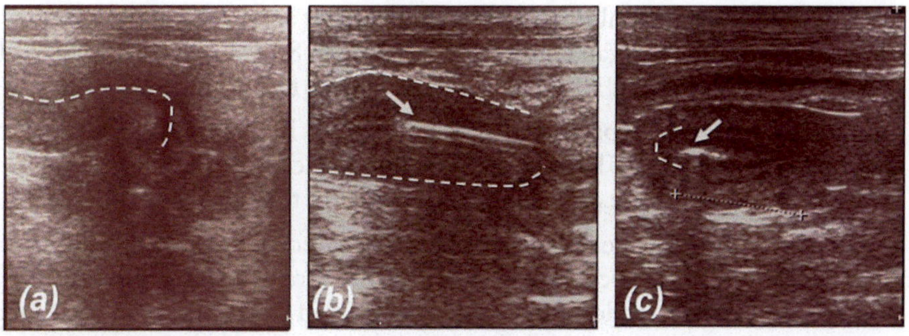

Fig. 6 Ultrasound images of (**a**) cervical path (---), (**b**) stylet (arrow) placed 1.5 cm from the uterine fundus. Uterus outlined (---), and (**c**) embryos after transfer (arrow) to the uterine fundus (---)

13. Transfer embryos to the equilibrated 50 μl drops of CULTURE and return to the incubator.

3.14 Nonsurgical Embryo Transfer

In most instances, the transfer of frozen-thawed embryos is undertaken as recipient synchronization is a limiting factor. Once a female has entered a transfer cycle and ovulation has been determined, an appropriate time for embryo transfer can be estimated. This depends on the stage of the embryo and the female's cycle length and history. Early cleavage stage embryos can be transferred surgically to the Fallopian tubes of the recipient female [3]. Nonsurgical, intrauterine transfer of later stage embryos is possible with ultrasound guidance (Fig. 6), as well as modifications to the catheter and stylet.

Macaca fascicularis do not show predictable physical changes, such as sex skin breakdown, as an indication of ovulation. As a result, the monitoring of previous cyclicity and the current transfer cycle is very important. Conscious blood sampling for daily hormone analysis is the best way to determine the time of ovulation. If this is not available, other measures include:

Ultrasound: *M. fascicularis* ovaries in a natural cycle undergo a 35% increase in size 24–36 h prior to ovulation (as measured by the LH surge). These changes can be monitored by visualization and measurement (long and short axis) of the ovary. A high-powered probe can also allow for visualization of a dominant follicle and possible corpus luteum formation.

Day of ovulation-to-cycle length ratio: The day of ovulation-to-cycle length ratio (DO/CL) is described by Dukelow and Brüggemann [6] and can be used as the basis for determining presumptive ovulation. For the purpose of embryo transfer, the day of ovulation is assumed to be the average cycle length × 0.40. Blastocyst stage embryos are transferred 3 or 4 days after presumptive ovulation. All dishes and media should be prewarmed or equilibrated overnight.

1. Embryos selected for transfer are placed in the center well of an organ culture dish filled with 800 μl of CULTURE and 2–3 ml CULTURE in the outer rim to minimize evaporation.
2. A prewarmed, 1 ml syringe is aseptically loaded with CULTURE and returned to the incubator. This is to be used to flush the catheter prior to loading the embryos.
3. The recipient female is sedated, a blood sample taken for retrospective measurement of estradiol and progesterone, and the abdominal region shaved and peritoneal region cleaned.
4. For transcervical embryo transfer, first insert the nasal speculum into the vagina to better visualize the vaginal vault and identify the external cervical os.
5. Using sterile swabs and BENCH clean the vaginal vault to remove excess mucus and debris.
6. Gently introduce the stylet into the cervix. Once placed, remove the speculum, and apply the ultrasound probe to identify the stylet. Under ultrasound guidance transverse the cervix to the cervical-uterine junction and enter the uterus.
7. Advance the stylet to within 1.5–2 cm of the fundus.
8. With the stylet in place prepare the transfer catheter.
9. Using aseptic technique attach the syringe containing CULTURE securely to the transfer catheter. Free the tip of the catheter from the packaging and expel the contents of the syringe through the catheter to the 0.2 ml mark.
10. To load embryos, remove the embryo dish from the incubator. Using a stereomicroscope, draw the embryos into the catheter in a minimal amount of medium (10–15 μl).
11. Carefully insert the catheter into the stylet and under ultrasound guidance advance the tip of the catheter to within 0.5 cm of the fundus.
12. Gently expel the embryo/s in 20–30 μl of medium. Wait about 10 s then remove the catheter, keeping the stylet in place.
13. The catheter is returned to the laboratory, where the remaining fluid is expelled into a dish containing BENCH. The dish and the tip of the catheter are examined for the presence of embryos. If any embryos remain then the procedure should be repeated with fresh media and catheters.
14. To monitor for pregnancy continue daily swabs to detect signs of menses. Note that cynomolgus macaques frequently display implantation bleeding [3] that is difficult to distinguish from menses. To confirm pregnancy perform an ultrasound assessment to detect the presence of a gestational sac at gestation day 30 (GD30).

4 Notes

1. We advise using BSA for all medium used during sperm preparation and insemination as other sources of protein (e.g., FCS, HSA) leads to rapid sperm agglutination.
2. While several options for embryo culture are available we have achieved consistently high rates of blastocyst development (>50% blastocyst rate per zygote) using Global Total culture medium (Life Global).
3. Due to the complexity of the *M. fascicularis* cervical canal we produced our own transfer stylet consisting of a metal luer of around 22 G, a section of metal hypodermic tube approx. 6 in. in length and a ball bearing that has been hollowed out consistent with the 22 G and soldered on at a slight angle. The gauge needs to large enough to accommodate a transfer catheter (1–1.6 mm outer diameter).

Acknowledgments

E.C.C. and E.S.H. would like to thank the veterinary and animal care staff at the Washington National Primate Research Center, particularly Cliff Astley for surgery support and Clayton Ferrier for assistance in working with the macaques for cycle detection, gamete collection, and embryo transfer. The authors thank the WaNPRC's Tissue Distribution Program for their assistance with the collection of reproductive tissues. Studies at the WaNPRC were supported by NIH grant P51 OD010425.

References

1. Curnow E, Ryan J, Saunders D, Hayes E (2010) In vitro developmental potential of macaque oocytes, derived from unstimulated ovaries, following maturation in the presence of glutathione ethyl ester. Hum Reprod 25:2465–2474
2. Curnow EC, Ryan JP, Saunders DM, Hayes ES (2011) Primate model of metaphase I oocyte in vitro maturation and the effects of a novel glutathione donor on maturation, fertilization, and blastocyst development. Fertil Steril 95:1235–1240
3. Hayes E, Curnow E, Trounson A, Danielson L, Unemori E (2004) Implantation and pregnancy following in vitro fertilization and the effect of recombinant human relaxin administration in *Macaca fascicularis*. Biol Reprod 71:1591–1597
4. Curnow E, Kuleshova L, Shaw J, Hayes E (2002) Comparison of slow-and rapid-cooling protocols for early-cleavage-stage *Macaca fascicularis* embryos. Am J Primatol 58:169–174
5. Balmaceda JP, Heitman TO, Garcia MR, Pauerstein CJ, Pool TB (1986) Embryo cryopreservation in cynomolgus monkeys. Fertil Steril 45:403–406
6. Dukelow WR, Brüggemann S (1979) Characteristics of the menstrual cycle in nonhuman primates. II. Ovulation and optimal mating times in macaques. J Med Primatol 8(2):79–87

Chapter 23

In Vitro Culture of Rhesus Macaque (*Macaca mulatta*) Embryos

Cathy Ramsey and Carol Hanna

Abstract

Production of nonhuman primate (NHP) embryos in vitro begins with recovery of gametes. Females undergo a controlled ovary stimulation to produce multiple preovulatory follicles from which oocytes may be recovered. Mature ova are subjected to in vitro fertilization (IVF) and presumptive zygotes are cultured to the intended stage of development. Essential to this practice is the culture medium unique to each step in the process. Here we describe medium preparation, oocyte recovery, in vitro fertilization, and in vitro culture (IVC) of embryos in the rhesus macaque model.

Key words Rhesus macaque, Controlled ovarian stimulation, Fertilization, Embryo culture, Nonhuman primate

1 1 Introduction

Rhesus macaques represent a valuable resource in biomedical programs and conservation initiatives. Their close phylogenetic relationship to humans makes them an ideal species for studies investigating infectious disease and chronic degenerative disorders [1, 2], as well as reproductive function—gametogenesis, tract biology, pregnancy, and assisted reproductive technologies (ART) [3–7]. The advent of gene editing technology has made possible the generation of precise macaque disease models by injecting zygote stage embryos with editing material (e.g., CRISPR-Cas9 or TALENs) followed by transfer of cultured embryos into recipient dams [8, 9]. Additionally, rhesus monkeys provide a foundation for conservation strategies of endangered macaque species, such as the Celebes Crested or Pagai Island macaques, through development of macaque specific in vitro ART and banking of gametes and embryos [10–17]. The use of rhesus macaques for all of these techniques requires, at some level, the ability to produce and culture embryos in vitro so that optimized protocols and validated medium become critical for success.

Jason R. Herrick (ed.), *Comparative Embryo Culture: Methods and Protocols*, Methods in Molecular Biology, vol. 2006,
https://doi.org/10.1007/978-1-4939-9566-0_23,

In this chapter, methods are sequentially provided to walk through the steps required to obtain the biological materials necessary for creation and culture of rhesus macaque embryos. Protocols include (1) media preparation for gamete collection and culture, (2) ovarian stimulation with recombinant gonadotropins and a GnRH inhibitor for production of multiple preovulatory follicles, (3) oocyte recovery from follicle aspirates and grading of meiosis, and (4) IVF followed by IVC and the stages of embryo development. At the beginning of the methods section is a graphical timeline indicating the sequence of events beginning with initiation of the ovarian stimulation protocol and ending with embryo production. As with all United States Department of Agriculture (USDA) regulated laboratory species, appropriate Institutional Animal Care and Use Committee (IACUC) approval must be acquired prior to initiating any NHP studies.

2 Materials

Wear gloves during medium preparation and if necessary, spray with 70% ethanol (EtOH) to remove latex powder (toxic to embryos) before handling materials. Use ultrapure water (≥18 MΩ) unless otherwise specified and prepare medium in clean glassware free of residue and debris (*see* **Note 1**).

2.1 Preparation of Culture Media

1. *TALP-HEPES:* 114 mM sodium chloride, 3.2 mM potassium chloride, 0.34 mM sodium phosphate dibasic, 5.0 mM D-(+)-glucose, 10.0 mM sodium DL-lactate solution 60% (w/w), 28.2 mM phenol red, 2.0 mM sodium bicarbonate, 0.2 mM gentamicin sulfate, 11.0 mM HEPES, 0.5 mM sodium pyruvate, 0.5 mM magnesium chloride hexahydrate, and 2.0 mM calcium chloride dihydrate. On a magnetic stir plate with an appropriately sized stir bar, place a 1-L glass beaker filled with ~700 mL ultrapure water and adjust rotational speed to a "medium" setting. In the order listed in Table 1, weigh out and add chemicals allowing each to dissolve into solution before adding the next (*see* **Note 2**).

   Once all chemicals are thoroughly mixed and in solution, adjust the pH to 7.3–7.5 with either NaOH or HCl. Add sufficient ultrapure water to bring the final volume to 1000 mL. Measure and confirm that the osmolality is between 275 and 290 mOsm/kg. If the osmolality is greater than 290 mOsm/kg, an adjustment can be made by gradually adding more water and measuring again. If the osmolality is below the target range, a new batch of medium must be prepared. Filter medium through a 0.2 μm filter unit and store at 4 °C for up to 1 month. Prior to use add 3 mg/mL of BSA, ensuring all crystals are completely dissolved and pass through a 0.2 μm

Table 1
Composition of HEPES-buffered Tyrode's albumin[a], lactate, and pyruvate medium (TALP-HEPES)

Compound	MW (g/mol)	Amount per L	Concentration (mM)
Sodium chloride	58.44	6.660 g	114.0
Potassium chloride	74.55	0.240 g	3.2
Sodium phosphate	141.96	0.048 g	0.3
D-(+)-glucose	180.16	0.900 g	5.0
Sodium D,L-lactate solution 60% (w/w)	112.06	1.870 mL	10.0
Phenol red	354.38	0.010 g	28.2
Sodium bicarbonate	84.01	0.168 g	2.0
Gentamicin sulfate	202.27	0.050 g	0.2
HEPES	238.30	2.600 g	11.0
Sodium pyruvate	110.04	0.060 g	0.5
Magnesium chloride hexahydrate	203.30	0.102 g	0.5
Calcium chloride dihydrate	147.01	0.294 g	2.0

[a]3.0 mg/mL bovine serum albumin added prior to use

filter. TALP-HEPES can be stored at 4 °C for up to 7 days. Warm medium to 37 °C prior to use.

2. *TALP Stock Solutions*: 157 mM sodium chloride stock solution, 166 mM potassium chloride stock solution, 120 mM calcium chloride dihydrate stock solution, 120 mM magnesium chloride hexahydrate stock solution, 295 mM D-(+)-glucose stock solution, 167 mM sodium bicarbonate stock solution, and 1.7% sodium DL-lactate stock solution. Prepare stock solutions by weighing and adding the indicated chemicals to the specified volume of ultrapure water listed in Table 2.
3. *TALP Sodium Phosphate Stock (20 mM):* Add 140 mg of sodium phosphate monobasic monohydrate to 50 mL of glucose stock solution.
4. *TALP Medium*: On a magnetic stir plate with an appropriately sized stir bar, place a 1000 mL glass beaker filled with ~300 mL of sodium chloride stock solution and adjust rotational speed to a "medium" setting. In the order listed in Table 3, weigh out or measure the indicated chemicals allowing each to dissolve into solution before adding the next. Adjust the pH to 7.3–7.5 and bring the final volume to 500 mL with sodium chloride stock solution. The osmolality should be between 285 and 295 mOsm/kg. Filter medium through a 0.2 μm filter unit and store at 4 °C for up to 1 month. Prior to use add 3 mg/mL of BSA and 0.006 g/100 mL sodium pyruvate (0.5 mM),

Table 2
Composition of stock solutions used to make Tyrode's albumin, lactate, and pyruvate (TALP) medium

Compound	Amount	MW (g/mol)	Ultrapure water	Stock concentration
Sodium chloride	9.2 g	58.44	1000 mL	157 mM
Potassium chloride	2.48 g	74.55	200 mL	166 mM
Calcium chloride dihydrate	3.52 g	147.01	200 mL	120 mM
Magnesium chloride hexahydrate	4.88 g	203.30	200 mL	120 mM
D-(+)-glucose	26.55 g	180.16	500 mL	295 mM
Sodium bicarbonate	2.8 g	84.01	200 mL	167 mM
Sodium DL-lactate solution, 60% (w/w)	2.0 mL	112.06	70 mL	1.7% (149 mM)

Table 3
Preparation of Tyrode's albumin[a], lactate, and pyruvate[a] (TALP) medium

Component	Amount per 500 mL	Concentration (mM)
Sodium chloride stock	300 mL	114.0
Phenol red	0.005 g	28.2
Sodium D,L-lactate stock	33.5 mL	10.0
Gentamicin sulfate	25 mg	0.2
Potassium chloride stock	9.5 mL	3.2
Calcium chloride dihydrate stock	8.5 mL	2.0
Magnesium chloride hexahydrate stock	2.05 mL	0.5
Sodium bicarbonate	75 mL	25.0
Sodium phosphate stock[b]	8.5 mL	0.3

[a]Add 3 mg/mL BSA and 0.006 g/100 mL (0.5 mM) sodium pyruvate prior to use
[b]Prepared in glucose stock solution for a final concentration of 5.0 mM glucose

ensuring all crystals are completely dissolved before passing through a 0.2 μm filter. TALP can be stored at 4 °C for up to 7 days. Medium intended for culture should be equilibrated at 37 °C in 5% CO_2 in humidified air for at least 2 h before use.

5. *Sperm Activation Medium*: 2 mg (10 mM) caffeine, 5 mg (10 mM) dibutyryl cAMP, and 1 mL of sterile saline. Mix well. Make fresh prior to use for in vitro fertilization and maintain at 37 °C.

6. *Embryo Culture Medium*: Commercially available single step, or single phase, culture medium better supports development of rhesus embryos from zygote to the blastocyst stage than sequential culture medium [18]. Examples of medium include

those formulated for human embryo culture (e.g., Global® by the LifeGlobal® Group) or similar (e.g., BO-IVC from IVF Bioscience). All media should contain 4 mg/mL protein (*see* **Note 3**). Culture medium should be equilibrated for at least 2 h before use in an atmosphere of 6% CO_2, 5% O_2, and 89% N_2 at 37 °C in humidified air to achieve a pH of approximately 7.3 (range 7.2–7.4). Altitude and characteristics of individual incubators may alter the concentration of CO_2 required to maintain the desired pH of the culture medium.

7. *Blastocyst Cryopreservation*: Commercial cryopreservation kits marketed for use with human blastocysts are suitable for rhesus macaque embryos.

2.2 Controlled Ovarian Stimulation (COS)

1. Recombinant human follicle stimulating hormone (r-hFSH).
2. Recombinant human luteinizing hormone (r-hLH).
3. Human chorionic gonadotropin (hCG): available commercially either as a recombinant product or purified from the urine of pregnant women.
4. Gonadotropin Releasing Hormone (GnRH) antagonist.

3 Methods

Both indoor and outdoor housed rhesus macaques undergo a seasonal breeding cycle where ovulation typically initiates in September and continues until May. The months of June–August constitute the off-season, where animals may still experience menses but ovulation of mature ova may not occur [19]. When selecting animals to use for in vitro embryo production, it is best to avoid the off-season months as oocyte quality and embryo development can be affected. Additionally, using regularly cycling females between the ages of 6 and 12 years is ideal for providing developmentally competent ova. Characterization of the ovarian cycle can be determined by measuring estradiol-17β (E_2) and progesterone (P_4) in blood serum and has been well described [20–23]. The following methods describe the steps necessary to produce and culture rhesus macaque embryos in vitro, including ovarian stimulation to recover multiple mature ova, sperm preparation, IVF, IVC, and embryo cryopreservation (Fig. 1).

3.1 Controlled Ovarian Stimulation (COS)

For this 10-day protocol, r-hFSH and r-hLH are used to stimulate development of multiple ovarian follicles. GnRH antagonist is used to suppress the endogenous LH surge. The timed bolus of hCG initiates the ovulatory process inducing the signaling cascade for meiotic resumption resulting in mature ova (Fig. 2).

1. Begin the COS protocol within the first 4 days after the onset of menses (beginning of the follicular phase). Blood samples to

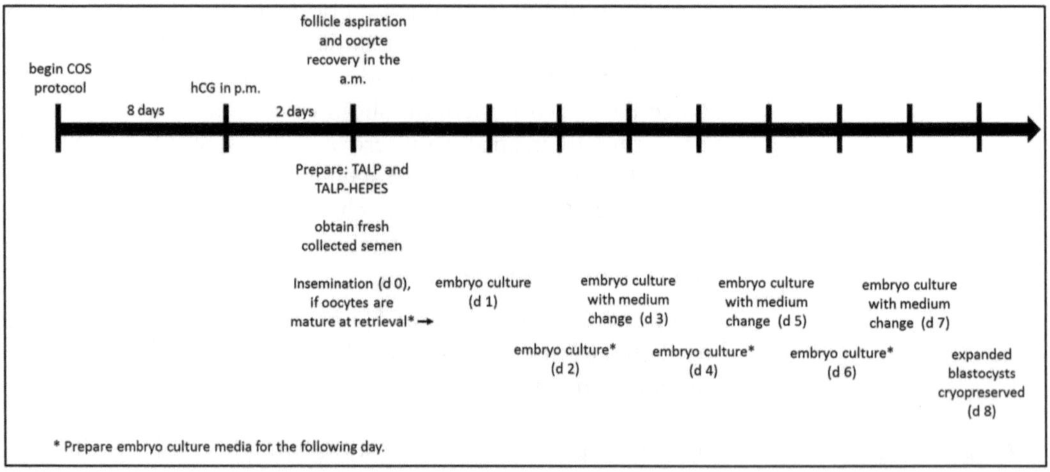

Fig. 1 Timeline of events for in vitro production of rhesus macaque embryos

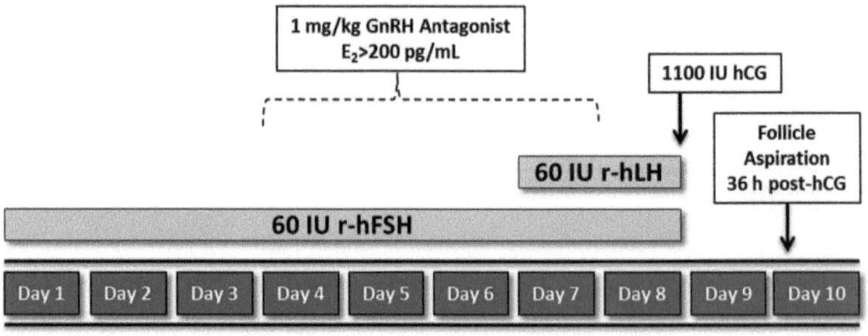

Fig. 2 Schematic representation of the controlled ovarian stimulation (COS) protocol used to promote development of multiple antral follicles in rhesus macaques. Protocol should be initiated within the first 4 days after the onset of menses

measure serum estradiol (E_2) and progesterone (P4) should be taken every 1–2 days prior to morning administration of gonadotropins. On day 1 (first day of injections) of the protocol, E_2 concentrations should be 30–60 pg/mL and P_4 concentrations should be approximately 0 (<0.05 ng/mL) (*see* **Note 4**).

2. Administer 30 IU of r-hFSH intramuscularly (IM), in a.m. and p.m., with at least 8 h between doses on days 1–8.
3. Similar to r-hFSH, administer 30 IU of r-hLH IM, a.m. and p.m., with at least 8 h between doses on days 7 and 8 (*see* **Note 5**).
4. When serum E_2 concentrations are >200 pg/mL, administer a single dose of 1 mg/kg GnRH antagonist subcutaneously in the morning of the following day. If P_4 begins to rise prior to the hCG, a second dose of 1 mg/kg GnRH antagonist may be given.

5. On day 8, administer 1100 IU of hCG IM in the p.m., timed so follicle aspiration occurs 36 h later. For example, an 8:00 p.m. dose on day 8 accommodates an 8:00 a.m. follicle aspiration on day 10. Serum P_4 should measure approximately 1 ng/mL or higher the morning following hCG administration. If P_4 remains low following hCG, it is presumed the animal did not respond to the bolus and oocyte maturation may not have been initiated. In this case, the COS protocol is considered unsuccessful and retrieval for ova should not be attempted.
6. Aspirate follicles 36 h following hCG to ensure collection of mature ova. Follicular aspirates should be collected into warmed TALP-HEPES with 5 IU/mL heparin added to prevent blood clot formation. If collection time is delayed beyond 37 h, follicles may undergo ovulation diminishing the number of recoverable ova.
7. Isolate oocytes into warmed TALP-HEPES and wash free of any residual blood and follicle cells. Remove cumulus granulosa cells surrounding the oocyte by gentle pipetting to visualize the cytoplasm and assess meiotic stage (Fig. 3). Metaphase II (MII) oocytes with an extruded polar body are considered mature and able to undergo fertilization. Oocytes that have not initiated meiotic resumption and have a visible germinal vesicle (GV) or those undergoing meiosis (GVBD or metaphase I, MI) are considered immature and may require additional culture time in TALP medium.
8. Wash ova selected for fertilization into fresh, equilibrated TALP. Add five ova per 100 μL drop (covered with oil and equilibrated) and maintain at 37 °C in 5% CO_2 humidified air until insemination—this will later increase the total volume of each drop to 111 μL.

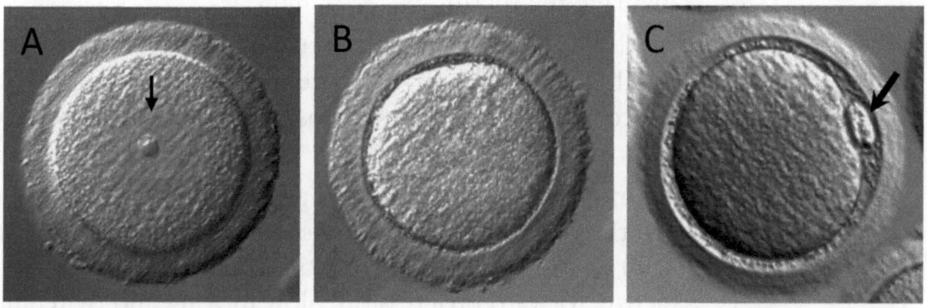

Fig. 3 Stages of meiosis in recovered rhesus macaque oocytes. (**a**) Germinal vesicle (GV) intact (arrow), oocyte has not yet resumed meiosis and is considered immature; (**b**) Metaphase I (MI), oocyte has resumed meiosis but not yet considered mature; (**c**) Metaphase II (MII), ovum has completed meiosis, extruded the first polar body (arrow), and is now considered mature

3.2 Collection and Processing of Spermatozoa for In Vitro Fertilization (IVF)

When performing IVF, fresh semen samples are preferred. They can be easily washed by centrifugation and typical samples maintain motility throughout culture. This protocol describes the process to wash the sperm from the seminal plasma, assess motility, and determine optimal concentration for IVF. Appropriate personal protection equipment (PPE) should always be worn when handling biological samples from nonhuman primates.

1. Semen collected by electroejaculation with penile stimulation will form a seminal coagulum that would otherwise function as a postcoital plug during natural breeding [24]. The coagulum forms quickly at the time of semen collection, trapping spermatozoa in the plug. To recover the spermatozoa, samples should be maintained undisturbed for 30 min at ambient temperature to allow liquefaction of the outer layer of the coagulum and release of motile spermatozoa.
2. Transfer the liquefied portion of the sample to a 15 mL conical tube and record the volume.
3. Use 1 mL of warm (37 °C) TALP-HEPES to rinse the coagulum and collection container. Add the wash to the same conical tube containing the liquefied sample and repeat the wash twice more. Add enough warmed TALP-HEPES into the conical tube to bring the final volume to 10 mL.
4. Centrifuge the sample for 8 min at 1400 rpm (no more than $400 \times g$) at room temperature.
5. Discard the supernatant, being careful not to disturb the medium close to the pellet that may contain live sperm. Suspend the sperm pellet with 10 mL of warmed TALP-HEPES.
6. Centrifuge the sample for another 8 min at room temperature.
7. Discard enough supernatant to leave a final volume of 1 mL. Mix gently, avoiding rapid pipetting as the shear force may damage the sperm.
8. Add 10 μL of the washed sperm to 190 μL of warmed TALP-HEPES in a 0.5 mL microcentrifuge tube. Mix gently then place 10 μL on a glass microscope slide and place a coverslip over the sample.
9. Using an upright compound phase contrast microscope, focus on the center of the sample to score a group of approximately 100 sperm for motility. Motile sperm are those that exhibit active, forward progression. Slow moving and twitching sperm should be scored as nonmotile. Record the percentage of motile sperm.
10. Determine the concentration of the washed sperm sample and adjust the final concentration to 20 million sperm/mL in 1 mL of TALP. Maintain the sperm sample at 37 °C in 5% CO_2 humidified air until ready to perform IVF.

3.3 In Vitro Fertilization (IVF)

Once oocytes have been recovered and confirmed to have completed meiosis, insemination should occur within 4 h. This allows time to obtain and evaluate the semen sample for suitability and acquire additional samples if necessary. Sperm cells must undergo hyperactivation to initiate sperm capacitation, observed as increased motility characterized by pronounced flagellar movements, a nonlinear trajectory, and acquisition of the ability to undergo the acrosome reaction upon zona binding before introduction to the ova (insemination) [25]. This protocol describes the use of gametes intended to be used on the same day as they were collected, although oocytes that have undergone in vitro maturation (IVM) would be inseminated similarly but require a source of fetuin during IVM to prevent zona hardening (*see* **Note 6**). All incubations are performed at 37 °C in 5% CO_2 humidified air.

1. To activate the sperm, combine 900 μL of the 20 million/mL sperm sample in TALP with 100 μL of the sperm activation solution and maintain in the incubator for 15 min. Check for activation by observing a small drop of sperm under the microscope for increased, rapid movement.
2. Add 10 μL of warmed sperm activation solution to each fertilization drop containing oocytes, then follow by adding 1 μL of activated sperm (*see* **Note 7**). It is best to observe the addition of sperm under the microscope to ensure the full microliter is delivered. Note the insemination time and incubate the oocytes with spermatozoa for 14–16 h.
3. Upon completion of gamete coincubation, remove the ova from the TALP drops and wash in TALP-HEPES. Use a micropipettor with a 125 μm diameter tip to gently remove remaining granulosa cells and attached sperm, which should dislodge easily from the zona without enzymatic treatment. Transfer the oocytes to a fresh well containing 500 μL of TALP-HEPES and note the presence of polar bodies (PB) and/or pronuclei (PN) (Fig. 4a, b).
4. When evaluations are complete, transfer presumptive zygotes to equilibrated medium for culture.

3.4 In Vitro Embryo Culture (IVC)

While performing embryo culture, the amount of time the embryos are out of the incubator should always be minimized to avoid excessive changes in temperature and pH. When observing embryos for developmental progress, also assess the culture medium on multiple focal planes for signs of bacterial or fungal contamination.

1. Day 0—On the day of IVF, prepare 2 60 mm culture plates each containing four drops (100 μL/drop) of embryo culture medium layered with enough mineral oil to completely cover the medium—approximately 7 mL. One plate will serve to

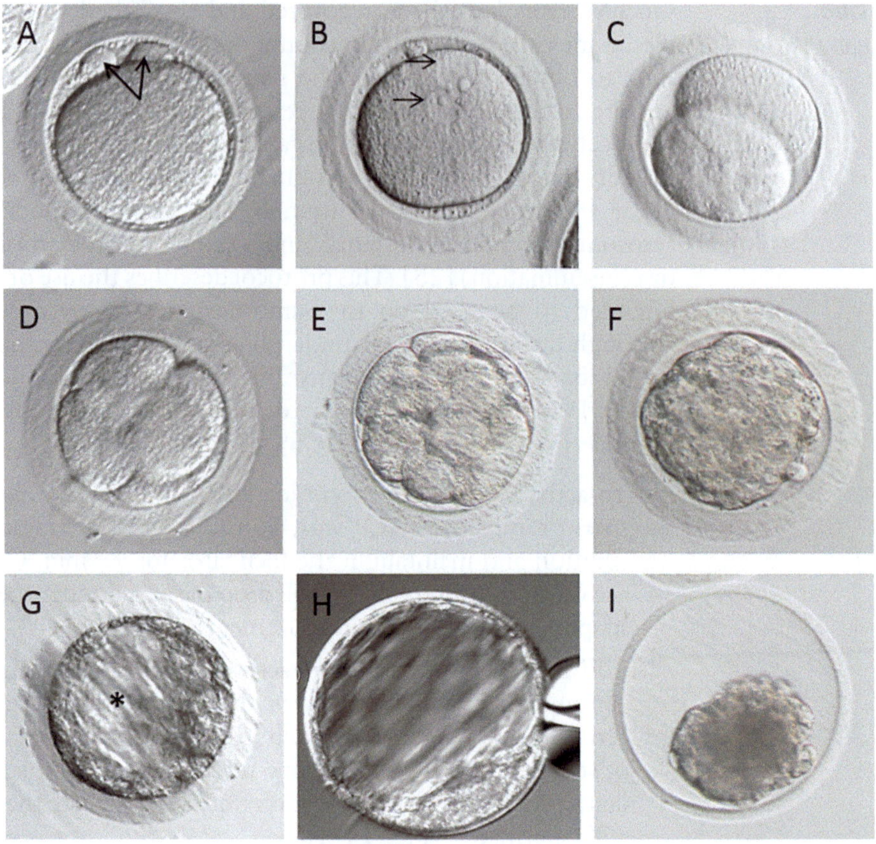

Fig. 4 Stages of embryo development. (**a**) Zygote with two polar bodies (PB) in the perivitelline space (arrows); (**b**) Zygote with two pronuclei (PN) present (arrows); (**c**) 2-cell; (d) 4-cell; (**e**) 12-cell; (**f**) compacted morula; (**g**) blastocyst with clearly forming blastocoel (asterisk); (**h**) expanded blastocyst; (**i**) collapsed blastocyst

wash embryos out of TALP-HEPES from **step 4** above and the other will be used to culture embryos. Allow the plates to equilibrate at 37 °C in an atmosphere of 6% CO_2, 5% O_2, and 89% N_2 with humidified air overnight.

2. Day 1—Wash presumptive zygotes recovered from the IVF culture through at least three drops of embryo culture medium in the wash plate before transferring to a fresh drop of culture medium, maintaining no more than five embryos per drop.

3. Day 2—Most embryos will begin to cleave by the second day in culture. Observe individual embryos for developmental quality characteristics. The nuclei should be clearly visible and the blastomeres should have homogenous cytoplasm with little to no retraction (clearing) around the periphery. There should be no vacuole formation or fragmentation of blastomeres. Any zygotes that have not cleaved or are obviously degenerating, should be removed.

4. Day 3—Developing embryos should be actively cleaving and have reached the 8–12 cell stage. Wash embryos through at least three drops of fresh, preequilibrated (18–24 h), culture medium and transfer to a new plate containing 100 μL drops culture medium covered in oil for continued culture.
5. Day 4—All embryos in culture should have developed to the 12–16 cell stage.
6. Day 5—Healthy embryos should begin to show signs of compaction, where individual blastomeres become indistinguishable (Fig. 4f). All viable embryos should be washed and moved to fresh culture medium as before—100 μL drops, preequilibrated for 18–24 h, and covered in oil.
7. Day 6—Early stage blastocysts should begin developing cavities that will form the blastocoel.
8. Day 7—Blastocysts should have a well-formed blastocoel with minimal cellular fragmentation (Fig. 4g). Embryos that continue development are washed and moved to fresh, preequilibrated culture medium—100 μL drops, preequilibrated for 18–24 h, and covered in oil.
9. Day 8—Blastocysts may undergo expansion with a noticeable increase in size and a thinning of the zona pellucida (Fig. 4h). Some embryos may collapse after expansion and have a similar appearance to a compacted morula (Fig. 4i).
10. Day 9—Expanded blastocysts of good quality may hatch and may either be partially or completely extruded from the zona pellucida.

3.5 Blastocyst Cryopreservation

Blastocysts intended for embryo transfer can be cryopreserved until a recipient is properly synchronized [20]. Protocols developed for use with human blastocysts are effective when followed as directed by manufacturer instructions. While the longevity of embryos stored in liquid nitrogen has yet to be determined, they can be frozen and subsequently thawed at the time of embryo transfer to produce viable offspring [23].

4 Notes

1. For optimal results, it is best to use glassware dedicated to medium preparation only. Never use soap or detergents to clean. Always rinse out glassware at least seven times with ultrapure water immediately following use to remove residue before allowing to air dry.
2. Be sure to add the calcium chloride dihydrate last to prevent precipitation out of solution.

3. In place of defined protein supplements such as human or bovine albumin, 10% fetal bovine serum (FBS) may be used during embryo culture. Medium should be passed through a 0.2 μm filter after adding protein and before use.
4. While the ideal progesterone concentration during a COS protocol, prior to hCG administration, is less than 0.05 ng/mL, it is not uncommon to see levels fluctuate reaching 0.15 ng/mL without disruption to follicular growth.
5. The r-hFSH and r-hLH doses can be administered in combination in the same injection syringe to limit the number of treatments to the animal.
6. For oocytes that must first undergo IVM, add 5% fetal bovine serum or 1 mg/mL fetuin to prevent hardening of the zona pellucida [26].
7. Some metaphase I oocytes will complete meiosis during the overnight IVF culture and can be fertilized. However, metaphase I oocytes should be kept separate from the MII during IVF, so that accurate rates of fertilization success may be calculated.

Acknowledgments

C.H. and C.R. are grateful for the hard work and dedication of the surgeons, veterinarians, and animal care staff in the Division of Comparative Medicine, Behavioral Services Unit, and Surgical Services Unit at the Oregon National Primate Research Center (ONPRC) without which none of the animal work described here would have been possible. They would also like to thank the Assisted Reproductive Technology Core and the Endocrine Services Laboratory at the ONPRC for developing and optimizing the protocols presented in this chapter. Support at the ONPRC is provided by funding from the NIH grant P51 OD011092.

References

1. Sibal LR, Samson KJ (2001) Nonhuman primates: a critical role in current disease research. ILAR J 42(2):74–84
2. Yang S-H, Cheng P-H, Banta H, Piotrowska-Nitsche K, Yang J-J, Cheng ECH, Snyder B, Larkin K, Liu J, Orkin J, Fang Z-H, Smith Y, Bachevalier J, Zola SM, Li S-H, Li X-J, Chan AWS (2008) Towards a transgenic model of Huntington's disease in a non-human primate. Nature 453:921
3. Appt SE, Ethun KF (2010) Reproductive aging and risk for chronic disease: Insights from studies of nonhuman primates. Maturitas 67 (1):7–14
4. Kelleher MA, Liu Z, Wang X, Kroenke CD, Houser LA, Dozier BL, Martin LD, Waites KB, McEvoy C, Schelonka RL, Grigsby PL (2017) Beyond the uterine environment: a nonhuman primate model to investigate maternal-fetal and neonatal outcomes following chronic intrauterine infection. Pediatr Res 82(2):244–252
5. Schramm R, Bavister B (1999) A macaque model for studying mechanisms controlling

oocyte development and maturation in human and non-human primates. Hum Reprod 14 (10):2544–2555

6. Simerly C, Schatten G (2012) Utility of animal models for human embryo culture: nonhuman primates. Methods Mol Biol 912:39–59
7. Lee DM, Yeoman RR, Battaglia DE, Stouffer RL, Zelinski-Wooten MB, Fanton JW, Wolf DP (2004) Live birth after ovarian tissue transplant. Nature 428(6979):137–138
8. Xin L, Min L, Bing S (2016) Application of the genome editing tool CRISPR/Cas9 in non-human primates. Zool Res 37(4):241
9. Liu H, Chen Y, Niu Y, Zhang K, Kang Y, Ge W, Liu X, Zhao E, Wang C, Lin S, Jing B, Si C, Lin Q, Chen X, Lin H, Pu X, Wang Y, Qin B, Wang F, Wang H, Si W, Zhou J, Tan T, Li T, Ji S, Xue Z, Luo Y, Cheng L, Zhou Q, Li S, Sun Yi E, Ji W (2014) TALEN-mediated gene mutagenesis in rhesus and cynomolgus monkeys. Cell Stem Cell 14(3):323–328
10. Engelhardt A, Muniz L, Perwitasari-Farajallah-D, Widdig A (2017) Highly polymorphic microsatellite markers for the assessment of male reproductive skew and genetic variation in critically endangered crested macaques (*Macaca nigra*). Int J Primatol 38(4):672–691
11. Roos C, Boonratana R, Supriatna J, Fellowes JR, Groves CP, Nash SD, Rylands AB, Mittermeier RA (2014) An updated taxonomy and conservation status review of Asian primates. Asian Primates J 4(1):2–38
12. Bavister BD, Boatman DE, Collins K, Dierschke DJ, Eisele SG (1984) Birth of rhesus monkey infant after in vitro fertilization and nonsurgical embryo transfer. Proc Natl Acad Sci 81(7):2218–2222
13. Gabriel Sanchez-Partida L, Maginnis G, Dominko T, Martinovich C, McVay B, Fanton J, Schatten G (2000) Live rhesus offspring by artificial insemination using fresh sperm and cryopreserved sperm. Biol Reprod 63(4):1092–1097
14. Lanzendorf SE, Gliessman PM, Archibong AE, Alexander M, Wolf DP (1990) Collection and quality of rhesus monkey semen. Mol Reprod Dev 25(1):61–66
15. Meng L, Ely JJ, Stouffer RL, Wolf DP (1997) Rhesus monkeys produced by nuclear transfer. Biol Reprod 57(2):454–459
16. Chen Y, Niu Y, Yang S, He X, Ji S, Si W, Tang X, Xie Y, Wang H, Lu Y, Zhou Q, Ji W (2012) The available time window for embryo transfer in the rhesus monkey (*Macaca mulatta*). Am J Primatol 74(2):165–173
17. Gangrade BK (2013) Cryopreservation of testicular and epididymal sperm: techniques and clinical outcomes of assisted conception. Clinics 68:131–140
18. Hanna C, Ramsey C, Hennebold J (2016) 100 Sequential versus single-step medium for rhesus macaque embryo culture. Reprod Fertil Dev 28(2):180–180
19. Walker ML, Gordon TP, Wilson ME (1983) Menstrual cycle characteristics of seasonally breeding rhesus monkeys. Biol Reprod 29 (4):841–848
20. Wolf D, Thormahlen S, Ramsey C, Yeoman R, Fanton J, Mitalipov S (2004) Use of assisted reproductive technologies in the propagation of rhesus macaque offspring. Biol Reprod 71 (2):486–493
21. Wolf DP (2004) Assisted reproductive technologies in rhesus macaques. Reprod Biol Endocrinol 2(1):37
22. Wolf DP, Thomson JA, Zelinski-Wooten MB, Stouffer RL (1990) In vitro fertilization-embryo transfer in nonhuman primates: the technique and its applications. Mol Reprod Dev 27(3):261–280
23. Wolf DP, Vandevoort CA, Meyer-Haas GR, Zelinski-Wooten MB, Hess DL, Baughman WL, Stouffer RL (1989) In vitro fertilization and embryo transfer in the rhesus monkey. Biol Reprod 41(2):335–346
24. Danzy J, Gutierrez V, Pampush J, Campbell B (2009) Factors affecting the distribution of copulatory plugs in rhesus monkeys (*Macaca mulatta*) on Cayo Santiago. Folia Primatol 80 (4):264–274
25. Boatman DE, Bavister BD (1984) Stimulation of rhesus monkey sperm capacitation by cyclic nucleotide mediators. J Reprod Fertil 71 (2):357–366
26. Schroeder AC, Schultz RM, Kopf GS, Taylor FR, Becker RB, Eppig JJ (1990) Fetuin inhibits zona pellucida hardening and conversion of ZP2 to ZPZ during spontaneous mouse oocyte maturation in vitro in the absence of serum. Biol Reprod 43(5):891–897

Chapter 24

Culture of Human Preimplantation Embryos in a Clinical ART Setting

Rebecca L. Krisher and Terry Schlenker

Abstract

In vitro fertilization (IVF) and embryo culture in the human is a unique endeavor. Human assisted reproductive technology (ART) is practiced clinically to treat infertility. Due to the obvious ethical considerations of ART as applied to human medicine, only rarely is embryo culture undertaken for research purposes. As most IVF clinics are for profit businesses, a robust industry has developed to supply embryologists with quality control-tested media, equipment, and supplies necessary to support human IVF laboratory operation. Moreover, commercial items are preferred for efficiency and consistency, and strict quality control is required by accrediting organizations. As such, very little manual formulation or preparation of culture medium is typically required. Although human embryo culture is performed clinically, there is a surprising degree of variability in the details of the techniques used. In this chapter, we describe state-of-the-art procedures for gamete collection, in vitro fertilization, embryo culture, and embryo transfer that result in excellent blastocyst development and pregnancy rates for patients seeking treatment for infertility.

Key words Sequential media, Single step media, Blastocyst, In vitro embryo culture, IVF

1 Introduction

Humans are perhaps the only species in which embryo culture is not only a scientific endeavor but also a highly lucrative commercial enterprise. In the USA alone, over 72,000 babies are born from IVF cycles each year [1]. Human ART, including embryo culture, is a highly relevant clinical treatment, and as such the protocols for patient management, embryo culture, genetic testing, embryo transfer, and pregnancy support are intensively studied and widely practiced [2]. The human menstrual cycle can be tightly controlled, ovarian stimulation protocols are well established, and transvaginal oocyte recovery is easily accomplished. Due to the medical treatment aspect of human embryo culture, the technologies used in the human ART laboratory are tightly regulated, quality control standards are high, and outcomes are reported to the public [3, 4]. However, this does not mean that we have reached the

Jason R. Herrick (ed.), *Comparative Embryo Culture: Methods and Protocols*, Methods in Molecular Biology, vol. 2006, https://doi.org/10.1007/978-1-4939-9566-0_24, © Springer Science+Business Media, LLC, part of Springer Nature 2019

pinnacle of embryo culture success. Overall, only 23% of initiated cycles, and 37% of embryo transfers, result in a live birth [1]. This picture is a bit more optimistic when only women under 35 years of age are considered; in this case 41% of embryos transferred implant, and 33% of all cycles result in a live birth [1]. This still means that even in young patients, which typically have the best oocyte quality, almost six out of ten embryos that develop in vitro to a staged deemed suitable for transfer are unable to support a pregnancy to term. One must consider, however, when scrutinizing these numbers that these are infertile patients with a variety of diagnoses and as such may not truly represent the optimal blastocyst development and viability success rate in vitro in the human. An examination of donor oocyte ART cycles likely provides a more accurate picture. Indeed, transfer of a fresh embryo from a fresh donor egg results in ~55% implantation and live birth rate [1]. Thus, odds are improved in the best clinical scenario, although still only about one of every two embryos is capable of supporting pregnancy. Although one cannot rule out uterine factors contributing to embryo loss, certainly these numbers suggest that there is room for improvement in the quality of embryos produced in human ART.

Constantly in the pursuit of improved embryo development, a host of companies manufacture and market a variety of culture medium specifically for human embryos. The content of this medium is unknown to the end user. Thus, comparisons between media often reflect individual laboratory performance and do not contribute to our knowledge of embryo physiology nor do they suggest how to improve culture conditions for human embryos in the broader sense. Embryo culture medium is certainly one of the critical factors affecting success of an ART laboratory. Because of this role, culture medium is often touted as the secret to high pregnancy rates but can just as easily be the first suspected culprit when trouble arises.

One outcome not currently reported by fertility clinics is embryo development success, also called blastocyst conversion rate. This describes the percentage of normally fertilized zygotes that develop into a good quality blastocyst acceptable for transfer. Each clinic closely monitors its own developmental parameters, but there is no universal average readily available. A recent consensus report from ESHRE Special Interest Group of Embryology suggests that good quality (transferrable) blastocyst development is recommended to be 30% or greater on day 5 for competency, and 40% or greater as a benchmark [3]. For most laboratories, these would be excellent outcomes. Of course, patient population can dramatically affect the success of embryo culture, particularly the percentage of patients that have failed previous cycles or are of advanced maternal age. In general, an estimate of total blastocyst development (days 5, 6, and 7) of 60–70% of normally fertilized zygotes, and 50–60% good quality blastocysts can be expected.

Compared to some of the domestic species, this appears to be good developmental success. One of the most important predictors of IVF cycle success is the number of mature oocytes retrieved [5], as this leads to more blastocysts produced, resulting in the most attempts possible for a successful embryo transfer. If one looks at the overall picture, though, when half of the fertilized eggs become good quality blastocysts on average, and half of the blastocysts transferred are capable of implantation in the best case scenario, this represents a high rate of attrition for most patients. Although the widespread implementation of preimplantation testing for aneuploidies (PGT-A) to detect aneuploidy embryos, thereby transferring only chromosomally normal embryos, has resulted in improved implantation rates, there are still many euploid embryos that do not implant. Refining our ability to support both the quantity and quality of blastocysts produced in vitro is therefore critical to the successful treatment of infertility for our patients.

2 Materials

1. 1 cc tuberculin syringe with attached 27 G needle.
2. Automated hand pipette aid.
3. Centrifuge.
4. Disposable, sterile plasticware: 35, 60 and 100 mm dishes; center well organ culture dishes, 15 mL centrifuge tubes; 14 and 5 mL snap-cap culture tubes; 1, 5, 10, 25 mL serological pipettes; 150 mL sterile plastic receiver unit, specimen cup (*see* **Note 1**).
5. Fertilization and embryo culture media (*see* **Note 2**).
6. Gas and temperature-controlled incubator set to 37C, 6% CO_2 and 5% O_2 (*see* **Note 3**).
7. Gas supply (CO_2 and N_2); in line gas filters.
8. Hemacytometer or Makler chamber.
9. Holding micropipette, ICSI micropipette.
10. Hyaluronidase.
11. Inverted microscope equipped with heated stage, Hoffman optics, microinjectors, and laser.
12. IVF isolette chamber.
13. Laminar flow hood.
14. Micropipette with tips of various sizes (130–275 μm).
15. Mineral or paraffin oil (*see* **Note 4**).
16. pH meter.

17. Pipettes (20, 100, 200, 1000 μL) and low-retention aerosol filter tips.
18. Protein supplement (*see* **Note 5**).
19. PVP ICSI medium.
20. Sperm wash medium.
21. Stereomicroscope with heated stage.
22. Sterile borosilicate glass pipettes with reusable bulb.
23. Sterile colloidal silica suspension in an isotonic salt solution.
24. Warming blocks.
25. Warming plate.

3 Methods (*See* Note 6)

3.1 Preparation of Dishes

Media used for human IVF and embryo culture is commercially purchased (*see* **Note 7**). It is required that companies provide a certificate of analysis with each shipment of medium that contains information regarding lot-specific quality control (QC) information. Media also come with instructions for use, which may contain a product description with a list of components although the concentrations are not usually provided. All media containers should be clearly labeled with contents, storage conditions, date of production, and date of expiration.

Care must be taken to prepare dishes for use in human IVF in a sterile environment, so all dish preparation is done in a laminar flow hood. Each piece of plastic ware used is labeled with a sticker containing the specific patient's information, including the patient's name, date of birth and ID number, and the type of medium in the dish or tube (*see* **Note 8**). Each patient is also assigned a specific location in an incubator, and that location is noted on the patient's chart.

When preparing microdrop dishes for embryo culture, it is important to minimize evaporation prior to covering the media droplets with oil [6]. To prepare dishes for group embryo culture, the patient sticker is placed on the dish lid and the bottom of the dish is marked with the patient's last name and the medium used. A pipette tip, or alternatively a sterile glass pipette, is rinsed (aspirate sterile media and discard) and then used to place 30 μL drops of culture medium in a 35 or 60 mm dish, and promptly covered with oil. The oil must completely cover the top of the media drops; this typically requires 3 mL of oil in a 35 mm dish, or 7–8 mL in a 60 mm dish. If center well dishes are needed, 800 μL of medium is added to the center well, and the moat filled with 2 mL of medium. Medium in the moat of the dish is used for rinsing tips prior to handling embryos. The center well may or may not be covered with oil (0.5 mL). Dishes are prepared the day prior to use, and are used within 24 h of preparation. No dish is immediately discarded. After a dish has been used, it is placed into a designated incubator and discarded several days later to ensure that no material is inadvertently discarded.

3.2 Collection of Mature Oocytes

Following ovarian stimulation, oocytes are recovered from patients via ultrasound guided aspiration of ovarian follicles 34 h after the administration of hCG or GnRH agonist, used to trigger final maturation of the follicle and eggs [7]. The isolette is turned on approximately 30 min prior to retrieval to allow it to warm (37 °C) and gas (5% CO_2). To prepare for oocyte retrieval, gather several 100 mm plastic dishes, two glass pipettes with a bulb, two sterile 14 mL culture tubes, an empty 150 mL receiver unit, and two 1-cc tuberculin syringes with 27 G needles in the isolette. The isolette should contain a heating block; place the 14 mL tubes in the heating block. Remove five 35 mm petri dishes filled with 3 mL culture medium, a 150 mL plastic receiver unit containing 50 mL of rinse medium (this medium may contain heparin), and a 14 mL tube with 1 mL of culture medium from the incubator, and place them into the isolette. Rinse the glass pipettes with the medium in the 14 mL tube and place them into a sterile tube in the heating rack, one pipette per tube. The identity of the patient and the number of oocytes expected is verified by the embryologist in the operating room (OR) prior to beginning the procedure. The isolette is placed either in the OR or in the embryology lab connected to the OR by a pass-through window. The receiver unit containing warmed, equilibrated rinse medium is given to the OR nurse to fill the syringe used for aspiration.

Once follicle aspiration has begun, the embryologist will receive culture tubes containing the follicular aspirates. Tubes are searched immediately, or held in a heating block in the isolette for a short period until they can be searched. The contents of the tube are poured into a 100 mm dish on the microscope stage and the contents examined under the stereoscope for the presence of cumulus–oocyte complexes (*see* **Note 9**). Once identified, oocytes are picked up using the glass pipette and moved into one of the 35 mm dishes containing culture medium. The recovered eggs are divided between at least two 35 mm dishes to prevent accumulation of follicular fluid and blood; if one of these dishes does become bloody a new dish should be used. Once the 100 mm dish has been thoroughly searched, the contents are poured into the empty 150 mL receiver unit. After all the tubes have been examined and eggs removed, the fluid collected in the receiver unit is rescreened to be sure no oocytes were missed. The cumulus mass is partially trimmed away from the oocyte using the 1-cc syringes with needle. If any blood clots are contained in the expanded cumulus mass, they should be excised using this method.

The trimmed oocytes are then moved from the 35 mm dishes into two center well dishes. For conventional insemination, these dishes contain fertilization medium; for ICSI cases these dishes contain culture medium (step one medium when using a sequential culture system). The identity of the patient in the retrieval isolette should be checked against the name on the center well dishes to be sure it matches; this should be witnessed by a second embryologist (*see* **Note 10**). The center well dishes containing the eggs are then returned to the incubator. Empty dishes and tubes, used pipettes and syringes are discarded from the isolette appropriately (biohazard trash and sharps container, respectively), and any spills in the isolette cleaned. After each egg retrieval the appropriate paperwork [4] is completed detailing the time of oocyte collection, and the number of oocytes collected (*see* **Note 11**). After the last retrieval of the day, or following retrieval from any patient with a communicable disease, the isolette is cleaned with 70% ethanol.

3.3 Collection and Preparation of Sperm for In Vitro Fertilization

Collection of sperm is scheduled for the same time as the egg retrieval. The male patient's identity is verified by the embryologist by checking a standard form of photo ID (driver's license or passport), and he is requested to thoroughly wash his hands and penis using the soap provided (*see* **Note 12**). A specimen cup obtained from the IVF laboratory is used to collect the ejaculate following masturbation (*see* **Note 13**). The patient is asked to completely fill out the label provided with his name and date of birth, his partner's name, and the number of days since the previous ejaculation, and then place the label on the specimen cup (*see* **Note 14**). The labeled specimen cup must then be given by the patient directly to an embryologist from the IVF lab, and he must sign the paperwork verifying his identity and that an embryologist has received the specimen provided. Alternatively, sperm may be obtained surgically or by electroejaculation.

In the IVF laboratory, the sperm specimen is held at room temperature until processing [8, 9]. More than one specimen is never processed simultaneously by a single embryologist. All sperm processing is performed in a sterile laminar flow hood. The specimen will liquefy at room temperature typically within 5–10 min; processing begins as soon as liquefaction is complete. During liquefaction, the density gradient is prepared. The colloidal silica suspension (at room temperature) is diluted to 90% and 45% using the sperm wash media. To prepare the gradient, 1 mL of the 45% suspension is layered over 1 mL of the 90% suspension in a 15 mL centrifuge tube. One gradient is prepared for each 1 mL of sample. If less than ten million motile sperm or less than 1 mL total volume are expected based upon previous semen analysis, 0.3–0.5 mL of each density suspension is used (rather than 1 mL) to make the density gradient. After liquefaction, 1 mL of semen is layered over the top of the density gradient. One gradient tube is prepared for each mL of semen collected. The gradient tubes are then centrifuged for 20 min at 300 × *g*. During centrifugation, sperm count, motility, and forward progression from the initial sample is assessed. Volume and viscosity are recorded, and any agglutination or debris in the sample is noted. After centrifugation, the supernatant above the pellets is aspirated and discarded. Multiple pellets are combined and then placed into one or two tubes at this time, depending on the amount of sperm recovered. To wash the pellet, 3 mL of sperm wash media is added to each tube with gentle mixing. The samples are then centrifuged for 5 min at 300 × *g* and the supernatant discarded, two times. Following centrifugation through the density gradient and washing, a swim-up is performed. Each pellet is gently layered under 1 mL sperm wash medium in a round bottom snap-cap tube. If the sperm concentration is low, less medium may be used. Motility and forward progression are checked again and recorded post gradient density centrifugation. The patient identification on the centrifuge tubes and the swim-up tubes are confirmed by a second embryologist to be sure that it matches, and the swim-up tubes are placed into the incubator until the time of ICSI or insemination.

3.4 In Vitro Fertilization

3.4.1 Insemination

At this time, the oocytes are in labeled center well organ culture dishes and the sperm in labeled swim-up tubes, both in the incubator. Typically, there are between 5 and 10 mature oocytes per dish. Oocytes are not assessed for maturity prior to insemination. Four to seven hours after the scheduled retrieval time, the sperm tubes are removed from the incubator and placed into an isolette. Then, 1 mL is removed from the top of the sperm tubes and sperm concentration and motility is determined. A sperm dilution of 3.3 million motile sperm/mL is prepared with equilibrated sperm wash medium. All sperm assessments and calculations are recorded. If motility is low then the physician is notified and the patient may be converted from an insemination to an ICSI protocol for IVF. The oocytes are removed from the incubator and placed into the same isolette. A second embryologist confirms that the patient ID on the oocyte dishes and sperm tubes matches. Thirty microliter of the sperm dilution (99,000 motile sperm) is added to each dish (800 μL) containing oocytes, and the dish examined under the microscope to verify that motile sperm are present [10] (*see* **Note 15**). Dishes containing oocytes and sperm are returned to the incubator and coincubated for 15–17 h.

3.4.2 Intracytoplasmic Sperm Injection (ICSI)

Intracytoplasmic sperm injection [11, 12], or ICSI, revolutionized assisted fertilization in the late twentieth century, significantly increasing the effectiveness of treating male infertility [13–17]. Three to four hours after retrieval, hyaluronidase (*see* **Note 16**) is used to remove cumulus cells from the oocytes. For this procedure, several micropipettes with multiple sized diameter tips are readied in the isolette (*see* **Note 17**). One half of the oocytes are placed into the equilibrated hyaluronidase solution, and gently pulled in and out of the large and then mid-sized micropipette tips until the majority of the cumulus cells have been removed. The time of oocyte exposure to hyaluronidase should be limited. The oocytes are then placed back into the original dish, rinsed, and the smaller diameter tips are used to remove any remaining cumulus or corona cells. Oocytes are then washed through multiple drops or wells of fresh culture medium to rinse away all traces of the hyaluronidase and placed into a dish of fresh culture medium. The process is then repeated for the remaining oocytes. If the resultant embryos are scheduled for biopsy and genetic testing, it is imperative that all cells be completely removed from the zona. Once all oocytes have been denuded, they are examined for maturity by visualization of the first polar body. Immature oocytes cannot undergo ICSI. If immature oocytes are recovered, they are often discarded from the treatment cycle at this point, but may be retained for day 2 (rescue) ICSI. This decision is typically made on a patient by patient basis depending on history and current retrieval success. Immature oocytes have not responded appropriately to hCG and are generally regarded as lower quality. However, they may mature in vitro at which time they can undergo ICSI. These in vitro matured oocytes are capable of successful fertilization and development into normal blastocysts, and do result in pregnancies following transfer although at a lower frequency.

The ICSI dish is now prepared using equilibrated culture medium. Several 6 μL drops are placed in the lid of a 60 mm sterile disposable tissue culture dish and covered with 8 mL oil. The drop in the center of the dish is then removed and replaced with PVP solution. The dish is labeled with the patient's name, the media drops are numbered, and a circle is drawn around the center PVP drop. Sperm (1 μL) is placed into the PVP drop (*see* **Note 18**). A second embryologist must verify that the name on the sperm tube, the ICSI dish and the egg dishes match.

ICSI is performed no earlier than 4 h post-egg retrieval. Denuded oocytes are placed individually into the drops of medium in the ICSI dish. No more than a few eggs should be placed into the dish at one time. The ICSI dish is moved onto the heated stage of the inverted microscope, where a holding and ICSI micropipette have been previously set up (*see* **Note 19**).

The ICSI micropipette is lowered into the PVP drop, PVP aspirated into the micropipette, and a motile sperm located in the PVP drop (*see* **Note 20**). The ICSI tool is used to immobilize the sperm by lowering the tool onto the sperm midpiece. The immobilized sperm is then aspirated into the ICSI micropipette tail first. The dish is moved so that a drop with an egg is in the field of view. The holding pipette is lowered into the drop and aspiration suction is used to hold the egg on the pipette. The egg is rotated until the polar body is located at either the 6 or 12 o'clock position. The ICSI pipette is then placed at the 3 o'clock position, in focus and touching the zona pellucida. The sperm is moved to the tip of the ICSI pipette. The ICSI pipette is advanced through the zona and into the egg, until it reaches the center of the ooplasm. Cytoplasm is then aspirated into the ICSI pipette until the ooplasmic membrane breaks. The sperm and the cytoplasm inside the ICSI pipette are expelled into the egg between the center of the egg and the holding pipette, and the ICSI pipette is withdrawn. The egg is then released from the holding pipette. This procedure is repeated for all oocytes in the ICSI dish currently, and then for all remaining oocytes. When all eggs in the dish have been injected with sperm, the eggs are removed from the ICSI dish, rinsed several times in fresh equilibrated culture medium, and placed into 30 μL microdrops of culture medium under oil for embryo culture, 2–4 eggs per drop. The lid of the dish is labeled with the number of eggs in each of the drops. After each batch of eggs has been injected, the media in the ICSI drops should be replaced with new, equilibrated medium before other eggs are added to the dish for injection.

3.5 Assessing Fertilization Success

The success of IVF is assessed 15–17 h after insemination or ICSI [18]. In the isolette, presumptive zygotes are quickly examined in the insemination dish or post ICSI culture dish. If the eggs have been inseminated, they will need to be stripped of any remaining cumulus cells at this time. After insemination, cumulus cells can be removed in the insemination dish without using hyaluronidase by passing them in and out of the small micropipette tips. Eggs that have undergone ICSI will already be free of cumulus cells. The oocytes are gently rolled with the micropipette to visualize pronuclei (PN) under increased magnification. The number of PN in each zygote is noted, and those with two PN are separated into a new center well wash dish of equilibrated culture medium. Embryos with three or more PN are discarded. Embryos with one PN after insemination are typically retained and cultured separately; after ICSI embryos with only 1 PN are discarded. If an egg has no PN after insemination or ICSI it is cultured separately from normally fertilized zygotes; if a blastocyst results it is tested for aneuploidy prior to transfer. This is done to be sure than PN formation and fading did not occur faster than anticipated and was simply missed. All zygotes with two PN are washed and moved into a fresh 30 μL microdrop for culture, with 2–4 embryos per drop [19]. Any presumptive/potential zygotes with 0 or 1 PN are also washed and cultured separately from those zygotes with two PN. A second embryologist again checks the names on the insemination/post-ICSI dish and the culture dish to be sure that they match, and dishes are returned to the incubator. This is considered day 1 of culture. All information is recorded, the disposition of each oocyte is clearly noted, and the fertilization results are reported to the patient.

3.6 Embryo Assessment

3.6.1 Day 3 of Culture

On the morning of day 3 of culture, the microdrop culture dishes are quickly examined on the inverted microscope. At this time, human embryos should contain ≥4 cells. The number of cells in the embryo is recorded, as well as the embryo grade. The grade of day three embryos is based primarily on the degree of fragmentation observed, from a grade 4 embryo with even size blastomeres and no fragmentation to a grade 1 embryo, in which individual blastomeres are indistinguishable. A plus or minus designation can be used to differentiate within grade. Embryos are then washed thoroughly and transferred into a new, equilibrated dish of culture medium, patient name on the old and new dish verified by a second embryologist, and returned to the incubator.

3.6.2 Day 5 and 6 of Culture

On the morning of day 5 of culture, embryos are again assessed on the inverted microscope. The Gardner and Schoolcraft grading system is used [20]. In this system, grade reflects the stage of blastocoel expansion, from grade 1 (early blastocyst) to grade 6 (completely hatched blastocyst). For blastocysts that have reached grade 3 (blastocyst) or higher, a letter score (A, B, or C) is also assigned to reflect the quality of both the inner cell mass (ICM) and the trophectoderm (TE). For example, a 4AB blastocyst would be an expanded blastocyst with a large, compact ICM and a TE with a lesser number of cells forming a loose epithelium. Each embryo is examined and a score recorded. Embryos of 3BB or better are considered good quality embryos, and may be either transferred or cryopreserved at this time. Remaining embryos are moved to a fresh, equilibrated dish of culture medium (matching names on both dishes are again verified by a second embryologist) and reassessed 24 h later, when any additional embryos of grade 3BB or better are cryopreserved. Embryos may occasionally be continued in culture until day 7. If an embryo does not develop into a blastocyst, it is discarded at that time. In human ART, it is typical to measure blastocyst conversion rate, or the percentage of blastocysts developing of normally fertilized (2 PN) zygotes.

3.7 Assisted Hatching

Often assisted hatching, or an artificial breaching of the zona pellucida, is used to ensure that the embryo is able to successfully escape from the zona prior to implantation, as well as to provide a means for trophectoderm biopsy if necessary [13, 21, 22]. The embryologist, in consultation with the physician and patient, will determine if and how many embryos will be hatched for any given patient. The majority of IVF clinics now use a laser for assisted hatching, and several options are commercially available [23]. It is safest to perform assisted hatching on D3 of embryo development, as there is reduced danger of damaging the embryonic cells that lie beneath the zona before blastulation. After quality assessment of the day 3 cleavage stage embryo is complete, the dish of microdrops containing the embryos to be hatched are moved from the isolette to the stage of an inverted microscope equipped with a laser objective. The embryos to be hatched are centered in the field of view using a 4× or 10× objective, then the objective is changed to 20×. A location for hatching is identified at a gap between blastomeres, or at an area of fragmentation. The identified location for hatching is brought into focus between 12 and 4 o'clock. The laser objective is rotated into place and the view refocused on the embryo. The settings for the laser will vary depending on the manufacturer. In general, the pulse length of the laser should be set to result in a diameter of 10–12 μm. A smaller diameter opening may be required for subsequent biopsy at the blastocyst stage than is used for assisted hatching alone. The embryo is placed so that the laser target, which is visible on the computer screen, appears over the location in the zona chosen for hatching. The laser is started using the computer mouse or a foot pedal if available, and the pulse of the laser creates a channel and breaks the membrane on the inner surface of the zona. After hatching, the dish containing the embryos is moved back into the isolette, and the embryos placed into a fresh dish of second step culture medium. The number of embryos and the number of embryos that were hatched contained within each dish is written on the dish lid, and dishes are replaced into the culture incubator. All information regarding hatching is recorded in the patient's medical records.

3.8 Embryo Transfer

The transfer of viable embryos into the uterus resulting in a healthy pregnancy to term is the goal of human ART. The procedure for transferring embryos into the uterus is critical for a successful IVF cycle, as damage to the embryo(s) or the endometrium must be avoided. Historically zygotes, as well as day 2 and day 3 cleavage stage embryos have been transferred, although currently blastocysts are the most common stage selected for transfer [24]. This is due in part to our improved ability to culture human embryos to the blastocyst stage in recent years, as well as the ability to obtain more information about which embryos are of the highest quality at this stage [25]. The number of embryos to be transferred is decided by the physician and patient (*see* **Note 21**). All personnel in the embryo transfer room should wear surgical scrubs, shoe covers, and hair covers. All procedures are performed using sterile technique.

When the blastocyst(s) to be transferred is selected (*see* **Note 22**), either fresh or vitrified and warmed, it is placed into a center well organ culture dish (*see* **Note 23**) filled with 0.8 mL of preequilibrated commercial medium designed for embryo transfer that contains an elevated hyaluronic acid content to increase viscosity; no oil overlay is used. Two milliliters of this medium is placed in the moat of the dish. Alternatively, culture medium may be used for transfer. The embryos are then placed back into the incubator. When the physician and patient are ready, the center well dish containing only the embryo(s) to be transferred is placed into the isolette and the isolette rolled into the transfer room. Once in the room, the embryologist conducts an extensive identity verification to ensure the correct embryo is being transferred to the correct patient. The embryologist checks the patient's wristband for their identifying information as well as asking the patient their name and date of birth. This information is then checked against the information on the patient's chart and on the embryo transfer dish to be sure everything matches. This is done in the presence of another witness. The name on the dish is also verified by the physician. In addition, the number of embryos to be transferred is verified with both the patient and physician at this time. Then, the embryo is placed under the stereomicroscope within the isolette (*see* **Note 24**). The transfer catheter (*see* **Note 25**) is opened, attached to the 1 cc syringe, and then removed from its packaging maintaining sterility. The catheter is then loaded with a series of air and medium columns from the center well, with the embryo(s) drawn into the catheter in the final column of medium. The total volume of medium surrounding the embryo(s) should be approximately 20 μL. The catheter is then handed to the physician, who will use ultrasound guidance to place the catheter into the uterine lumen and inject the embryo(s). The physician then returns the catheter to the embryologist, who checks to be sure there are no embryos remaining in the catheter. If so, the procedure is repeated with a new catheter. At the completion of the procedure, all transfer details are recorded on the patient's chart.

4 Notes

1. Each lot of disposable plastic ware that contacts gametes or embryos, or holds medium that will come into contact with gametes or embryos, should be tested for toxicity using a mouse embryo assay or a sperm survival assay. Lot numbers should be carefully tracked.
2. Commercially available media are used for all the steps of human IVF [26]. These media are quality controlled by the manufacturers, including a one-cell mouse embryo bioassay (MEA). Most human IVF labs do not have the capability to produce a similar quality or the required quantity of media. Unfortunately, the chemical formulations for these media are not released by the producing companies, although some do provide a list of ingredients. Most media are available from the manufacturer with or without protein added.

3. There are a variety of incubators available for human embryo culture. The more traditional "big box" incubators have been used historically and are the most economical. There are also smaller chamber sizes available in this style now. However, the recent trend is to use benchtop incubators for human embryo culture. They are more expensive, but have better temperature and gas control and recovery time, and save significant space in the laboratory. However, they provide a nonhumidified environment so an oil overlay must always be used. Even with oil, evaporation occurs over time so osmolality should be closely monitored. In addition, it is important for laboratories at higher elevations to appropriately adjust incubator CO_2 and O_2 settings to compensate for the reduced atmospheric pressure (mmHg) and corresponding gas ppm.
4. Oil is purchased commercially, and should be quality control tested for use in clinical IVF. Oil is available in light and heavy forms, although any difference in preventing evaporation is anecdotal at this time. Some laboratories "wash" oil before use by mixing a 10% volume of sterile medium with the oil and allowing it to separate before use, in an attempt to sequester any water-soluble toxic substances present in the oil. If quality control has been adequately performed this step should be unnecessary, and evidence suggesting that washing oil results in better results is lacking.
5. Protein supplements are also purchased commercially. These include complex protein supplements that consist of biologically derived human serum albumin (HSA) with alpha and beta-globulins added, and both biological and recombinant forms of HSA.
6. All procedures are carried out by trained and certified embryologists wearing surgical scrubs, gloves, shoe covers, and head covers. Universal precautions are used despite the fact that all patients are tested for HIV, hepatitis B and C, and syphilis. Each IVF laboratory is required to undergo external audits by the American College of Pathologists semiannually for quality assurance.
7. Culture systems vary widely lab to lab, depending on the culture media used and the preferences of the laboratory director [27]. There are two culture media systems currently in use for human embryos, single step [28] or sequential [29], and different companies market their own variations within each of these two types. Single step medium consists of one medium formulation that is used throughout in vitro preimplantation embryo culture. Medium can be replenished one or two times during the culture period, meaning that embryos are moved to fresh drops of the same medium. Alternatively, a single step

medium can be used without any media change for the entire culture period. A sequential medium system is composed of two separate media formulations, one for the early cleavage stages (days 1–3) and one for blastocyst development (days 4–6 or 7). In a sequential system, embryos are always moved on day 3 to a new medium, and are sometimes moved on day 5 to a fresh drop of the same medium. To date, there is no overarching scientific agreement that one medium or system is superior for human embryo culture than another [30, 31], due in part to the large role of factors other than the culture medium that affect embryo development and that vary widely between laboratories.

8. All contact materials such as tubes, culture dishes, and center well dishes are rinsed with commercially available bicarbonate buffered salt solution prior to use with embryos. Medium is added to dishes or tubes, aspirated, and then discarded immediately before the dish is prepared. Dishes used to make microdrops for culture are not rinsed.
9. Aspirates may contain follicular fluid from one or more follicles, so there are often multiple oocytes per tube. Some physicians will rinse the follicle by injecting fluid into the follicle and aspirating this fluid following aspiration of the follicular fluid itself to increase recovery rate.
10. There is now electronic witnessing software available that can take the place of manual double witnessing in the IVF laboratory. Electronic witnessing systems can reduce the risk of human error when gametes and embryos are moved and mixed during the IVF cycle. These systems may consist of some type of handheld Wi-Fi scanner that reads bar codes on culture dishes and tubes, or radio frequency identification reader and tags. These systems download information directly into electronic medical records to ensure traceability.
11. Any broken zonas, atretic oocytes, or other unusual occurrences are also noted.
12. Some common hand soaps are toxic to sperm.
13. Instruct the patient to avoid touching the inside of the specimen cup with the penis, and do not use lubricants. It is also important to note if any of the ejaculate was not captured in the specimen cup.
14. The recommended time between the last ejaculation and sperm collection for IVF is 2–7 days.
15. More than 30 μL of sperm may be added to reach the desired concentration in the dish if the sperm concentration is low.
16. Hyaluronidase, an enzyme used for the removal of the expanded cumulus complex from the mature oocyte, is

available lyophilized or in solution as single use, ready to use vials from various commercial vendors. Hyaluronidase is typically supplied as raw, purified material of bovine or ovine origin. A recombinant human hyaluronidase is now commercially available, also reconstituted in single use vials. Either of these products can be used for removing cumulus cells from human oocytes.

17. Each micropipette tip should be rinsed prior to use.
18. PVP solution is commercially available for this purpose. More or less sperm may be added to the PVP drop depending on sperm concentration.
19. Micropipettes for all human gamete and embryo micromanipulations are commercially available for purchase, in varying sizes and angles of bend.
20. Pentoxyfylline (1.5 mM) treatment may be used immediately prior to ICSI to improve sperm motility if necessary. In addition, eggs may be incubated for 15 min in calcium ionophore (1 mg/mL) after the ICSI procedure to improve egg activation in cases of previous ICSI fertilization failure.
21. Typically, it is recommended to transfer only one embryo at the blastocyst stage, particularly if the embryo has undergone preimplantation genetic testing and is known to be euploid, in an effort to avoid multiple births and the associated complications. However, based on each individual patient's diagnosis and history, two embryos may be transferred. It is unusual and considered risky to transfer more than two embryos at the blastocyst stage.
22. Blastocysts are typically selected for transfer based upon morphology. Embryos developing to the blastocyst stage on D5 are typically considered more viable than those developing to blastocyst on day 6. However, pregnancies routinely result from day 6 embryos, so they are still considered a viable option for transfer [1]. Embryos graded as stage 4, 5, or 6 and quality AA on day 5 are considered the best candidates and are the highest priority for transfer, if multiple embryos are available. If genetic testing has been done (PGT-A, preimplantation genetic testing for aneuploidy; or PGT-M, preimplantation genetic testing for monogenic/single gene diseases) then aneuploid or affected embryos are removed from consideration for transfer.
23. When embryos are moved from the culture to the transfer dish, there must be a second person present as a witness to verify that the name on the two dishes match.
24. The stereoscope may be equipped with a camera connected to a monitor on the top of the isolette, so that the patient and their partner can visualize the embryo(s) to be transferred.

25. Various types of catheters are sold commercially for human embryo transfer. They are typically made of a soft, flexible, smooth plastic, a pliable outer sheath or guide catheter, markings to show depth, and have an echogenic tip for visualization on the ultrasound during the transfer procedure. They are available with or without a curve. Some catheters come with a trial transfer catheter to allow positioning of the outer sheath before introducing the catheter containing the embryo.

References

1. Centers for Disease Control and Prevention (CDC), and American Society for Reproductive Medicine (ASRM) Society for Assisted Reproductive Technology (SART) (2017) 2015 Assisted reproductive technology national summary report. US Department of Health and Human Services, Atlanta, GA
2. Adamson GD, de Mouzon J, Chambers GM, Zegers-Hochschild F, Mansour R, Ishihara O, Banker M, Dyer S (2018) International Committee for Monitoring Assisted Reproductive Technology: world report on assisted reproductive technology, 2011. Fertil Steril 110:1067–1080. https://doi.org/10.1016/j.fertnstert.2018.06.039
3. ESHRE Special Interest Group of Embryology and Alpha Scientists in Reproductive Medicine (2017) The Vienna consensus: report of an expert meeting on the development of ART laboratory performance indicators. Reprod Biomed Online 35:494–510. https://doi.org/10.1016/j.rbmo.2017.06.015
4. The Practice Committee of the American Society for Reproductive Medicine (ASRM) and the Practice Committee of the Society for Assisted Reproductive Technologies (SART) (2008) Revised guidelines for human embryology and andrology laboratories. Fertil Steril 90:S45–S59. https://doi.org/10.1016/j.fertnstert.2008.08.099
5. van Loendersloot LL, van Wely M, Limpens J, Bossuyt PMM, Repping S, van der Veen F (2010) Predictive factors in in vitro fertilization (IVF): a systematic review and meta-analysis. Hum Reprod Update 16:577–589. https://doi.org/10.1093/humupd/dmq015
6. Morbeck DE, Leonard PH (2012) Culture systems: mineral oil overlay. Methods Mol Biol 912:325–331. https://doi.org/10.1007/978-1-61779-971-6_18
7. Edwards RG, Steptoe PC, Purdy JM (1980) Establishing full term human pregnancies using cleaving embryos grown in vitro. BJOG Int J Obstet Gynaecol 87:737–756. https://doi.org/10.1111/j.1471-0528.1980.tb04610.x
8. Berger T, Marrs RP, Moyer DL (1985) Comparison of techniques for selection of motile spermatozoa. Fertil Steril 43:268–273
9. Hyne RV, Stojanoff A, Clarke GN, Lopata A, Johnston WI (1986) Pregnancy from in vitro fertilization of human eggs after separation of motile spermatozoa by density gradient centrifugation. Fertil Steril 45:93–96
10. Veeck L (ed) (1986) Atlas of the human oocyte and early conceptus, vol 1. Williams & Wilkins, Baltimore, MD. https://doi.org/10.1002/mrd.1120200110
11. Kimura Y, Yanagimachi R (1995) Intracytoplasmic sperm injection in the mouse. Biol Reprod 52:709–720
12. Yanagimachi R (1995) Is an animal model needed for intracytoplasmic sperm injection (ICSI) and other assisted reproduction technologies? Hum Reprod 10:2525–2526
13. Alikani M, Cohen J (1992) Advances in clinical micromanipulation of gametes and embryos. Assisted fertilization and hatching. Arch Pathol Lab Med 116:373–378
14. Alikani M, Cohen J, Palermo GD (1995) Enhancement of fertilization by micromanipulation. Curr Opin Obstet Gynecol 7:182–187
15. Mansour R (1998) Intracytoplasmic sperm injection: a state of the art technique. Hum Reprod Update 4:43–56
16. Tesarik J, Mendoza C (1999) In vitro fertilization by intracytoplasmic sperm injection. BioEssays 21:791–801. https://doi.org/10.1002/(sici)1521-1878(199909)21:9<791::aid-bies11>3.0.co;2-z
17. Tesarik J, Mendoza C, Greco E (1998) Treatment of severe male infertility by micromanipulation-assisted fertilization: news and views. Front Biosci 3:E238–E246
18. Meldrum DR, Chetkowski R, Steingold KA, de Ziegler D, Cedars MI, Hamilton M (1987)

Evolution of a highly successful in vitro fertilization-embryo transfer program. Fertil Steril 48:86–93

19. Reed ML (2012) Culture systems: embryo density. Methods Mol Biol 912:273–312. https://doi.org/10.1007/978-1-61779-971-6_16
20. Gardner DK, Lane M, Stevens J, Schlenker T, Schoolcraft WB (2000b) Blastocyst score affects implantation and pregnancy outcome: towards a single blastocyst transfer. Fertil Steril 73:1155–1158
21. Cohen J (1991) Assisted hatching of human embryos. J In Vitro Fert Embryo Transf 8:179–190
22. Cohen J, Alikani M, Trowbridge J, Rosenwaks Z (1992) Implantation enhancement by selective assisted hatching using zona drilling of human embryos with poor prognosis. Hum Reprod 7:685–691
23. Alteri A, Vigano P, Maizar AA, Jovine L, Giacomini E, Rubino P (2018) Revisiting embryo assisted hatching approaches: a systematic review of the current protocols. J Assist Reprod Genet 35:367–391. https://doi.org/10.1007/s10815-018-1118-4
24. Gardner DK, Schoolcraft WB (1999) Culture and transfer of human blastocysts. Curr Opin Obstet Gynecol 11:307–311
25. Gardner DK, Lane M, Schoolcraft WB (2000a) Culture and transfer of viable blastocysts: a feasible proposition for human IVF. Hum Reprod 15(Suppl 6):9–23
26. Cohen J, Rieger D (2012) Historical background of gamete and embryo culture. Methods Mol Biol 912:1–18. https://doi.org/10.1007/978-1-61779-971-6_1
27. Quinn P (2014) Culture media, solutions, and systems in human ART. Cambridge University Press, Cambridge. https://doi.org/10.1017/CBO9781139059053
28. Machtinger R, Racowsky C (2012) Culture systems: single step. Methods Mol Biol 912:199–209. https://doi.org/10.1007/978-1-61779-971-6_12
29. Quinn P (2012) Culture systems: sequential. Methods Mol Biol 912:211–230. https://doi.org/10.1007/978-1-61779-971-6_13
30. Youssef MM, Mantikou E, van Wely M, van der Veen F, Al-Inany HG, Repping S, Mastenbroek S (2015) Culture media for human pre-implantation embryos in assisted reproductive technology cycles. Cochrane Database Syst Rev 11:CD007876. https://doi.org/10.1002/14651858.CD007876.pub2
31. Dieamant F, Petersen CG, Mauri AL, Comar V, Mattila M, Vagnini LD, Renzi A, Petersen B, Ricci J, Oliveira JBA, Baruffi RLR, Franco JG Jr (2017) Single versus sequential culture medium: which is better at improving ongoing pregnancy rates? A systematic review and meta-analysis. JBRA Assist Reprod 21:240–246. https://doi.org/10.5935/1518-0557.20170045

Chapter 25

Isolation and Culture of Periimplantation and Early Postimplantation Mouse Embryos

Niraimathi Govindasamy and Ivan Bedzhov

Abstract

Key developmental processes of cell fate decisions and morphogenetic transformations take place during the periimplantation and early postimplantation stages of mouse embryogenesis. However analysing these fundamental events relies on direct observations of cultured embryos, which are challenging to obtain. To address this challenge, here we provide a detailed protocol describing a workflow for isolating early implanted embryos, removing of redundant extraembryonic tissues and describing the culture conditions that support further embryo development in vitro.

Key words Mouse embryo, Implantation, Periimplantation, Postimplantation, Blastocyst, Egg cylinder, In vitro culture, Isolation, Uterus, Decidua, E4.75, E5.0, E5.5, E6.5

1 Introduction

The implantation of the mouse embryo is a critical developmental process of connecting the conceptus to the maternal tissues during the early stages of pregnancy. This process starts as the mature blastocyst hatches from its glycoprotein envelope, the zona pellucida, and attaches to the uterine epithelium. The cells of the mural trophectoderm (TE) mediate the first direct contact with the mother and penetrate into the underlying stroma [1] (Fig. 1a). In turn, the stromal cells rapidly proliferate and differentiate into polyploid cells that form the decidua [2]. This is a specialized tissue that supports embryo development by mediating nutrient and gas exchange, and it also restricts entry of immune system cells, enabling immune tolerance of the fetomaternal unit [3]. By E5.0 the decidua completely engulfs and conceals the embryo. Shortly after that, during E5.5–E6.5, patterning of the epiblast (EPI) into anterior and posterior regions is initiated, primordial germ cells (PGCs) are specified and subtypes of the trophoblast lineage emerge [4, 5].

Jason R. Herrick (ed.), *Comparative Embryo Culture: Methods and Protocols*, Methods in Molecular Biology, vol. 2006,
https://doi.org/10.1007/978-1-4939-9566-0_25,

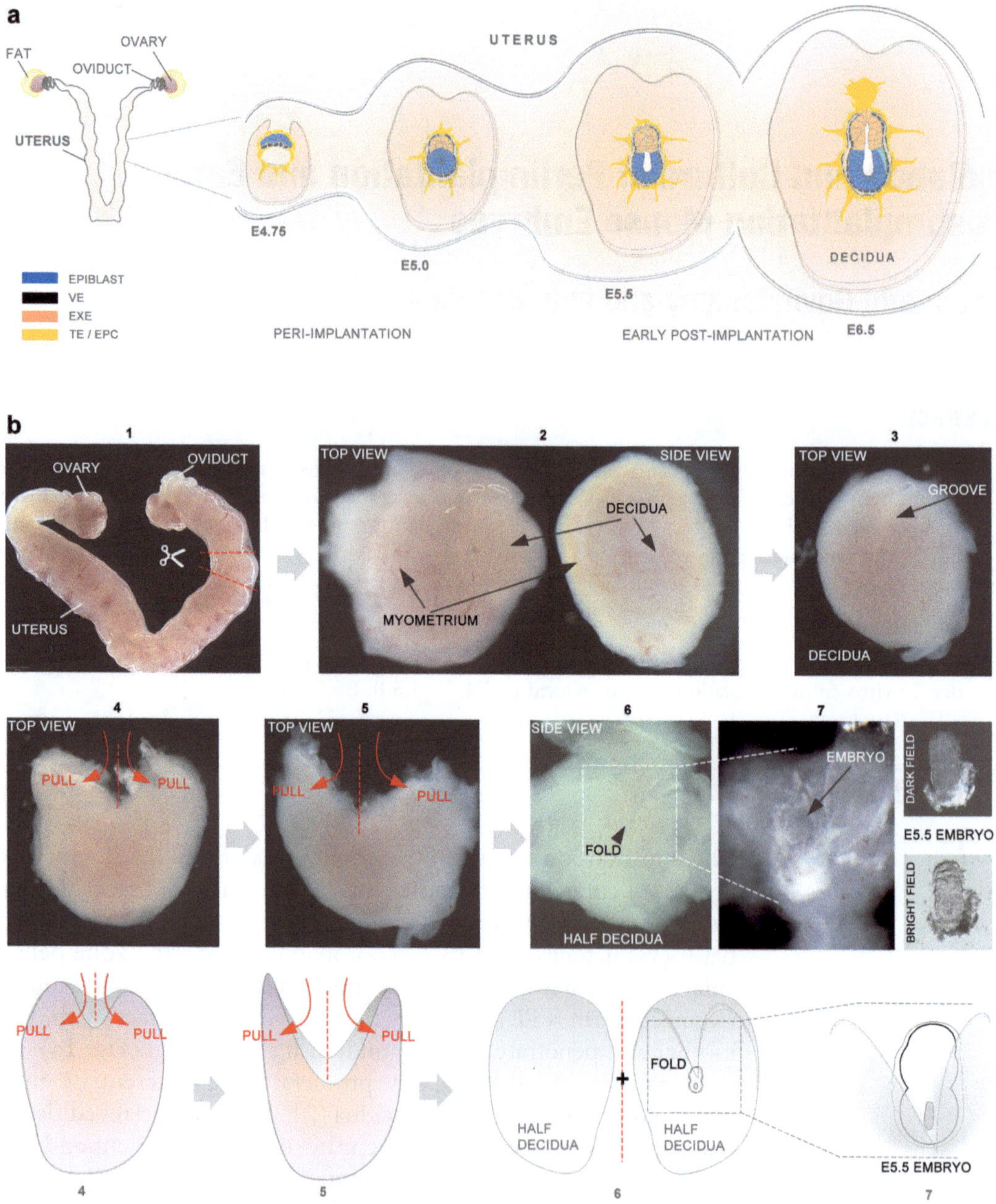

Fig. 1 Isolation of early mouse embryos. (**a**) Schematic overview of the mouse female reproductive tract. The peri-implantation (E4.75–E5.0) and early post-implantation (E5.5–E6.5) embryos are located in the uterus. The E4.75 embryo attaches to and penetrates the uterine wall triggering the formation of the decidua that engulfs the conceptus from E5.0 onward. The embryonic and extraembryonic lineages during the consecutive stages are annotated using a color code: EPI—blue, VE—black, ExE—orange, TE and EPC—yellow. (**b**) Isolation of E5.5 embryos. (1) The uterus is cut between individual implantation sites. (2) The outer layers of the uterus (perimetrium and endometrium) are cut off to expose the decidua. (3) The decidua forms a natural groove as it engulfs the embryo. (4 and 5) Using this groove as guidance, the decidua is gradually opened. (6) As the cut is usually not precisely in the middle, one half of the decidua shows a characteristic fold (implantation site) that

Understanding of these key developmental events requires extensive analysis of periimplantation (E4.75–E5.0) and early post-implantation (E5.5–E6.5) embryos. However, working with these stages is challenging, as the embryos are hidden within the decidua and are also relatively small and fragile. To help overcome this issues, here we provide a step-by-step protocol for isolation, micro-dissection, and culture of such embryos. Mastering the isolation procedure requires a period of practice to recognize certain features of the decidua that can be used as signposts for finding the implanted embryo. Once the embryos are recovered from the maternal tissues, the mural TE of E4.75–E5.0 and the Reichert's membrane (RM) of E5.5–E6.5 embryos have to be removed to allow for proper growth in vitro. While the mural TE has to be manually cut off, the RM can be removed via enzymatic digestion [6]. This is an elegant approach that circumvents the risk of mechanically damaging the embryo. To safely pick individual embryos, mouth pipetting is recommended, as it enables very precise control over the collection and the transfer of embryos between different wells.

Embryos that are free of mural TE or RM can be cultured in vitro for a few hours to several days in IVC1 and/or IVC2 media depending on the experimental setup [7, 8]. The IVC1 medium contains fetal calf serum (FCS), which is required for activating the proliferation of E4.75 embryos. Testing several batches of high-quality (embryonic stem cell grade) FCS might be necessary to fine-tune the culture conditions [9]. E5.0, as well as E5.5–E6.5 embryos, can be cultured directly in serum-free IVC2 medium. This medium is completely synthetic, providing a consistent and reproducible embryo culture environment [10].

2 Materials

2.1 Dissection Tools

1. Stereomicroscope equipped with transillumination base with rotating mirror and connected to a light source.
2. Bunsen burner.
3. Glass Pasteur pipettes.
4. Rubber tubing.
5. Iris scissors.
6. Fine scissors.

Fig. 1 (continued) contains the embryo. (7) The embryo can be directly scooped out of this fold, or the whole tissue with the fold can be dissected out for better visibility. For the final steps of the dissection procedure bight field microscopy provides better contrast for easier identification of the implantation site and the embryo itself

7. Fine forceps (Dumont #5).
8. Lapping film (3 M, AL2O5 1 U 262 × 1 MIC × 8 1/2″ × 11″).
9. 1- ml syringes.
10. 27- G needles.
11. Petri dishes, 35 × 10 mm.
12. 4-well plates.
13. Mouth pipette assembly. Exchange the provided mouthpiece of the rubber tubing with a 200-μl filter tip. On the other side of the tubing, attach 1000-μl tip without a filter. After pulling the glass Pasteur pipette over the flame of the Bunsen burner, attach the Pasteur pipette to the 1000-μl tip. The open end of the glass pipette should have a diameter that allows for free movement of the embryo, taking into account the developmental stage.
14. Sharpening of the forceps. Sharp forceps are absolutely essential for the successful isolation of periimplantation and early postimplantation embryos. Although the Dumont #5 forceps have fine tips, further sharpening is required. Using a stereomicroscope, place the Lapping film on the stage, press to hold the forceps closed and focus on the tips. Continuously slide the forceps over the film to sharpen the tips. As the tips are easily damaged, the forceps have to be sharpened every time before starting the embryo isolation procedure.

2.2 Stock solutions

1. Advanced DMEM/F-12 (Life Technologies, cat. no. 12634-010).
2. Fetal Calf Serum (FCS), embryonic stem cell-grade. Heat-inactivate the serum by incubating it at 56 °C for 30 min.
3. Knock-out Serum Replacement—KSR (Life Technologies, cat. no. 10828010).
4. L-Glutamine Stock: 200 mM in ddH_2O.
5. Penicillin–Streptomycin Stock: 5000 U/ml penicillin and 5000 μg/ml streptomycin.
6. ITS-X Stock: insulin 1 g/l, transferrin 0.55 g/l, sodium selenite 0.00067 g/l, ethanolamine 0.20 g/l.
7. β-Estradiol Stock: 10 μM solution in DMSO. Store at −20 °C for up to 2 months.
8. Progesterone Stock: 1 mg/ml solution in DMSO. Store at −20 °C for up to 2 months.
9. *N*-Acetyl-L-cysteine (NAC) Stock: 50 mM solution in ddH_2O. Store at −20 °C for up to 2 months.

2.3 Media

1. M2 medium [11]: Warm (37 °C) before use.
2. IVC1 Medium: Supplement Advanced DMEM/F12 with 20% (vol/vol) FCS, 2 mM L-glutamine, penicillin (25 U/ml), streptomycin (25 μg/ml), 1× ITS-X (10 mg/l insulin, 5.5 mg/l transferrin, 0.0067 mg/l sodium selenite, and 2 mg/l ethanolamine), 8 nM β-estradiol, 200 ng/ml progesterone, and 25 μM NAC. The medium can be kept at −20 °C for 3 months or at 4 °C for up to 2 weeks. Warm the medium (37 °C) and equilibrate in a humidified air atmosphere of 5% CO_2 for at least half an hour before use.
3. IVC2 Medium: Supplement Advanced DMEM/F12 with 30% (vol/vol) KSR, 2 mM L-glutamine, penicillin (25 U/ml), streptomycin (25 μg/ml), 1× ITS-X (10 mg/l insulin, 5.5 mg/l transferrin, 0.0067 mg/l sodium selenite, and 2 mg/l ethanolamine), 8 nM β-estradiol, 200 ng/ml progesterone, and 25 μM NAC. The medium can be kept at −20 °C for 3 months or at 4 °C for up to 2 weeks. Warm the medium (37 °C) and equilibrate in a humidified air atmosphere of 5% CO_2 for at least half an hour before use.

2.4 Enzymes for removal of the Reichert's Membrane

1. Collagenase, Type IV.
2. Dispase II.
3. Hyaluronidase, Type IV-S.

Prepare 10 mg/ml stock solutions of each enzyme in Ca^{2+}/Mg^{2+}-free phosphate buffered saline (PBS). Store aliquots (100 μl) at −20 °C for up to 6 months. Avoid repetitive freezing and thawing.

3 Methods

Isolation of embryos from the uterus is performed at room temperature. After that, the embryo culture must be carried out at 37 °C and 5% CO_2 in air in a humidified incubator.

1. Humanely euthanize pregnant female mice at the respective day postcoitum (d.p.c.) corresponding to the desired periimplantation or early postimplantation stage. Cut the skin and the peritoneum and push away the abdominal organs to locate the reproductive tract. Cut between the cervix and the uterus. After that dissect and collect the uterus, the oviduct, and the ovary in one piece. Place the organs in warm M2 medium.
2. Transfer the tissue into the inverted lid of a 35-mm dish filled with M2 medium. It is more convenient to use the lid instead of the plate for the dissection procedure, as the wall of the lid is

lower and provides better access to the sample. Place the lid under a stereomicroscope and cut off the ovary, oviduct and fat attached to the uterus. For isolation of E4.75 embryos, follow **step 3**, for all later stages go to **step 4**.

3. At E4.75, some of the embryos are still loosely attached to the uterine wall and can be flushed out using a syringe filled with M2 medium and attached to a 27-G needle. Place the needle into the lumen of one of the uterine horns and flush at least three times with 500 μl of M2 medium. Repeat the same procedure with the other horn. Using the mouth pipette assembly, collect and transfer the embryos into a 4-well plate filled with 250 μl M2 medium with a mouth pipette. To isolate stably attached embryos, cut the uterus lengthwise to convert the uterine horn into a flat sheet. Using forceps, scratch the tissue following the endometrial folds. Carefully examine the tissue debris to locate detached E4.75 embryos (*see* **Note 1**).
4. After E4.75, the embryos are completely engulfed by the decidua. Individual deciduas are visible as bulges of the uterus (Fig. 1b). Cut between the bulges and dissect the outer layers of the uterus (perimetrium and myometrium) to expose the decidua (*see* **Note 2**).
5. As it engulfs the embryo, the decidua forms a groove that is visible as a brighter line in the middle of the tissue (Fig. 1b). Use one pair of forceps to keep the position of the decidua fixed and another pair to open the groove. Hold the forceps closed and insert the tip into the groove then slowly open the forceps. This motion will cut the tissue in the middle by pulling apart the opposite sides (*see* **Note 3**).
6. Keep on cutting the decidua until you locate a characteristic fold that is slightly darker than the rest of the tissue. This is the implantation site. If you notice the embryo do not try to grab it; instead, use only one end of the open forceps to gently scoop it out.
7. If the embryo is not visible, completely slice the decidua in half. As the cut is usually not precisely in the middle, only one half of the decidua will contain the implantation site (the tissue fold). Cut out the fold with a little bit of the surrounding decidua. This piece of tissue is much smaller, providing better visibility for identifying the embryo. Carefully dissect the fold and scoop out the embryo (*see* **Note 4**). Transfer the embryos into a 4-well plate filled with 250 μl M2 medium using a mouth pipette.
8. The mural TE of the periimplantation embryo and the RM of the early postimplantation embryos have to be removed before in vitro culture. For dissection of the mural TE follow step 9, for enzymatic removal of the RM go to **step 10**.

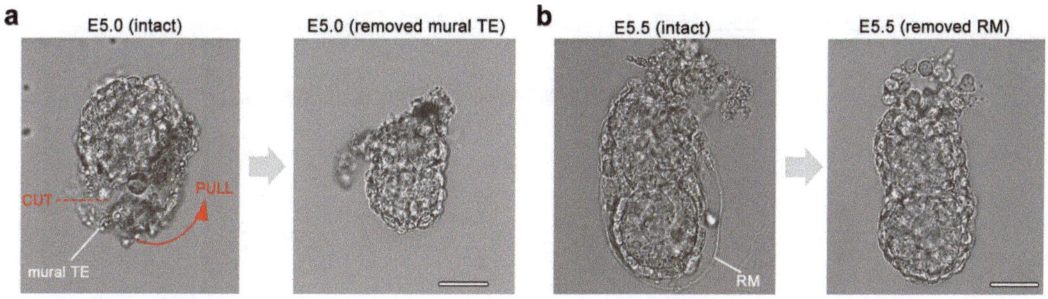

Fig. 2 Removal of the mural TE and RM. (**a**) The mural TE of the E5.0 embryo is cut off using sharp forceps. (**b**) The RM of the E5.5 embryo is removed via enzymatic digestion. Scale bars—50 μm

9. The mural TE of E4.75 and E5.0 embryos appears as dark, rough tissue at the distal pole of the embryo. Place the embryo into the inverted lid of a 35-mm plate filled with M2 medium. Keeping the forceps open, use one tip of the forceps as a scalpel, slide and cut the mural TE, pulling the tissue aside (Fig. 2a) (*see* **Note 5**).

10. The RM of the early postimplantation embryos can be dissected manually, following the same procedure as the removal of the mural TE from earlier stages. However, the space between the RM and the visceral endoderm (VE) is narrow, and the likelihood of damaging the egg cylinder is high. Enzymatic digestion of the RM is a convenient and fast alternative method [6]. Prepare a solution of 1 mg/ml collagenase type IV, 1 mg/ml dispase and 0.3 mg/ml hyaluronidase in Ca^{2+}/Mg^{2+}-free PBS. Place 250 μl of the warm (37 °C) triple-enzyme solution in one well of a 4-well plate and 250 μl of warm M2 medium in another well. Using a mouth pipette, transfer embryos to the triple enzyme solution. Pipette the embryos up and down a few times while observing the disappearance of the RM (approximately 2–3 min for E5.5 embryos). Immediately transfer the RM-free embryos to the M2 medium (Fig. 2b).

11. After dissection of the mural TE or the RM, the embryos are ready for culture. E4.75 embryos require 1 day of culture in IVC1 medium after that the medium is replaced with IVC2. All later stages are cultured only in IVC2 medium (Fig. 3a). Using a mouth pipette, transfer embryos from the M2 medium to wells containing 250 μl warm IVC media (IVC1 for E4.75 and IVC2 for E5.0, E5.5, and E6.5 embryos) (*see* **Note 6**).

12. Observe the development of the embryos daily. As the whole period of culture in IVC2 medium takes 1–2 days, it is not necessary to refresh the medium. It is expected that at each day of the culture the embryos will double their length, while

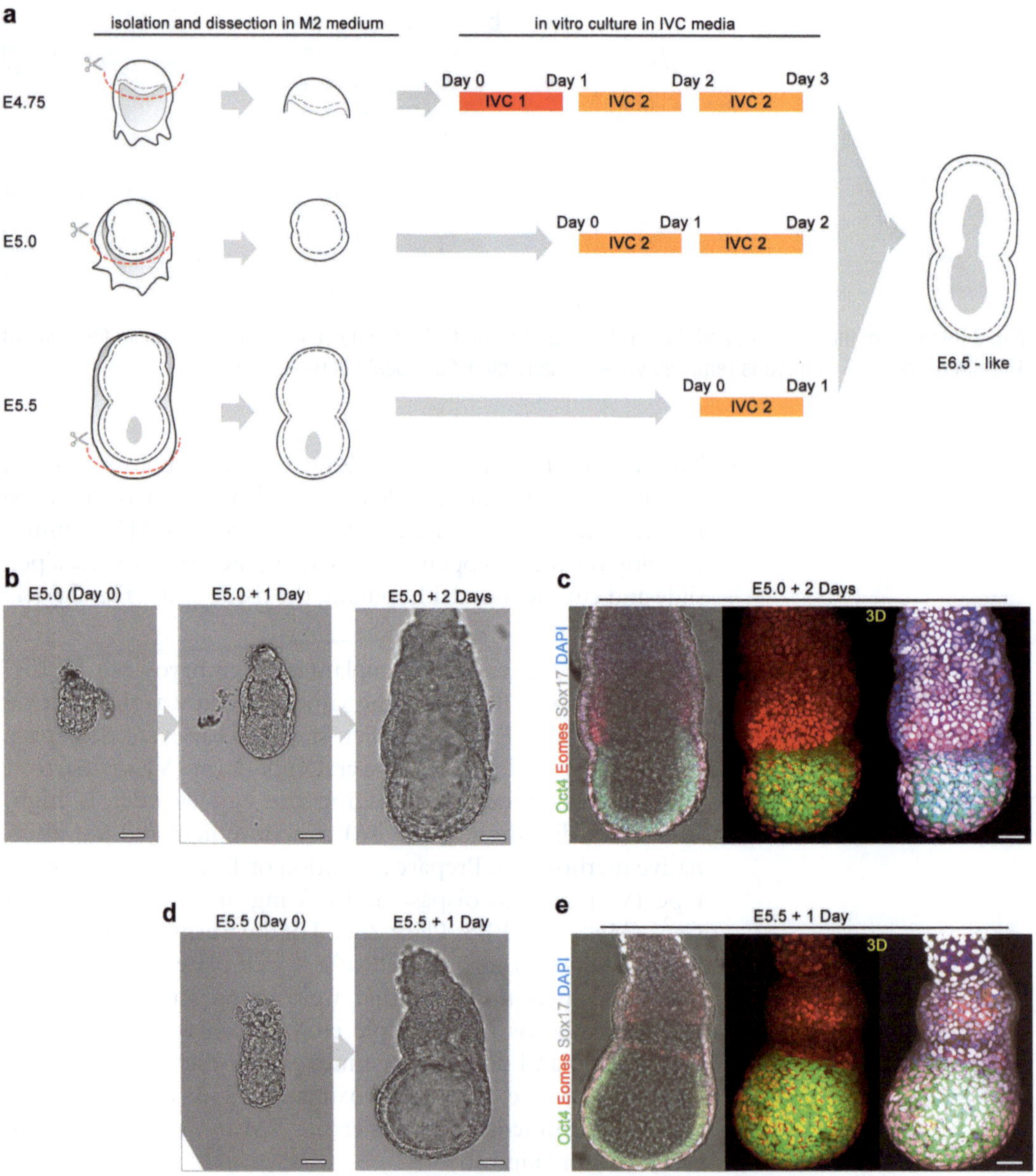

Fig. 3 In vitro culture of periimplantation and early postimplantation embryos. (**a**) Schematic overview of the procedure. After removal of the mural TE (E4.75–E5.0) or the RM (E5.5), the embryos are ready for culturing. E4.75 embryos are cultured for 1 day in IVC1 medium and then transferred to IVC2 medium. From E5.0 onward the embryos are cultured only in IVC2 medium. (**b**) Snapshot capture of E5.0 embryo cultured for 2 consecutive days in IVC2 medium. (**c**) The same embryo is fixed and stained for Oct4 (EPI), Eomes (ExE and VE), Sox17 (VE) and DAPI (nuclei). (**d**) Snapshot capture of E5.5 embryo cultured for 1 day in vitro. (**e**) At the end of the culture, the embryo is fixed and stained for Oct4 (EPI), Eomes (ExE and VE), Sox17 (VE), and DAPI (nuclei). Scale bars—50 μm

expressing proper cell lineage markers. For example, E5.0 embryos should form a visible proamniotic cavity after 1 day of culture. The embryos further elongate and the cavity expands during the second day (Fig. 3b, c). E5.5 embryos reach similar size after only 1 day of culture in IVC2 medium (Fig. 3d, e).

4 Notes

1. The E4.75 embryos have no zona pellucida, and they appear flat with differentiated (darker) mural TE and compacted (shiny) inner cell mass (ICM).
2. The decidua is very fragile, try not to tear it apart while removing the myometrium. Use fine scissors to cut out the outer layers of the uterus. Keeping the decidua intact helps to identify the groove and to more precisely separate the decidua in half.
3. It is essential to use very sharp forceps. *See* Subheading 2 about how to sharpen the forceps.
4. Use brightfield illumination of the stereomicroscope for better contrast.
5. It is critical to keep the ICM intact, while cutting off the mural TE. Even slight mechanical damage of the ICM will result in failure of further embryonic development.
6. Culture embryos on glass-bottom or optical plastic plates for live imaging experiments.

Acknowledgments

This work was supported by the German Research Foundation (DFG) Emmy Noether grant to I.B. and the Cells-in-Motion Cluster of Excellence/International Max Planck Research School (CiM/IMPRS) Ph.D. fellowship of N.G.

References

1. Bedzhov I, Graham SJ, Leung CY, Zernicka-Goetz M (2014) Developmental plasticity, cell fate specification and morphogenesis in the early mouse embryo. Philos Trans R Soc Lond Ser B Biol Sci 369(1657):pii:20130538. https://doi.org/10.1098/rstb.2013.0538
2. Mori M, Bogdan A, Balassa T, Csabai T, Szekeres-Bartho J (2016) The decidua-the maternal bed embracing the embryo-maintains the pregnancy. Semin Immunopathol 38(6): 635–649. https://doi.org/10.1007/s00281-016-0574-0
3. Nancy P, Tagliani E, Tay CS, Asp P, Levy DE, Erlebacher A (2012) Chemokine gene silencing in decidual stromal cells limits T cell access to the maternal-fetal interface. Science 336 (6086):1317–1321. https://doi.org/10.1126/science.1220030
4. Bedzhov I, Zernicka-Goetz M (2015) Cell death and morphogenesis during early mouse

development: are they interconnected? BioEssays 37(4):372–378. https://doi.org/10.1002/bies.201400147

5. Takaoka K, Hamada H (2012) Cell fate decisions and axis determination in the early mouse embryo. Development 139(1):3–14. https://doi.org/10.1242/dev.060095
6. Rivera-Perez JA, Diefes H, Magnuson T (2007) A simple enzymatic method for parietal yolk sac removal in early postimplantation mouse embryos. Dev Dyn 236(2):489–493. https://doi.org/10.1002/dvdy.21034
7. Bedzhov I, Bialecka M, Zielinska A, Kosalka J, Antonica F, Thompson AJ, Franze K, Zernicka-Goetz M (2015) Development of the anterior-posterior axis is a self-organizing process in the absence of maternal cues in the mouse embryo. Cell Res 25(12):1368–1371. https://doi.org/10.1038/cr.2015.104
8. Bedzhov I, Zernicka-Goetz M (2014) Self-organizing properties of mouse pluripotent cells initiate morphogenesis upon implantation. Cell 156(5):1032–1044. https://doi.org/10.1016/j.cell.2014.01.023
9. Bedzhov I, Leung CY, Bialecka M, Zernicka-Goetz M (2014) In vitro culture of mouse blastocysts beyond the implantation stages. Nat Protoc 9(12):2732–2739. https://doi.org/10.1038/nprot.2014.186
10. Zernicka-Goetz M, Bedzhov I (2015) Media and methods for culturing embryos and stem cells. Patent WO2015022541
11. Behringer R, Gertsenstein M, Vintersten Nagy K, Nagy A (2014) Manipulating the mouse embryo: a laboratory manual, 4th edn. CSH Press, Cold Spring Harbor, NY

INDEX

Jason R. Herrick (ed.), *Comparative Embryo Culture: Methods and Protocols*, Methods in Molecular Biology, vol. 2006, https://doi.org/10.1007/978-1-4939-9566-0, © Springer Science+Business Media, LLC, part of Springer Nature 2019

C

D

E

F

G

H

I

J

K

L

M

N

O

V

W

Z

MIX
Papier aus verantwortungsvollen Quellen
Paper from responsible sources
FSC® C105338

If you have any concerns about our products,
you can contact us on
ProductSafety@springernature.com

In case Publisher is established outside the EU,
the EU authorized representative is:
Springer Nature Customer Service Center GmbH
Europaplatz 3, 69115 Heidelberg, Germany

Printed by Libri Plureos GmbH
in Hamburg, Germany